# EMERGING OPTICAL NETWORK TECHNOLOGIES

# *Recent Related Titles*

**Metropolitan Area WDM Networks: An AWG-Based Approach**
Martin Maier
ISBN 1-4020-7574-X, 2003
http://www.wkap.nl/prod/b/1-4020-7574-X

**Optical Networks: Architecture and Survivability**
Hussein T. Mouftah and Pin-Han Ho
ISBN 1-4020-7196-5, 2002
http://www.wkap.nl/prod/b/1-4020-7196-5

**WDM Mesh Networks: Management and Survivability**
Hui Zang
ISBN 1-4020-7355-0, 2002
http://www.wkap.nl/prod/b/1-4020-7355-0

# EMERGING OPTICAL NETWORK TECHNOLOGIES
## Architectures, Protocols and Performance

Edited by
KRISHNA M. SIVALINGAM
University of Maryland, Baltimore County

SURESH SUBRAMANIAM
George Washington University

Springer

Krishna M. Sivalingam
University of Maryland, Baltimore County
Dept. of CS and EE
1000 Hilltop Circle
Baltimore, MD 21250
krishna@cs.umbc.edu

Suresh Subramaniam
George Washington University
Dept. of EE and CS
801 22nd St. N.W.
Washington, DC 20052
suresh@gwu.edu

Emerging Optical Network Technologies: Architectures, Protocols and Performance

Library of Congress Cataloging-in-Publication Data

A C.I.P. Catalogue record for this book is available from the Library of Congress.

ISBN 0-387-22582-X e-ISBN 0-387-22584-6 Printed on acid-free paper.



Printed in the United States of America.

9 8 7 6 5 4 3 2 1 SPIN 11052630

springeronline.com

*This book is dedicated to our families.*

# Contents

# Contributing Authors

Vishal Anand, *State University of New York, College at Brockport*
Amitabha Banerjee, *University of California, Davis*
Eric Bouillet, *Tellium, Inc.*
Hakki Candan Cankaya, *Alcatel USA*
Xiaojun Cao, *State University of New York at Buffalo*
Yung Chen, *University of Maryland, Baltimore County*
Imrich Chlamtac, *University of Texas at Dallas*
Linda Cline, *Intel Corporation*
Sudhir Dixit, *Nokia Research Center*
Samrat Ganguly, *NEC Labs America Inc.*
Nasir Ghani, *Tennessee Technological University*
David Griffith, *NIST*
Aihua Guo, *University of Maryland, Baltimore County*
Jian-Qiang Hu, *Boston University*
Rauf Izmailov, *NEC Labs America Inc.*
Myoungki Jeong, *Alcatel USA*
Admela Jukan, *Vienna University of Technology*
Leonid Kazovsky, *Stanford University*
Glen Kramer, *Teknovus, Inc.*
Jean-Francois Labourdette, *Tellium, Inc.*
Christian Maciocco, *Intel Corporation*
Harsha Madhyastha, *University of Washington*
Carmen Mas, *AIT, Greece*
Manav Mishra, *Intel Corporation*
Eytan Modiano, *MIT*

Biswanath Mukherjee, *University of California, Davis*

Hung Nguyen, *EPFL, Switzerland*

Chris Olszewski, *Tellium, Inc.*

Chunming Qiao, *State University of New York at Buffalo*

Matt Rogge, *Stanford University*

George Rouskas, *North Carolina State University*

Gokhan Sahin

Ramakrishna Shenai, *University of Maryland, Baltimore County*

Kapil Shrikhande, *Stanford University*

Mahesh Sivakumar, *University of Maryland, Baltimore County*

Krishna Sivalingam, *University of Maryland, Baltimore County*

C. Siva Ram Murthy, *Indian Institute of Technology, Chennai (Madras)*

Arun Somani, *Iowa State University*

Suresh Subramaniam, *George Washington University*

Patrick Thiran, *EPFL, Switzerland*

Slobodanka Tomic, *Vienna University of Technology*

Ian White, *Stanford University / Sprint Advanced Technology Laboratories*

Yizhi Xiong, *State University of New York at Buffalo*

Lisong Xu, *North Carolina State University*

Yinghua Ye, *Nokia Research Center*

Zhonghua Zhu, *University of Maryland, Baltimore County*

# Preface

Optical networks have moved from laboratory settings and theoretical research to real-world deployment and service-oriented explorations. New technologies such as Ethernet PON and optical packet switching are being explored, and the landscape is continuously and rapidly evolving. Some of the key issues involving these new technologies are the architectural, protocol, and performance aspects.

The objective of this book is to present a collection of chapters from leading researchers in the field covering the above-mentioned aspects. Articles on various topics, , spanning a variety of technologies, were solicited from active researchers in both academia and industry. In any book on such a quickly growing field, it is nearly impossible to do full justice to all of the important aspects. Here, rather than attempting to cover a large ground with a limited treatment of each topic, we focus on a few key challenges and present a set of papers addressing each of them in detail. It is our hope that the papers will be found to have sufficient detail for the new entrant to the field, and at the same time be a reference book for the experienced researcher.

This book is aimed at a wide variety of readers. The potential audience includes those who are interested in a summary of recent research work that cannot be found in a single location; those interested in survey and tutorial articles on specific topics; and graduate students and others who want to start research in optical networking. We hope that readers gain insight into the ideas behind the new technologies presented herein, and are inspired to conduct their own research and aid in further advancing the field.

## Organization of the book

The book is divided into six parts, each dealing with a different aspect: network architectures, switching, signaling protocols, traffic grooming, protection and restoration, and testbeds. At least two chapters have been selected for each part, with three or more chapters for most parts.

Part I is on network architectures and contains four chapters. The first chapter by Cline, Maciocco and Mishra from Intel Labs takes a look into the services and architectures for next generation optical networks. The second chapter by researchers from NEC Labs and UT Dallas presents a hybrid hierarchical network architecture wherein both all-optical and OEO switching co-exist within a cross-connect. Chapter 3 summarizes recent developments in passive optical network (PON) architectures. This chapter is written by researchers from UC Davis, Teknovus, and Nokia Research. Chapter 4, by Nasir Ghani of Tennessee Technological University, presents a detailed survey of the recent activities in regional and metro network architectures.

Part II focuses on switching and consists of three chapters. The first chapter presents an overview of optical packet switching and is written by Rouskas and Xu of North Carolina State University. Chapter 6, by researchers from the SUNY at Buffalo and Brockport, presents waveband switching OXC architectures, and algorithms for grouping wavelengths into wavebands. The last chapter of Part II is on the third main switching paradigm, namely optical burst switching (OBS). The article, written by researchers from Alcatel and Samsung, reviews OBS concepts and describes the work on OBS done at Alcatel USA.

Signaling protocols are the subject of Part III. The first chapter, by Tomic and Jukan of the Vienna University of Technology, discusses the architecture and functionality of GMPLS-enabled exchange points. The second chapter by David Griffith of NIST presents the GMPLS protocol framework including RSVP-TE, OSPF-TE, and LMP. Chapter 10, authored by three researchers from Tellium, explains the benefits and operational aspects of mesh optical networks.

Part IV contains two chapters on traffic grooming. The first chapter by Hu and Modiano introduces a simple traffic grooming problem and then presents various modifications and solution techniques. The next chapter by Madhyastha and Murthy presents a specific architectural solution for efficient traffic grooming.

Part V is dedicated to protection and restoration. The first chapter by Sivakumar, Shenai, and Sivalingam presents a survey of survivability techniques. The next chapter by Somani focuses on routing "dependable" connections and presents a novel solution. The following chapter by Sahin and Subramaniam presents a new strategy of scheduling restoration control messages to provide quality of protection in mesh networks using capacity sharing. The last chapter in this part, written by Mas, Nguyen, and Thiran, discusses methods to locate failures in WDM networks.

The final part of the book consists of two chapters describing the testbeds built at UMBC and Stanford. In Chapter 17, a multi-layered GMPLS optical network testbed is described and Chapter 18 describes the HORNET packet switched metro network developed at Stanford.

We invite you to sit back and read about the recent research in optical networking presented in these chapters and hope that it stirs your creativity and imagination leading to further innovations and advances in the field.

## Acknowledgments

Naturally, this book would not have been possible without the time and effort of the contributing authors, and we are grateful to them. Each of the chapters selected were proofread by the editors and their graduate students who have also spent considerable time in taking care of the little details that make the book right. We also like to acknowledge the valuable assistance of Minal Mishra, Rama Shenai, Manoj Sivakumar, Mahesh Sivakumar and Sundar Subramani, graduate students at the University of Maryland, Baltimore County; and Tao Deng, Sunggy Koo, and Venkatraman Tamilraj at George Washington University.

We also gratefully acknowledge our research sponsors who provided partial support for this work. This includes DARPA under grant No. N66001-00-18949 (co-funded by NSA), National Science Foundation under grant Nos. ANI-0322959 and ANI-9973111, Cisco Systems and Intel Corporation.

We thank Kluwer Academic Publishers for the opportunity to publish this book. We are especially thankful to Alex Greene and Melissa Sullivan at Kluwer Academic Publishers for their constant help and patience, without which this book would not have been possible.

Krishna Sivalingam
*Associate Professor*
*University of Maryland, Baltimore County*
Email: krishna@umbc.edu

Suresh Subramaniam
*Associate Professor*
*George Washington University*
Email: suresh@gwu.edu

**May 2004**

## Book Editor Biographies

**Krishna M. Sivalingam** (ACM '93) is an Associate Professor in the Dept. of CSEE at University of Maryland, Baltimore County. Previously, he was with the School of EECS at Washington State University, Pullman from 1997 until 2002; and with the University of North Carolina Greensboro from 1994 until 1997. He has also conducted research at Lucent Technologies' Bell Labs in Murray Hill, NJ, and at AT&T Labs in Whippany, NJ. He received his Ph.D. and M.S. degrees in Computer Science from State University of New York at Buffalo in 1994 and 1990 respectively; and his B.E. degree in Computer Science and Engineering in 1988 from Anna University, Chennai (Madras), India. While at SUNY Buffalo, he was a Presidential Fellow from 1988 to 1991.

His research interests include wireless networks, optical wavelength division multiplexed networks, and performance evaluation. He holds three patents in wireless networks and has published several research articles including more than twenty-five journal publications. He has published an edited book titled "Wireless Sensor Networks" in 2004 and an edited book titled "Optical WDM networks" in 2000. He is serving as a Guest Co-Editor for a special issue of ACM MONET on "Wireless Sensor Networks" in 2004 and, in the past, served as Guest Co-Editor for a special issue of ACM MONET on "Wireless Sensor Networks" (2003) and an issue of IEEE Journal on Selected Areas in Communications on optical WDM networks (2000). He is co-recipient of the Best Paper Award at the IEEE International Conference on Networks 2000 held in Singapore. His work has been supported by several sources including AFOSR, NSF, Cisco, Intel and Laboratory for Telecommunication Sciences. He is a member of the Editorial Board for ACM Wireless Networks Journal, IEEE Transactions on Mobile Computing, and KICS Journal of Computer Networks.

He is serving as Steering Committee Co-Chair for the First International Conference on Broadband Networks 2004 (www.broadnets.org); and as Technical Program Co-Chair for the First IEEE Conference on Sensor and Ad Hoc Communications and Networks (SECON) to be held in Santa Clara, CA in 2004. He has served as General Co-Chair for SPIE Opticomm 2003 (Dallas, TX) and for ACM Intl. Workshop on Wireless Sensor Networks and Applications (WSNA) 2003 held on conjunction with ACM MobiCom 2003 at San Diego, CA. He served as Technical Program Co-Chair of OptiComm conference at Boston, MA in July 2002. He is a Senior Member of IEEE and a member of ACM.

**Suresh Subramaniam** received the Ph.D. degree in electrical engineering from the University of Washington, Seattle, in 1997. He is an Associate Professor in the Department of Electrical and Computer Engineering at the George Washington University, Washington, DC. He is interested in a variety of aspects of optical and wireless networks including performance analysis, algorithms, and design. His research has been supported by DARPA, DISA, NSA, and NSF.

Dr. Subramaniam is a co-editor of the book "Optical WDM Networks: Principles and Practice" published by Kluwer Academic Publishers in 2000. He has been on the program committees of several conferences including IEEE Infocom, IEEE ICC, and IEEE Globecom, and is TPC Co-Chair for the 2004 Broadband Optical Networking Symposium, part of the First Conference on Broadband Networks (www.broadnets.org). He serves on the editorial boards of Journal of Communications and Networks and IEEE Communications Surveys and Tutorials. He is a co-recipient of the Best Paper Award at the 1997 SPIE Conference on All-Optical Communication Systems.

# I

# NETWORK ARCHITECTURES

Chapter 1

# ENABLING ARCHITECTURES FOR NEXT GENERATION OPTICAL NETWORKS

Linda Cline, Christian Maciocco and Manav Mishra
*Intel Labs, Hillsboro OR 97124*
Email: linda.s.cline@intel.com, christian.maciocco@intel.com

**Abstract** As the demand grows for higher network access speeds, technologies such as optical fiber have begun to overtake traditional copper wire for data transport in short haul networks as well as long haul networks. Optical networking plays a growing role in next generation networks with new capabilities such as LCAS (Link Capacity Adjustment Scheme) and Virtual Concatenation (VC), and services such as dynamic provisioning and traffic grooming. While these emerging capabilities hold the promise of an intelligent optical network, there are still obstacles. Protocols and standards to support these capabilities are still evolving. In addition, in order to realize the new benefits, carriers and providers must invest in new optical equipment, as well as upgrades to existing equipment. In the current economic environment, a choice which leverages lower cost equipment with software which can provide advanced functionality is significantly more attractive than expensive alternatives. In addition, upgradeable software- based components provide future cost savings as well as flexibility in supporting new and changing protocols and standards. In this paper, we discuss each of these issues in detail and present a solution for optical services and applications, including Optical Burst Switching, using a network processor based platform to overcome the obstacles facing next generation optical networks.

**Keywords:** Optical Networking, SONET/SDH, Network Processors, GMPLS, UNI, Link Capacity Adjustment Scheme, Traffic Grooming, Optical Burst Switching.

## 1.1 Introduction

New capabilities and services for optical networks combined with optical fiber pushing toward the edge require continued investment in equipment and upgrades to support these new functions. This equipment needs to be flexible to support the networks of today as well as the capabilities for tomorrow. An architecture that is flexible enough to support this type of investment for the future is one that leverages software to augment less complex, and thus less expensive, hardware. Optical network nodes need to support changing network protocols and increased complexity in functionality. Use of a mass produced, inexpensive network processor that is optimized for network processing functions and completely programmable in software, provides an appropriate platform for these nodes. By implementing the complexity in software, there is increased adaptability to protocol upgrades for continued cost savings.

In this chapter, we discuss the problems and requirements of an intelligent optical network, and provide a solution describing the use of a software framework implemented on a network processor based optical platform.

In Section 1.2, we discuss several of the emerging optical services which are required by next generation optical networks, as well as some of the issues surrounding them. In Section 1.3, we provide an overview of network processors. Section 1.4 discusses the various software building blocks which can be used to implement the next generation optical services. In Section 1.5, we present a solution for Optical Burst Switching, which is a next generation optical application. Finally, Section 1.6 summarizes the choice of a network processor platform as an enabler for the continuously evolving optical networking technology.

## 1.2 Next-generation Optical Services

Next-generation optical services will support more customers and provide greater bandwidth in access networks. This capability requires new supporting services to be provided by the underlying networks. These services include automated optical provisioning, sophisticated traffic grooming, and services that ease management of networks with ever increasing complexity. These services are described in more detail in the subsequent sections.

### 1.2.1 Optical Provisioning

In current networks, setting up an optical connection to send SONET/SDH [11,12] frames from one location to another is a manual process. Typically, a Network Management System (NMS) is configured by one or more humans to add each new connection. It is not unusual for the turnaround time for a new connection to take up to six weeks to configure after the initial request has been submitted. Once a human has begun directly configuring the NMS software, the completion of the task may still take several minutes or hours. Provisioning that takes months or minutes may be acceptable, if not desirable, for setting up long haul connections which may be in place for long periods of time. However, as optical networking moves to the metro area network (MAN), this delay in provisioning connections becomes less acceptable. Access connections for the MAN have a finer granularity in bandwidth requirements and are more transient than long haul connections. Quantities of service connection or service modification requests will increase rapidly, which can swamp a provisioning system which is accomplished manually. Dynamic, automated provisioning is vital if service providers are going to meet the rigorous turnaround time and scalability requirements of MANs. Dynamic provisioning can also improve operational expenditures by reducing the need for human control, improving time to revenue for new services.

Support for dynamic provisioning is beginning to emerge, although today this is typically implemented using proprietary means. Such proprietary schemes make end to end automated provisioning not possible except where certain carriers control the complete paths. Efforts are underway in standards groups to define protocols for dynamic provisioning, which may solve the end to end problem eventually. Currently, these standards are moving targets, which magnifies the need for programmable network nodes which can easily be updated as new versions are defined or protocols modified. We talk about just a few of these protocols for illustration.

One aspect of automation in provisioning involves the configuration of end to end connections. In the past this has been primarily accomplished through manual means, but there are currently efforts underway to define standard signaling protocols such as the GMPLS (Generalized Multi-Protocol Label Switching) suite of protocols [1][2][3], to automate some of this process. One such standards effort is UNI (User-Network Interface) [4], defined at the Optical Internetworking Forum (OIF). In brief, UNI provides an interface by which a client may request services (i.e. establishment of connections) of an optical network. By supporting dynamic connection requests, end to end provisioning can be accomplished.

LCAS [8] is another area where efforts are being made in automation of provisioning. LCAS is a recent SONET based protocol that allows a particular connection to be resized (to adjust the capacity or bandwidth). It utilizes Virtual Concatenation (VC) [9], a method for providing SONET/SDH virtual connections in a variety of sizes, that supports flexibility as well as better bandwidth utilization. Combined, these two mechanisms can support dynamic changes to connections and their capacities, which allows new virtual connections to be easily integrated into the SONET/SDH multiplex, or existing connections to be given more or less bandwidth. Smaller granularities of bandwidth can be supported and increased dynamically, making SONET/SDH a viable alternative to Ethernet for metro carriers. Addition of bandwidth on demand will allow service providers to be much more responsive to transient customer bandwidth needs, enabling better utilization of empty fiber along with addition of premium services for short term bandwidth bursts.

Once connection provisioning can be automated, additional services can be developed that utilize this automation, such as intelligent protection and restoration schemes that do not rely on expensive hardware redundancy, and may provide better restoration by creating fall back routes which avoid points of failure. Network Management Systems (NMS) can take advantage of these services for more resilient and fine grained manageability of the optical network.

### 1.2.2 Traffic Grooming

Another service which has great importance in the next generation optical network, especially for access networks, is traffic grooming. Traffic grooming refers to efficient multiplexing at the ingress of a network. Typically, it is used to group lower-rate traffic flows into higher-rate traffic flows in such a manner that add/drop operations are minimized. Grooming is a composite solution employing various traffic pattern, engineering, topology and routing schemes. Grooming can be employed at MAN gateways to exhaustively utilize bandwidth in an intelligent manner. There are three main components of traffic grooming for next generation optical networks: admission control, traffic management, and LCAS/VC.

Admission control ensures that the customers adhere to their Traffic Conditioning Agreements as specified by their SLAs (Service Level Agreements). This helps to support Authentication, Authorization and Accounting (AAA) of the customers. It also supports policing of the customer traffic flows and enforcement of domain policies. If a customer's flow exceeds the SLA, then a back pressure message (i.e. Ethernet PAUSE flow control message) can be sent to the customer to initiate a slow-down in the rate of traffic.

Once traffic has been authenticated and authorized, traffic management deals with queuing and scheduling of the incoming traffic flows onto the various egress queues available. The scheduler usually doubles as a shaper as well and thereby ensures that the traffic is pumped onto the network based on a profile characteristic to the network.

Use of the LCAS/VC feature of next-generation SONET networks allows the service provider to over-provision bandwidth on existing channels, which ensures rapid provisioning of services to customers. This feature also enables the service provider to add new customers to its clientele without making fork-lift or cumbersome upgrades to the network infrastructure.

Figure 1.1 illustrates a deployment scenario for traffic grooming at a metro gateway where numerous gigabit Ethernet lines are aggregated and provisioned over an outgoing PoS (Packet over SONET) or EoS (Ethernet over SONET) line for transport across the core of the network. Unlike traffic engineering, which is end-to-end, traffic grooming is done primarily at the ingress of the

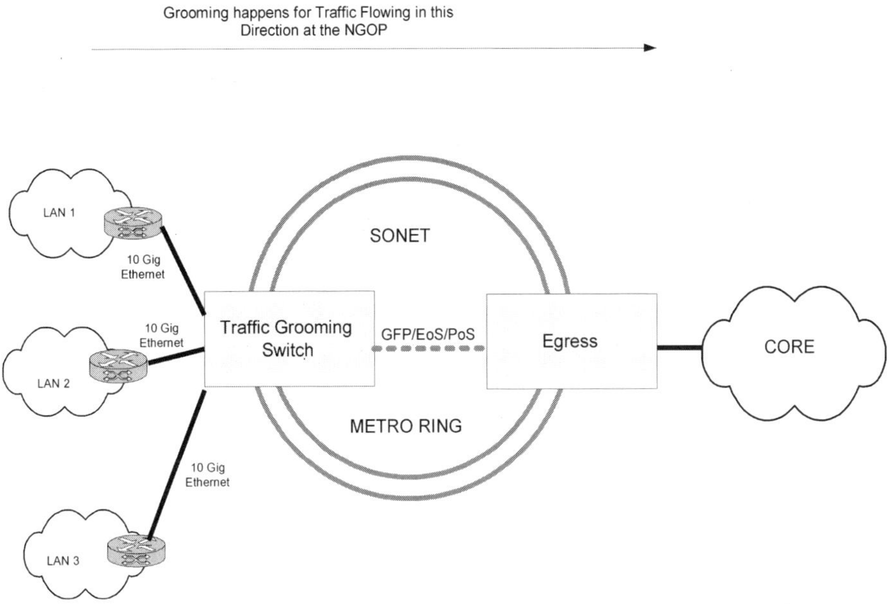


*Figure 1.1.* Traffic grooming switch.

MAN, as this is where a major aggregation of trunk lines and gigabit Ethernet lines happens.

### 1.2.3 Automated Device Control

Network management in optical networks has traditionally been implemented as a centralized control. As complexity in optical devices and networks increases, and the number of managed devices grows, it becomes an increasingly difficult management problem to centralize all functions. Network elements can off-load some of the NMS tasks, if they are capable of handling additional processing. This may include better statistics gathering and alarm/event correlation, support at the device level for some levels of automated provisioning, support for some policy administration at the device level, and higher level, easy to use interfaces for device configuration to ease the work of administrators.

One of the scalability problems with a large optical network is the sheer volume of statistics and events that must be analyzed and processed at the NMS. A single hardware failure can escalate into a large number of alarms which need to be handled with great efficiency to isolate the failure and select a solution or a workaround. A link failure can cause these alarm notifications to be generated from all affected network elements. As the size of the network grows and the number of elements increases, this can swamp a centralized management system. The network element can handle some correlation of multiple alarms and events if it can accommodate the analysis processing. Gathering of fine grained statistics and coherent summarization can be supported at the network element level and propagated in summary form to the NMS. This can relieve some of the obvious scalability problems.

While end to end path provisioning may be better served by a central NMS, some levels of provisioning as well as policy administration can be supported at the network element level. The NMS could delegate select policies for administration directly at the optical device level. This might include SLA information for traffic flows, or certain admission control policies. Local decisions about LCAS initiation and processing based on information provided by the NMS could support lower levels of automated provisioning directly at the device level.

Obviously, as the capabilities of optical network devices become more sophisticated, there is additional complexity in programming these devices. Remote administration, compatibility with existing as well as emerging management interface standards, and high level, easy to use functions with fine grained control are among the requirements of the interfaces supported by the optical network element.

All of these new capabilities and services that will be present in the next generation optical platform impose new demands on the network and the equipment used to support it. In addition, standards for the new protocols and interfaces to support these new services are still in development and subject to industry acceptance. Some of these standards are described in the next section.

### 1.2.4 Standards for Tomorrow's Optical Networks

We briefly describe some of the standards efforts that are geared toward improving automation of provisioning and adding intelligent capabilities to the optical network. The primary efforts in this space are GMPLS and UNI.

**GMPLS Overview.** GMPLS [1-3] is not a single protocol, but a collection of protocols being defined by the IETF, offering a consolidated control plane and extending topology awareness and bandwidth management across all network layers, thus enabling new, more efficient and cost effective core network architectures. The potential of GMPLS is that it makes possible the evolution to peer-based networks where all network elements have information about all other elements. The GMPLS suite of protocols is applicable to all types of traffic and provides mechanisms for data forwarding, signaling and routing on a variety of data plane interfaces. GMPLS enhances MPLS to additionally support Packet Switched Capable interfaces (PSC), Time Division Multiplexing Capable interfaces (TDMC), Lambda Switch Capable interfaces (LSC) and Fiber Switch Capable interfaces (FSC).

GMPLS extends IP technology to control and manage lower layers. In order to establish a connection, the GMPLS control plane is using a routing protocol, e.g. OSPF or IS-IS, to maintain route information, and also a signaling protocol, e.g. RSVP-TE or CR-LDP, to provide the messaging functionality. GMPLS also manages TE (Traffic Engineering) links, where combining multiple data links for routing purposes forms a single TE link, through the use of LMP (Link Management Protocol) running between neighboring nodes.

The intelligent optical network uses GMPLS to dynamically establish, provision, maintain / tear down, protect and restore, groom and shape traffic to make efficient use of SONET/SDH, WDM, or OTN networks. These features allow operators/service providers to offer new services over these networks such as bandwidth-on-demand, efficient traffic grooming, etc. GMPLS supports the establishment of labels with traffic engineering attributes between end-points. A GMPLS domain consists of two or more Label Edge Routers (LER) connected by Label Switched Routers (LSR). Label Switched Paths (LSPs) are established between pair of nodes along the path to transfer packets across the domain. An LSP consists of an ingress LER, one or more LSRs or OCX (Optical Cross-Connect) and an egress LER. For all packets/frames received at the ingress, the LER determines which packets should be mapped to which particular LSP based on packet classifications, i.e. destination address, source address, protocol port, etc. Multiple LSPs can be established between any ingress and egress LER pair. Figure 1.2 shows a GMPLS domain consisting of two LERs and two LSRs.

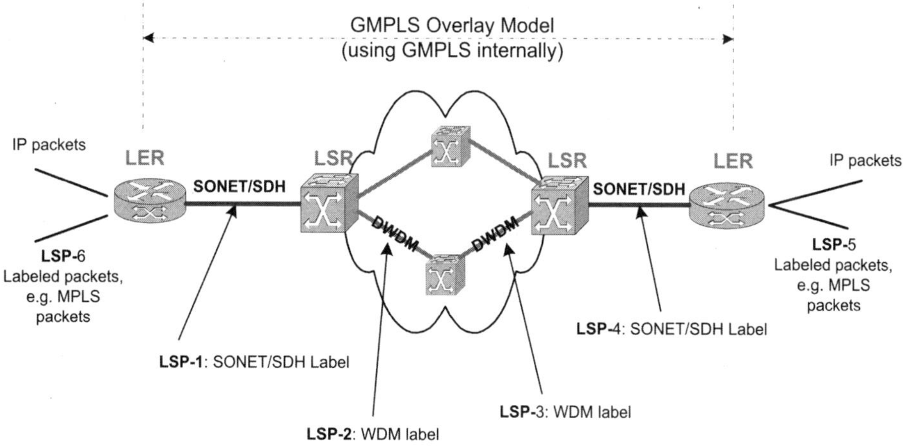


*Figure 1.2.* GMPLS Overlay Model (UNI).

**UNI overview.** New technologies in optical networking such as Dense Wave Division Multiplexing (DWDM) have evolved to provide a cost effective means to increase the bandwidth capacity of optical fiber networks as well as create a new optical network layer that can provide intelligent transport services to allow IP routers, MPLS label switching routers, ATM switches, etc., to interconnect using SONET/SDH or other future interfaces. To support provisioning of end to end connections across multi-vendor networks, a standard method of signaling to create these connections is required. The Optical Internetworking Forum (OIF) has been working toward defining such standards. The standard interface that provides a service control interface between the transport network and the client is called the User- Network Interface (UNI) [4]. The initial specification developed is Optical UNI 1.0.

UNI 1.0 allows a client from the user network to dynamically initiate and establish an optical connection with a remote node using GMPLS signaling. A neighbor discovery mechanism is defined, which allows a client (termed UNI-C) and the network node (termed UNI-N) which supports the client's request, to discover each other. In addition, a service discovery mechanism is included, which allows the client to discover the services offered by an optical network.

UNI 2.0 specification is underway and addresses features such as security, bandwidth modification, extension to physical layers such as Ethernet, and the ability to establish multiple connections with a single request.

While the work with UNI holds promise for paving the way to fully automated provisioning and advanced optical network services and management, it is still very much a work in progress. OIF and IETF are separate standards organizations, and thus may not be completely in agreement or consensus. IETF has not settled on a single signaling protocol for GMPLS, but OIF has selected CR-LDP as the signaling protocol for UNI. IETF supports a peer to peer model for GMPLS, whereas OIF, heavily guided by telecom companies, defines an overlay model. It can be difficult for equipment vendors to decide how to develop their implementations if they wish to become early adopters. The standards are currently moving targets, which means that any implementation of the protocols in these standards needs to accommodate frequent modifications and updates. The most flexible and cost-effective solution for this is to provide as much of the implementation in software as possible.

### 1.2.5 Optical Services: Advantages and Issues

In summary, emerging optical services provide the following advantages and features:

- Automated provisioning for finer grained and more efficient network management

- Traffic grooming for better traffic and customer management

- Network management capabilities to support scalable and more resilient and flexible networks

- Standards to allow inter-vendor interoperation

However, the following obstacles remain:

- Standards specifications are still unstable

- Implementions require increased complexity

- Upgrades to legacy optical equipment may be unavoidable

Next, we talk about network processors and the solution they provide for overcoming these obstacles.

## 1.3 Programmable Silicon and Network Processors

Historically, manufacturers have employed fixed-function ASICs (application specific integrated circuits) to perform packet processing in network devices at line rates. However, ASICs can be expensive to revise if adding or changing protocol functionality, and can be difficult to program. The complexity of required packet processing is directly related to the number and sophistication of the supported protocols. For IP networks, protocols such as IPv6 and MPLS are being added to the required lineup of supported protocols in network devices. For optical networks, the list grows with the addition of protocols such as LCAS, VC, GMPLS, CR-LDP, etc.

Instead of utilizing fixed-function ASICs to support growing and evolving packet processing functions, a network processor is a fast, but more flexible and programmable device that could constitute a better choice. Network processors are capable of processing packets at line rates, but provide users with the programmability of a generic processor. Use of commercial off-the-shelf network processors can eliminate development time for custom ASICs, and better support adaptation to changing customer requirements or evolving standards by allowing software updates [6].

A typical lineup of packet processing functions handled by a network processor could include header classification, deep packet inspection and analysis, packet processing, policing, statistics, and traffic management. The network processor could work in conjunction with a control plane processor to handle control and exception packets that are detected. Routing tables can be stored in SRAM, TCAM, or DRAM. Packets can be stored internally or in an external DRAM. Standard interfaces such as SPI-3, SPI- 4, CSIX [5], etc. can be used to connect network processors to framer or switch fabric devices, or to co-processors such as those used to perform encryption, etc.

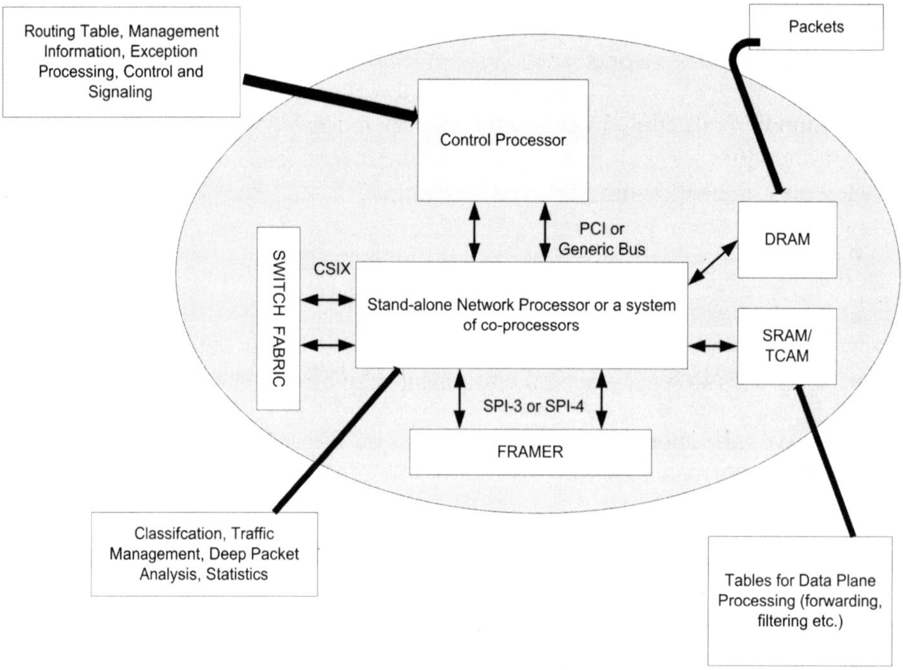


*Figure 1.3.* A generic network processor.

PCI or generic bus interfaces can be used to connect the network processor to a control plane processor. Figure 1.3 shows the architecture of a generic network processor and the functionality of the various components.

### 1.3.1 Network processor programming and software framework

With a software based platform, it becomes possible to create modular and reusable components that can form the basis for more sophisticated processing functions. These components become software building blocks which are aggregated in different ways to suit different and more sophisticated applications. This type of software development strategy can be followed and implemented in software development kits that are made available with network processors.

Such development kits emphasize network elements which support separate but interoperating control and data planes. This allows independent development of control and data path software. Typically, a host network processor, host operating system, higher-layer software, and client APIs run on a control plane processor. Software which runs on the control plane includes protocol processing for routing and signaling, exception handling and control protocols. Data plane software includes packet processing functions which need to handle data at line rates, also known as fast-path code. The fast-path code runs on a network processor. Many network nprocessor vendors supply reference fast-path code as part of their development kits.

Another model that is an integral part of the development strategy is the use of pipelined processing to handle packet data along the fast-path. An ingress software module receives packets from a hardware interface, and passes these along to one or more modules for classification, filtering, policing and shaping before passing to an egress module for transmission on hardware once again. As network processors gain in capability and speed, this type of fast-path handling could grow from simple functions such as filtering and forwarding, to deep packet analysis and more. If these processing modules represent different aspects of protocol handling, they can be replaced, or mixed and matched with future modules for updated or layered protocol handling. Vendors can take modules with standard protocol behaviors and insert their own special processing functions within a pipeline for value added functionality, without having to implement all of the standard behaviors themselves.

The software building block and pipelined processing model has the flexibility and potential to support a suite of network processing and protocol functions that can grow and adapt with the needs of future networks. We next discuss some specific examples of building blocks that can help solve some of the problems with development for next generation optical networks.

## 1.4 Software Building Blocks for the Intelligent Optical Network

The control and data plane separation and the pipelined functional building blocks models as described above provide an architectural platform implemented in software which utilizes the flexibility of a network processor and supports development for the services and applications of the next generation optical networks. Designed well, these functional blocks are modular and reusable, and can be used to build various combinations to suit a wide variety of networking functions. When protocols are updated to reflect changes in evolving standards, or to include value added vendor processing, only the affected blocks need to be updated. Common blocks can be provided with equipment in the form of libraries or reference code, and can be utilized as is, or modified for differences or additions in supported features. In this section, examples will be presented which illustrate in detail the way in which several next generation optical services can be implemented.

### 1.4.1 Traffic Grooming Data Plane Blocks

An example architecture of a traffic grooming switch is shown in Figure 1.4 below. The figure shows control plane based software modules which are re-

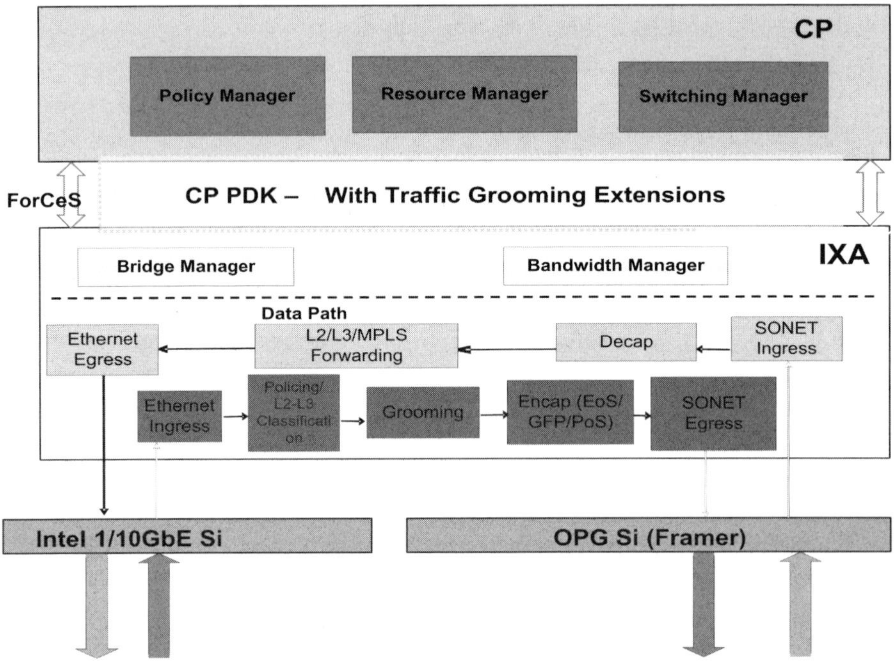


*Figure 1.4.* Traffic grooming architecture.

sponsible for policies, signaling, and management. These communicate with the data plane blocks through the ForCES [7] (Forwarding and Control Element Separation) protocol defined by IETF. The traffic grooming data plane blocks are responsible for authenticating, authorizing, accounting, provisioning of QoS of the incoming packet data, based on information downloaded from the control and management planes. The various data plane components for traffic grooming are shown in Figure 1.5. A brief overview on these blocks is provided below.

**Classifier.** A classifier is a functional data path element which consists of filters that select matching and non matching packets. Based on this selection, packets are forwarded along the appropriate data path within the gateway. Therefore, a classifier splits a single incoming traffic stream into multiple outgoing streams. This traffic grooming solution could employ an IEEE 802.1 p/q based classifier, or a Layer 3 classifier or a MPLS classifier. The IEEE 802.1 p/q classifier is employed for metro gateways with incoming Gigabit Ethernet or 10-Gigabit Ethernet links, as the VLAN ID in the header maps to the client's identity, and the priority bits map to the class of service for the flow.

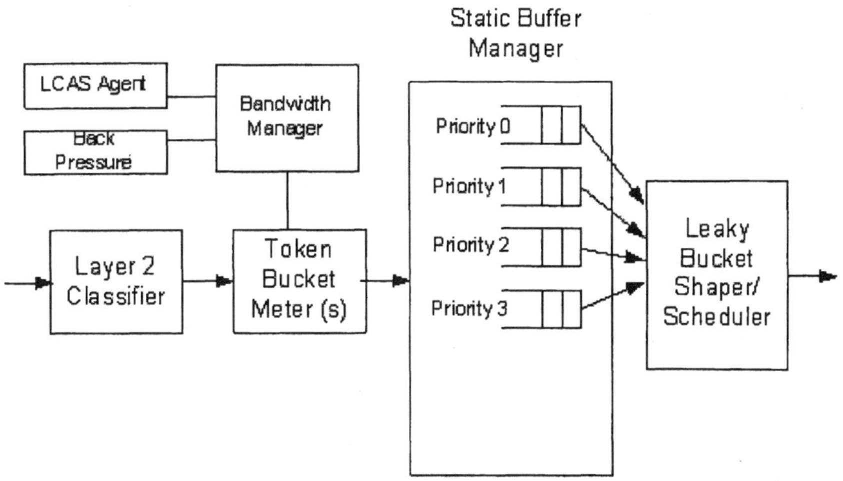


*Figure 1.5.* Traffic grooming data plane blocks.

**Meter.** A meter is a data path functional element that monitors the temporal characteristics of a flow and applies different actions to packets based on the configured temporal profile for that flow. A token bucket meter is used for the traffic grooming solution as it measures the conformance of a flow profile against the traffic profile specified by the SLA.

**Bandwidth Manager.** The bandwidth manager is a control block responsible for: Resource Management Admission Control through Ethernet back-pressure based throttling Provisioning of Bandwidth using LCAS, when over-subscription occurs and excess bandwidth is available.

**Buffer Manager.** A buffer manager is a queuing element which modulates the transmission of packets belonging to the different traffic streams and determines their ordering, possibly storing them temporarily or discarding them. Packets are usually stored either because there is a resource constraint (e.g., available bandwidth) which prevents immediate forwarding, or because the queuing block is being used to alter the temporal properties of a traffic stream (i.e., shaping). A simple, static buffer manager serves the purpose of traffic grooming.

**Scheduler.** A scheduler is an element which gates the departure of each packet that arrives at one of its inputs, based on a service discipline. It has one or more inputs and exactly one output. Each input has an upstream element to which it is connected, and a set of parameters that affects the scheduling of packets received at that input. The service discipline (also known as a scheduling algorithm) is an algorithm which might take any of the parameters such as relative priority associated with each of the scheduler's inputs (or) the absolute token bucket parameters for maximum (or) the minimum rates associated with each of the scheduler's inputs (or) the packet length or 802.1p QoS bits of the packet (or) the absolute time and/or local state as its input.

A leaky bucket based scheduler is employed as it implicitly supports traffic shaping on each outgoing queue according to a leaky bucket profile associated with that queue.

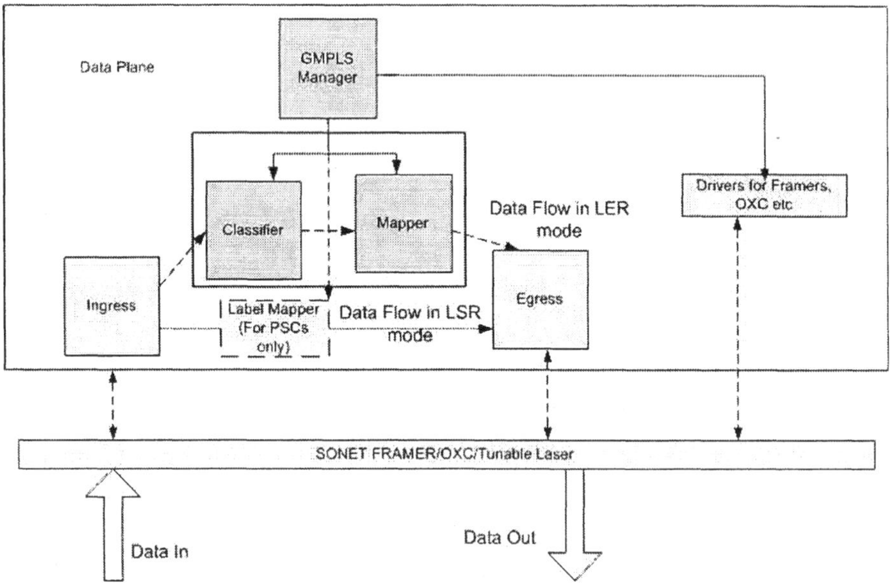


*Figure 1.6.* GMPLS/UNI data plane blocks.

## 1.4.2 UNI/GMPLS Example

An example architecture for a network element which supports UNI / GMPLS is described in this section. At present, UNI / GMPLS is a developing technology with continuously evolving specifications, and illustrates an application which would particularly benefit from a flexible software architecture which supports frequent protocol updates. GMPLS networks support three distinct functionalities, namely, ingress LER processing, LSR switching and egress LER processing. The various GMPLS data plane blocks used to manifest the GMPLS switching functionalities is shown in Figure 1.6. A brief discussion on the functionality of these data plane blocks is provided below.

**GMPLS Manager.** The GMPLS manager populates the FEC (Forward Equivalence Class), NHLFE (Next Hop Label Forwarding Entry) and Incoming Label Map (ILM) tables required for data processing. The GMPLS manager also configures and manages the SONET channels through the driver for the SONET Framer.

**Classifier.** The classifier classifies the packets to a FEC. The FEC could be the destination network prefix, source IP address or five-tuple based. The results of the classification are used for determining the NHLFE for the packet.

**Mapper.** The mapper maps the flow to an outgoing label and an outgoing interface. The outgoing label could be a physical label, a TDM slot or a wavelength. The mapper also provides the next-hop for the outgoing flow.

**Label Mapper.** This module is used in the LSR mode for PSC LSPs only. The label from the incoming packet is mapped to an entry in the ILM to generate the outgoing label which could be a physical label, a TDM slot or a wavelength and the outgoing interface.

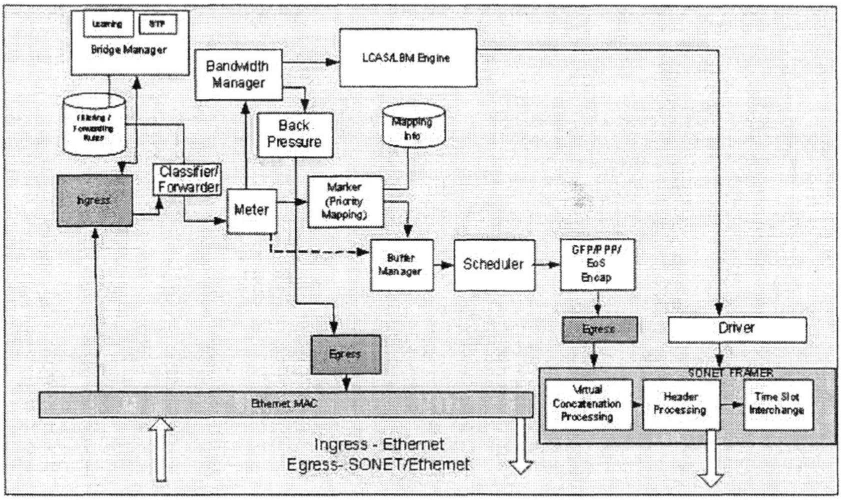


*Figure 1.7.* A programmable Next Generation Optical Platform.

## 1.4.3 Next Generation Optical Platform (NGOP)

A composite data plane solution for a multi-service provisioning platform that provides traffic grooming and multiple types of mappings such as GFP[10], PPP, EoS, etc., is illustrated in Figure 1.7. The figure shows blocks for the traffic grooming functionality in white boxes. Addition of the GMPLS LER building blocks shown in the dotted boxes illustrates easy integration of a new functionality with an existing functionality. This example serves to demonstrate how the network processing building blocks that implement various protocols can be combined into a single architecture. Additional services such as GMPLS based label switching can be easily incorporated into the architecture defined through the integration of software building blocks which implement those specific functionalities. This enables faster time to market of newer services and technologies.

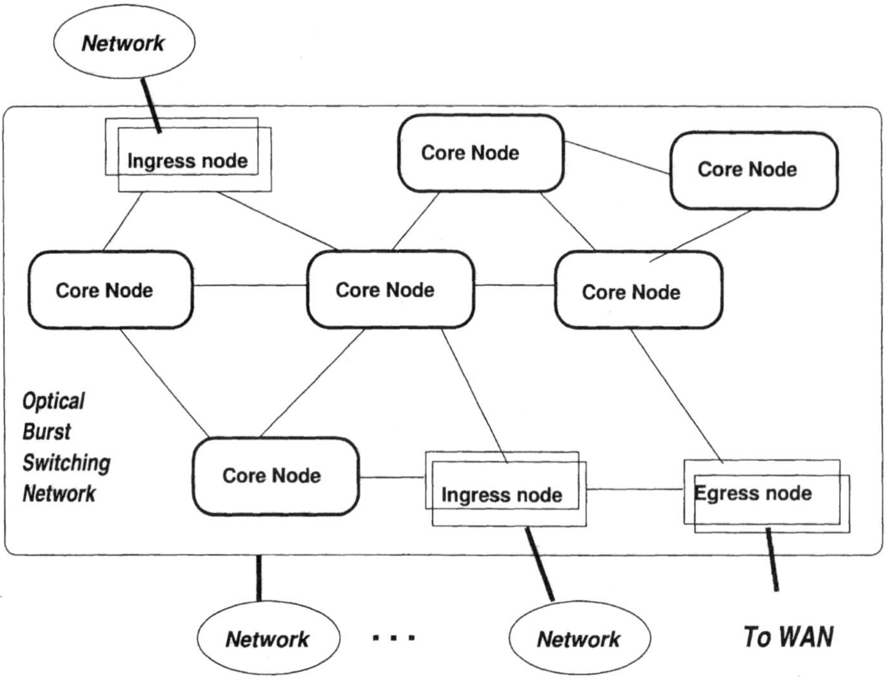


*Figure 1.8.* Optical burst switched network architecture.

## 1.5 Future Applications for the Intelligent Optical Network

In the future, optical network technologies will continue to evolve and gain the intelligence required to provide the level of services required by more sophisticated and demanding customers. The flexibility of a network processor architecture combined with optical hardware and software will continue to provide an appropriate platform for these technologies.

One such example is optical burst switching (OBS), a solution for high speed bursty data traffic over WDM optical networks, such as that which is typically observed around datacenters and in enterprise networks. OBS provides an intermediate granularity between existing optical circuit switching schemes and newer schemes based on optical packet switching. OBS switches on bursts of data of variable length, uses an optical switching fabric, and utilizes an end to end bandwidth reservation scheme [13][14]. Thus, it provides advantages of packet switching in the optical domain, but avoids some of the problems associated with optical packet switching, such as electrical/optical conversion at switching nodes. An OBS network has a combination of ingress and egress nodes at the edges, and switching nodes inside the OBS network, as shown in Figure 1.8. Ingress nodes classify data flows from LAN/WAN/MAN sources and form photonic control and data bursts. Control bursts include the signaling traffic for bandwidth reservation, using GMPLS signaling which has been extended to accommodate burst characteristics. Data bursts include groups of IP packets or Ethernet frames. Intermediate switching nodes and egress nodes process the control bursts to reserve bandwidth on the path.

Egress nodes perform the conversion from photonic bursts to LAN / WAN / MAN traffic once again. All of these nodes require some intelligence to perform control processing and conversion of the data flows into the appropriate traffic types. A network processor with appropriate software modules and hardware interfaces, performs tasks such as classification and other packet processing tasks, as well as burst assembly and framing, burst scheduling, and control processing. Control processing includes extraction of the routing information from the control burst (which can either be sent in-band or out-of-band with the data bursts) [15][16], reservation of the required switch resources on the local node, and formation of the outgoing control burst to the next switching node on the path to the egress node. An example high level diagram of an optical burst switching node is illustrated in Figure 1.9.

The control and data plane architecture and building blocks are shown in Figure 1.10.

In addition to optical burst switches, future optical devices will include smart optical framers with the capability to perform packet inspection or manipulation level processing for ingress or egress SONET data. Many future applications will require intelligence to be pushed toward the network devices. For these applications, software building blocks, network processors and emerging optical hardware can be combined to support the sophistication required in the future.

## 1.6 Enabling the Smart Optical Network

New technologies, standards and development methods are all moving toward a vision of an intelligent optical network. Technology needs to keep pace with demand, from increased bandwidth capabilities, to platforms and proces-

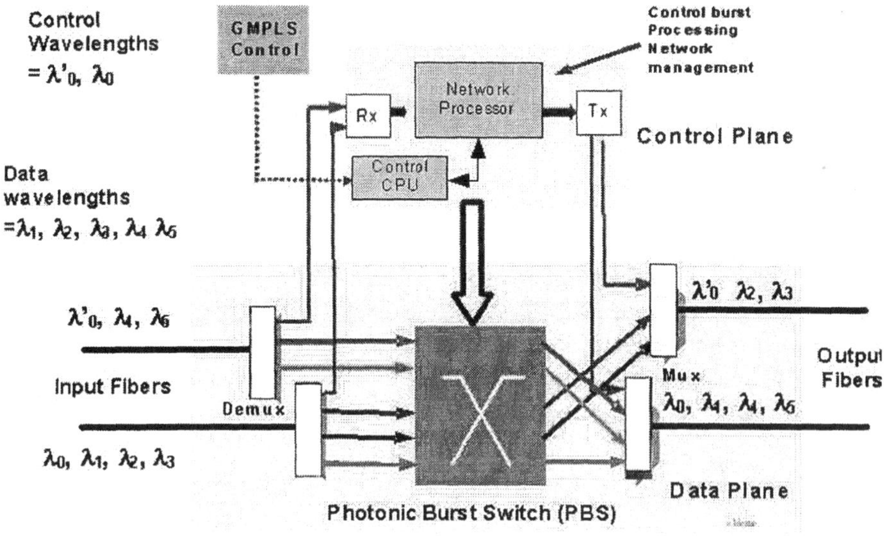


*Figure 1.9.* Block diagram of an OBS switching node.

sors which support more complex and sophisticated services and protocols. In addition, the sheer magnitude of the increase in network size, and quantities of devices and clients, requires better network management automation. At the same time, the current economic environment dictates that carriers and providers need to choose cost-effective and flexible equipment upgrades, that benefit them not only today, but grow to support tomorrow's environment as well. In this article, we have outlined some of the issues and problems which face the evolving optical network, and discussed possible solutions available for developers to choose today.

The obstacles include instability in standards, the need for implementations to support new and complex protocols, and the need to invest in new hardware that continues to provide cost savings as the networks change. We have shown how a network processor based platform can have advantages in cost of equipment, adaptability and flexibility for changing standards and a building block approach to development of software. It is evident that an optical network architecture which is derived by using network processors with software implemented protocols and services, has many advantages that can help ride the economic and technological waves to the next generation intelligent optical network.

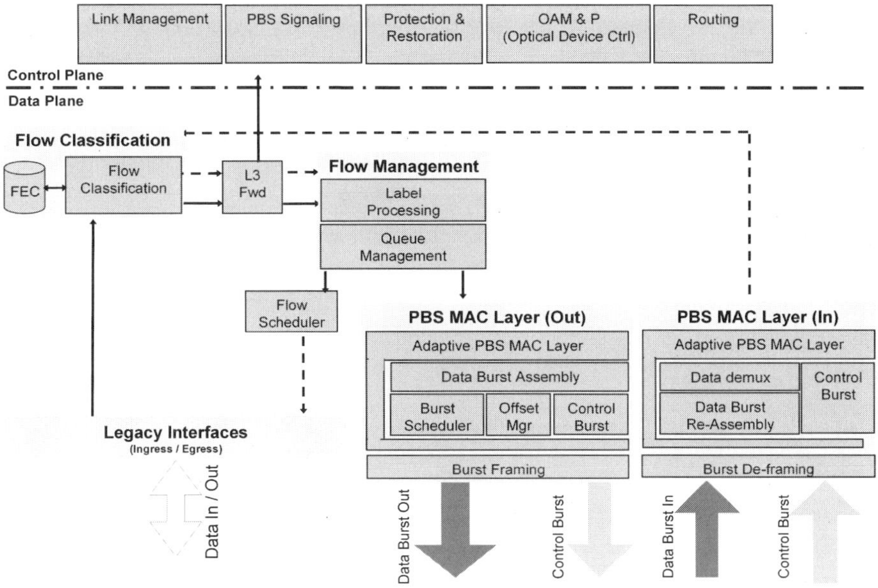


*Figure 1.10.* OBS software architecture and building blocks at ingress / egress nodes.

## Acknowledgements

The authors would like to acknowledge the support and help of the members of the Intelligent Optical Networking team at Intel Research and Development, Hillsboro, OR. The authors also thank Minal Mishra, a graduate student at UMBC for his help with parts of the document preparation.

## References

[1] E. Mannie, Generalized Multi-Protocol Label Switching (GMPLS) Architecture, draft-ietf-ccamp-gmpls-architecture-03.txt, IETF Draft, Work in Progress, Aug 2002.

[2] G. Bernstein, E. Mannie, V. Sharma, Framework for GMPLS-based Control of SDH/SONET Networks, draft-ietf-ccamp-sdhsonet-control-01.txt, IETF Draft, Work in Progress, May 2002.

[3] L. Berger, Generalized MPLS - Signaling Functional Description, draft-ietf-mpls-generalized-signaling-09.txt, IETF Draft, Work in Progress, Aug 2002.

[4] OIF User Network Interface (UNI) 1.0 Signaling Specification, Optical Internetworking Forum, Oct 2001.

[5] NPF CSIX-L1: Common Switch Interface Specification-L1 Specification, Network Processing Forum, Aug 2000.

[6] NPF Software API Framework Implementation Agreement, Network Processing Forum, Sep 2002.

[7] L. Yang, R. Dantu, T. Anderson, ForCES Architectural Framework, draft-ietf- forces-framework-03.txt, IETF Draft, Work in Progress, Oct 2002.

[8] ITU-T G.7042/Y.1305 LCAS for Virtually Concatenated Signals, International Telecommunication Union, Nov 2001.

[9] ITU-T G.707/Y.1322, Network Node Interface for the Synchronous Digital Hierarchy, International Telecommunication Union, Oct 2000.

[10] ITU-T G.7041/Y1303, Generic Framing Procedure, International Telecommunication Union, Jan 2002.

[11] ANSI T1.105-1995, Synchronous Optical Network (SONET) - Basic Description including Multiplex, American National Standards Institute, 1995.

[12] ITU-T G.707, Network Node Interface for the Synchronous Digital Hierarchy (SDH), International Telecommunication Union, Mar 1996.

[13] S. Ovadia, Christian Maciocco, and Mario Paniccia, Photonic Burst Switching Architecture for Hop- and Span- Constrained Optical Networks, Intel Corporation, IEEE Communication Magazine, Nov. 2003.

[14] C. Qiao, Myungsik Yoo, Optical Burst Switching (OBS) A New Paradigm for an Optical Internet, J. High Speed Networks, Vol. 8, No. 1, pp. 69-84, 1999.

[15] J. Wei, and R.I. McFarland Jr., Just-in-time signaling for WDM optical burst switching networks, IEEE/OSA Journal of Lightwave Technology, vol. 18, issue 12, pg. 2019-2037, Dec 2000.

[16] M. Yoo and C. Qiao, Just-Enough-Time (JET): a high speed protocol for bursty traffic in optical networks, Digest of the IEEE/LEOS Summer Topical Meetings, pp 26-27, Aug 1997.

Chapter 2

# HYBRID HIERARCHICAL OPTICAL NETWORKS

Samrat Ganguly[1], Rauf Izmailov[1] and Imrich Chlamtac[2]

[1]*NEC Laboratories America Inc, Princeton, NJ 08540*
[2]*University of Texas at Dallas, Richardson, TX 75083*

Email: samrat@nec-labs.com, rauf@nec-labs.com, chlamtac@utdallas.edu

**Abstract** The chapter presents a comprehensive discussion on the benefits of a hybrid optical WDM network solution while introducing the concept of waveband switching for port cost reduction. The chapter provides a detailed overview of different waveband routing and assignment algorithms that exploits the logical hierarchy in terms of wavebands and wavelengths. Cost effectiveness of the hybrid solution is also discussed through simple analytical and simulation results.

**Keywords:** Optical network, routing, paths, wavelength, waveband, optimization, aggregation, wavelength division multiplexing, optical cross- connects, demultiplexing, multiplexing, optical switches, packing.

## 2.1 Introduction

Telecommunication networks based on optical communication have been constantly evolving for the last decade following the changing industry landscape as shaped by the market conditions, technological innovations and regulatory decisions. Current optical networks [21] are expected to support the increasing network load by employing advanced transmission (wavelength division multiplexing (WDM)) and switching (optical switches and cross- connects (XC)) and routing [2,3] technologies. To cope with the growing data traffic, boosting line capacity is the most obvious strategy, albeit by itself not as attractive as a comprehensive long-term solution. While building trenches and laying multiple fibers helps to reduce the transportation cost, the key cost components and underlying complexity are being shifted to the bottleneck switching and regeneration nodes. In order to satisfy the growing bandwidth demands, more diverse and more intelligent management of capacity is thus required at the network nodes.

Hybrid hierarchical optical cross-connect has a potential to become one of the key elements for finding a comprehensive mid to long-term solution that will enable telecommunication carriers to create, maintain and evolve scalable networks in a cost effective and profitable way. This potential stems from significant capital expenditure savings that can be delivered by hybrid technology, which can replace a large part of expensive opto-electronic fabric with all-optical one [7]. The potential is augmented by hierarchical technology, which further reduces capital expenditure since the same optical port can process multiple wavelengths simultaneously [17]. Finally, the flexible structure of non-uniform wavebands reduces the capital expenditure even more by improving the optical throughput of the node [14].

The next section outlines the main optical technologies that enable hybrid hierarchical optical cross-connects. Section 2.2 describes the routing issues in hybrid hierarchical optical networks. Sections 2.3 and 2.4 describe the concepts of uniform and non-uniform wavebands, their applications and performance implications on node and network levels. Section 2.5 covers general architectural issues of hybrid hierarchical optical cross-connects.

## 2.2 Hybrid Hyerarchical Solution

The concept of hybrid solution is based on the combination of two promising technologies: all-optical (OOO) cross-connects and optical-electrical-optical (OEO) cross-connects. The concept of hierarchical solution includes different granularities of optical switching: from wavelengths to wavebands to fibers. These concepts are elaborated further in this section.

### 2.2.1 Hybrid Solution

All-optical cross-connects and optical-electrical-optical cross-connects are based on different technologies that determine their respective advantages and disadvantages.

*All optical OXC*: All-optical XC transparently switches the incoming optical signal through the switching fabric; the optical signal remains in optical domain when it emerges from the switching fabric. A variety of technologies can be used: arrays of tiny tilting mirrors (MEMS), liquid crystals, bubbles, holograms, and thermo- and acousto-optics.

All-optical XCs are less expensive than OEO-based XCs: they have a smaller footprint; consume less power and generate less heat. All-optical XCs are highly scalable: multiple wavelengths can be handled using the same port. Being naturally protocol-independent and bit-rate independent, they are able to carry services in their native format, providing for a future-proof alternative for OEO-based systems. However, the absence of optical 3R functions (required to clean up accumulated optical impairments) and wavelength conversion (required to resolve wavelength contention) restricts the capabilities of all-optical XCs. The lack of interoperability limits the deployment of all-optical XCs as well. Transparency of optical signals also makes it harder to monitor performance in all-optical XCs.

*OEO OXC*: OEO-based XC converts the incoming optical signal into electrical signal for subsequent switching and grooming; the electrical signal is regenerated as a new optical signal at the output port. OEO-based XCs are based on mature and reliable technology (O/E, Digital cross-connect, E/O). They can provide a variety of functions such as optical 3R (regeneration, reshaping, retiming), grooming and wavelength conversion. Each wavelength is thus processed individually, which becomes expensive as traffic volume grows. OEO-based XC can be controlled with a distributed control plane. The systems from different vendors can interoperate, and standards are already in place to extend the interoperability to the control plane. However, these functionalities need to be supported by expensive hardware, which, being protocol-dependent and bit-rate dependent cannot scale very well. OEO-based XCs are also characterized by large footprint and large power consumption, generating significant amount of heat.

*Hybrid OXC:* A hybrid cross-connect essentially consists of the all-optical (OOO) layer and the electronic OEO layer [20]. The benefits of hybrid solution are derived from leveraging the individual benefits of each of the above technologies [7]. Effectiveness of a hybrid cross-connect is based on the observation that most of the data passing through the optical nodes (as much as 75%) consists of transit traffic. A hybrid cross-connect can use its all-optical part to switch the transit traffic while the OEO part is used for grooming and adding/dropping the local traffic. Electronics is thus used to perform necessary and expensive processing of the traffic, while optics is used for inexpensive and transparent forwarding. In a WDM optical network, dominated largely by the switching cost, hybrid solution has the potential to reduce the average switching cost of wavelengths.

## 2.2.2 Hierarchical Solution

Main advantage of a hybrid cross-connect is based on the possibility of having waveband switching at the all-optical layer [18,26]. In waveband switching, multiple wavelengths forming a wavcband arc switched using single input port thus reducing the port complexity [8,19,4].

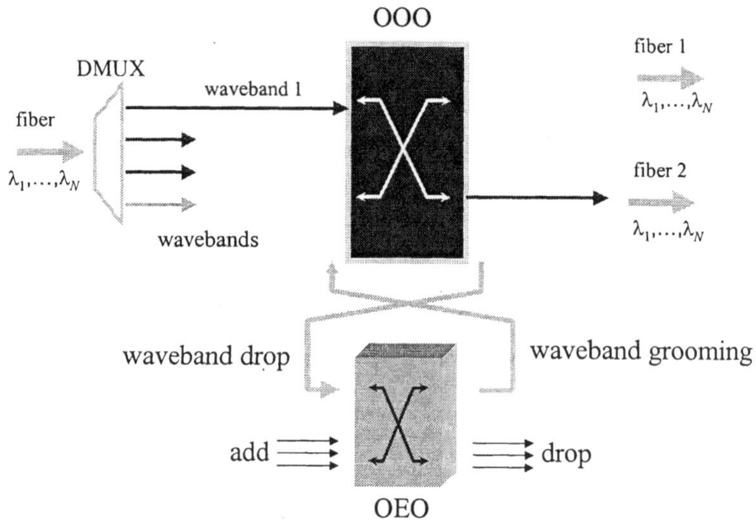


*Figure 2.1.* Waveband switching in hybrid cross-connect.

A simple architecture of a hybrid cross-connect with waveband switching is shown in Figure 2.1. The incoming wavelengths from an input fiber are grouped into a set of wavebands where each waveband consists of contiguous wavelengths. Each of these wavebands can be switched optically at the OOO layer to an output fiber (in Figure 2.1, waveband 1 from input fiber 1 is switched to output fiber 2). Transparent switching of a waveband from input fiber to output fiber is possible if all the individual wavelengths that are part of this waveband needs to be switched to the same output fiber (transit traffic). In the other case, the waveband is dropped to the OEO layer for individual wavelength processing/switching. The OEO layer apart from adding and dropping individual wavelengths also functions in grooming a waveband, which is then forwarded to the OOO layer for further switching.

The cost advantage of waveband aggregation is shown in Figure 2.2. Let each waveband be formed by aggregating $G$ wavelengths. Each waveband consists of contiguous wavelengths (the waveband number $m$ contains all the wavelengths with numbers from $(m-1)G+1$ to $mG$.

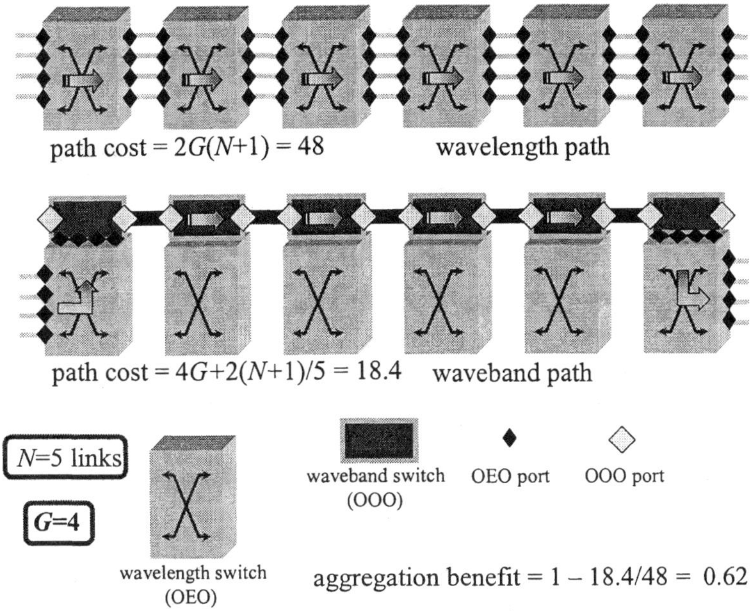


*Figure 2.2.* Aggregation benefit in a hybrid cross-connect.

Assume that the cost of an OEO port is five times the cost of an OOO port and define the total cost in terms of cost of a single OEO port. In a linear network of $N$ links or $N$+1 nodes, the cost of routing $G$ wavelength paths using OEO ports requires $2G(N+1)$ ports. In the same network, if the wavelengths are aggregated or groomed into a waveband and switched using OOO layer, the total cost is $4G+2(N+1)/5$. At the origination node, $2G$ OEO ports are required to aggregate the wavelengths into the waveband and similarly at the termination node resulting in $4G$ ports. Routing this waveband requires *two* OOO ports at each node resulting in the cost $2(N+1)/5$. The aggregation benefit is defined as the port reduction by employing waveband aggregation. In the example shown in Figure 2.2, the aggregation benefit is 62%.

### 2.2.3 Problems and Challenges

The waveband aggregation benefit is maximized if all wavelength paths can be aggregated into waveband at source node and switched through OOO layer at each intermediate node. In such a case, OEO ports are required only for adding and dropping wavelengths at termination nodes. Unfortunately, such an arrangement is difficult to attain for realistic traffic patterns where there may not exist demands in multiples of $G$ wavelengths for each source destination pair. Aggregation and waveband switching also may not be possible due to the waveband continuity (no waveband conversion is available at OOO layer). As a result, when there is a waveband contention at a node, one of the wavebands needs to be dropped to OEO. The amount of aggregation benefit then depends upon how the wavelength paths are routed (including primary and protection paths [25], where waveband aggregation is performed and what are the waveband assignment strategies. The main challenge, therefore, is in maximizing total aggregation benefit for a given network topology and traffic demand matrix based on efficient routing and provisioning strategies. The next two sections discuss these issues for uniform and non-uniform wavebands.

## 2.3 Hierarchical Routing on Optical Cross-Connects

Routing in WDM networks towards minimizing the cost of wavelength conversion has been extensively studied as in the landmark papers [3,22]. In an optical cross-connect that is equipped with multi-granularity switching [9] where wavelengths, wavebands and fibers can be switched. Such a switching leads to tunnels of various sizes between node pairs. Routing in such a network was studied in [9]. Detailed ILP formulation for optimization of the routing cost was proposed in [17,26]. The goal of this routing is reduction of the switching cost, which is accomplished by trying to accommodate traffic demands by using larger size tunnels as much as possible.

An online routing algorithm for dynamically changing traffic scenarios is proposed in [9]. In this algorithm, for a given traffic request between source and destination nodes, a combination of tunnels of different sizes is assigned (the cost of using a tunnel with larger size is lower since it uses less switching resources). Since creating new tunnels leads to reserving additional capacity along the tunnel, the routing of new lightpaths is directed towards using already established tunnels as much as possible. The problem of finding the optimal combination of routes has exponential complexity, so the proposed algorithm orders the tunnels based on their costs and then assigns tunnels in that order to the lightpaths while discarding other tunnels that conflict with the already assigned tunnel.

The offline version of the tunnel allocation assumes that the traffic matrix is known a priori. The offline version is therefore useful at the network planning stage. A simple heuristic algorithm was proposed in [17] where a two level hierarchy (wavelength and waveband) was considered. In this algorithm, lightpaths are first classified based on their destinations and then sorted based on their hop counts. Subsequently, for each class, $G$-1 lightpaths in the sorted order are grouped into a waveband based on a common segment and assigned a waveband number.

In another offline algorithm proposed in [9], multi-level hierarchy was considered. In this algorithm, each node is given two weights: potential and sink. The node with the largest *potential* has the highest priority to be the ingress of a tunnel, and the node with the largest *sink* has the highest priority to be the egress of a tunnel. Potential of a node is determined based on the total weights $W(l)$ of the outgoing links $l$ from the node. $W(l)$ for a link $l$ indicates the probability of using this link based on the traffic matrix and $k$ shortest paths between source and destination. For example, if a link belongs to $m$ alternate paths between a source and a destination pair with traffic load $T$, then $W(l)$ is $mT$.

Based on weights assigned to nodes, ingress-egress pairs of nodes are determined by sorting the nodes on weights. In creating such a pair, ingress node is taken as a node with a high potential and egress as a node with a high sink. Tunnels are then assigned between ingress and egress nodes of pairs in a sequential manner. After the tunnels are formed, lightpaths from the traffic matrix are assigned tunnels such that each tunnel is used to the fullest extent. Further details about the above-described algorithms can be found in [9]. Offline algorithm for hierarchical routing in multi-fiber case can be found in [26].

## 2.3.1 Hierarchical Routing on Hybrid Cross-Connects: Uniform Wavebands

Hybrid cross-connects with switching at two (OOO and OEO) layers create the opportunity to aggregate "express" optical wavelengths into wavebands (also called "fat pipes" or "super channels") that can be optically (transparently) switched for most part of their path through the network without being segregated into separate wavelengths at every node. The optical paths thus form a hierarchy in which higher-layer paths (waveband) consist of several segments of lower layer paths (wavelengths). The resulting hierarchy of wavelengths and wavebands is a mixed one: logical wavebands coexist on the same fiber with individual wavelengths. On the level of paths, one can view a wavelength path as a set of nodes traversed by the path and switching element used at each node (OOO or OEO). A waveband path can also be defined as set of nodes traversed using only OOO layer.

## 2.3.2 Routing Algorithms

A variety of waveband routing algorithms were explored in the literature [9,12,10]. Based on the sequence of routing and waveband aggregation, waveband routing algorithms can be classified into two models: *integrated routing* and *separate routing* [23]. Under the integrated routing model, the routes are computed for both the wavelength *and* waveband paths in the same module. The routing and resource allocation decision is based on the status of the advertised wavelength and waveband resources. Integrated routing can achieve a more cost effective use of resources, although it may require a more complex routing algorithm.

Under the separate routing model, there are two independent modules acting on wavelength and waveband layers, respectively. First, the wavelength module computes the routes for the traffic demands taking into account wavelength resources. Next, the waveband module determines the possibility of wavelength aggregation into wavebands taking into account waveband resources. If needed, wavelengths may have to be reassigned to create a new waveband path between intermediate nodes.

Depending on the number of hybrid cross-connect nodes, the network can be classified into two cases: *homogeneous* and *heterogeneous*. In the homogeneous case, all nodes are hybrid hierarchical OXCs, where a waveband path can be set up between any two nodes. In the heterogeneous case, some of the nodes are hierarchical while others are not. Since the latter nodes do not have waveband switches, a waveband path can only be set up between two hierarchical nodes.

Further, routing strategies can be explored for *offline* and *online* cases. Since the current wavelengths can carry a lot of aggregated data traffic, and traffic demands change infrequently, only the offline case is considered here. Routing strategies in the offline case assumes that the complete traffic demand is known a priori. The offline routing algorithm [24] works for general topologies in the following way.

First, a wavelength path along the shortest (in terms of hop count) route from the source node to the destination node is created. Then, starting from a large value *MaxHops* (the number of hops in the longest path) the algorithm tries to locate a common segment of length *MaxHops* belonging to *G* wavelength paths. If it is found, the corresponding waveband is created and the search is continued until no such segments could be found. In the subsequent steps, *MaxHops* is decreased by 1 and the procedure is repeated until *MaxHops* reaches the minimum value of 2. Since a waveband path formed over one link does not use any optical ports, the procedure is terminated for *MaxHops*=2.

Since the maximum aggregation cost benefit is created by longest wavebands, the offline algorithm attempts to maximize the cost benefits by locating the longest possible waveband paths first. The order of segments analyzed by the algorithm (longest segments first) facilitates the creation of longer wavebands. Figure 2.3 illustrates the algorithm for *G*=4. Wavelength path LP4 is set up after paths LP1,...,LP4 are computed. The longest waveband path is found for *MaxHops*=3 as shown by BP1 in Figure 2.3.

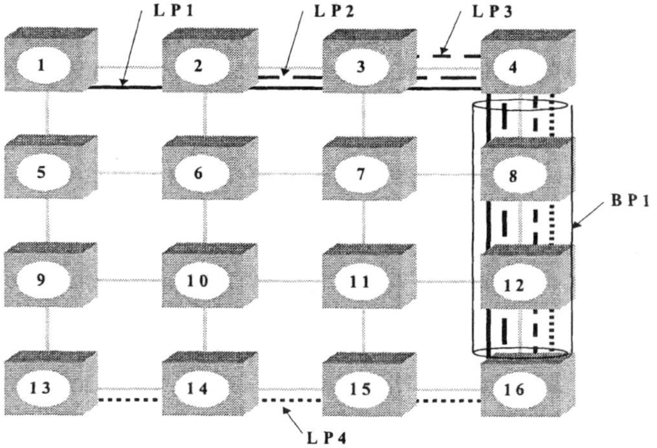


*Figure 2.3.* Offline waveband routing algorithm in mesh network.

### 2.3.3 Performance in Ring Networks

The performance of waveband routing and aggregation algorithm can be analyzed on a ring consisting of $2M+1$ nodes [23], one of which (node 0) functions as a hub, receiving and sending traffic to all other nodes (Figure 2.4). Suppose that each node of the ring establishes the same number $L$ of wavelength paths ($L$ is the load parameter) to the hub node 0. Then the traffic generated by the upper half of the ring (Figure 2.5) contains $L$ wavelength paths of $M$ links (which use $2(M+1)L$ ports), $L$ wavelength paths of $M-1$ links (which use $2ML$ ports), etc. In total, these paths utilize $D=LM(M+3)$ ports.

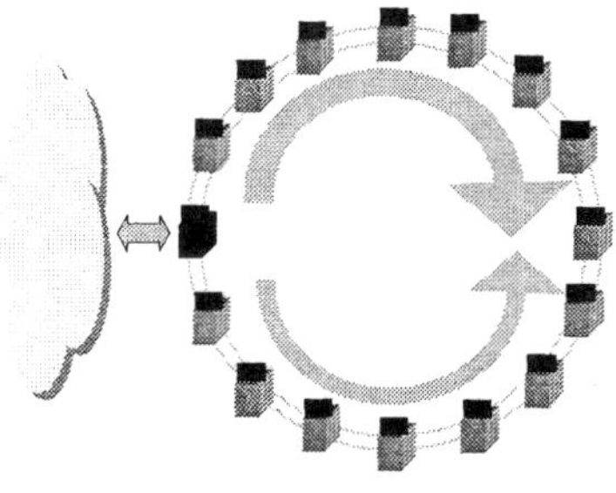

*Figure 2.4.* Ring network with a single hub.

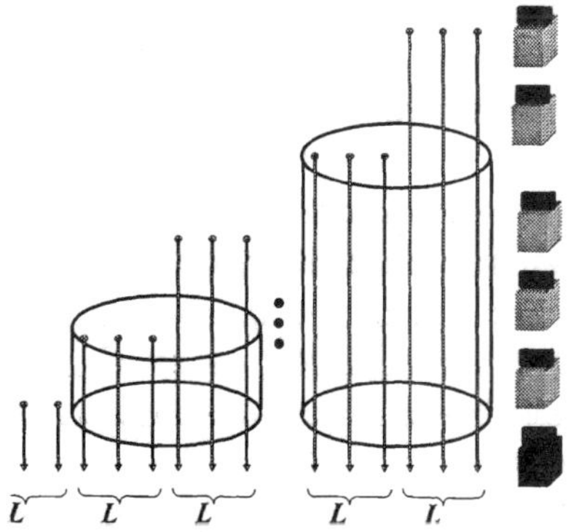


*Figure 2.5.* Wavelength and waveband paths in a ring.

For simplicity, suppose that $G$ is divisible by $L$: $G=kL$. Then, the aggregation of traffic flows from $k$ most remote nodes (from the hub) produces a waveband from node $M-k+1$ to the hub node 0. This waveband uses $2(M-k+1)+4G$ additional optical ports, while releasing $2G(M-k+1)$ OEO ports.

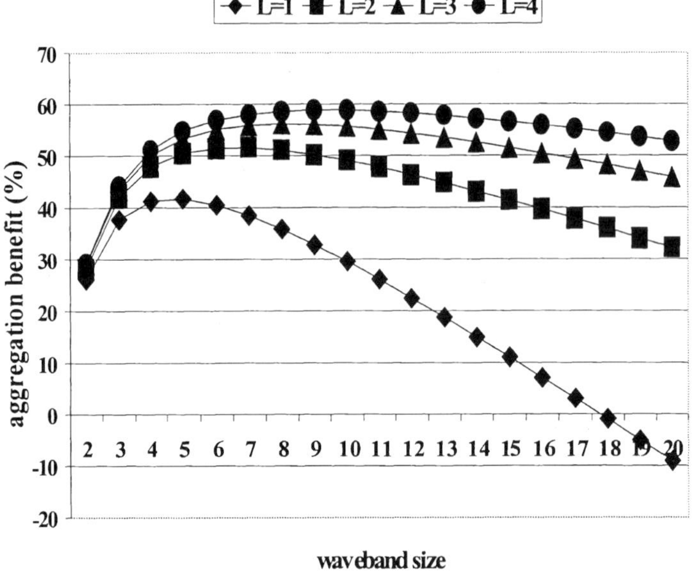


*Figure 2.6.* Aggregation benefit for various $G$ and $L$ ($M$=20).

Similarly, aggregation of flows from node $M-2k+1$ to the hub node 0, which uses $2(M-2k+1)+4G$ additional optical ports, releases $2G(M-2k+1)$ OEO ports, etc. In total, these $M/k$ wavebands use $(M/k)((M+2G+k+2)$ additional optical ports and release $(M/k)G(M-k+2)$ OEO ports. Since $k=G/L$, the reduction of number of ports is

$$R = M - GM - 2LM(1+1/G) + LM^2(1-1/G).$$

To compute the aggregation benefit, the reduction $R$ is divided by the default ports count $D$. The resulting aggregation benefit reaches its maximum

$$A_{\max} = \left((M-2)L + 1 - 2\sqrt{L(M+2)}\right)/\left((M+3)L\right)$$

for the optimum waveband size $G_{\max} = \sqrt{L(M+2)}$. Figure 2.6 shows the behavior of aggregation benefit $A$ for various $G$ and $L$ (for $M$=20).

As granularity $G$ increases to $G_{\max}$, the aggregation benefit first increases and then falls off. The optimal granularity $G_{\max}$ slowly increases (as $\sqrt{L}$) with the load $L$. It also slowly increases (as $\sqrt{M+2}$) with the size (equal to $2M+1$) of the ring. For rings containing 10, 20, and 40 nodes, where each node sends $L$=6 wavebands to the hub, the optimal waveband sizes $G_{\max}$ are 6, 8, and 11, respectively. If the load $L$ increases, the maximum benefit $A_{\max}$

converges to $(M-2)/(M+3)$. For rings containing 10, 20, and 40 nodes these benefits are 37%, 62%, and 78%, respectively.

## 2.3.4. Performance in Mesh Networks

Simulations of offline routing algorithm were carried out using the topology of European optical network (EON) shown in Figure 2.7. The results [23] show the cost performance benefit by employing the above offline algorithm on is shown in Figure 2.8.

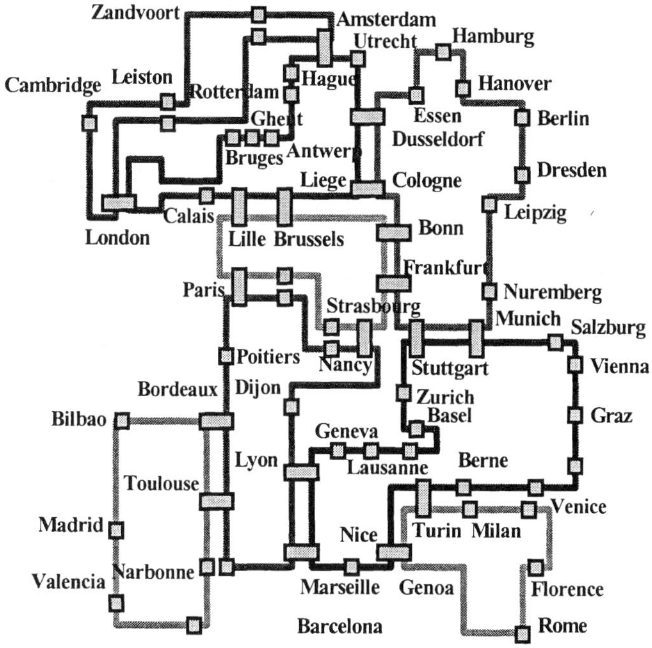


*Figure 2.7.* European optical network (EON).

Simulation scenario assumes that each link can carry up to 160 wavelengths and each hybrid hierarchical node had full wavelength conversion capability at OEO layer. Since the wavelengths could be aggregated into wavebands consisting of $G$ wavelengths each, the granularity $G$ determines the maximum number $160/G$ of wavebands on a link. The horizontal axis of Figure 2.8 shows different traffic load created by scaling the traffic demand.

The simulation results shows how aggregation benefit depends upon the size of the waveband $G$. As can be observed, at low load (60%), $G$=10 provides maximum aggregation benefit, while at higher load (90%), $G$=8 is better. At lower load, due to small number of wavelength paths, there are not enough wavelengths paths sharing same links that can be packed into a large size waveband. Thus at lower load, small waveband size is appropriate and the situation is reversed at higher load.

More recent studies [16] have been carried out for online routing algorithms in mesh topologies. The performance results are consistent with those obtained for offline routing.

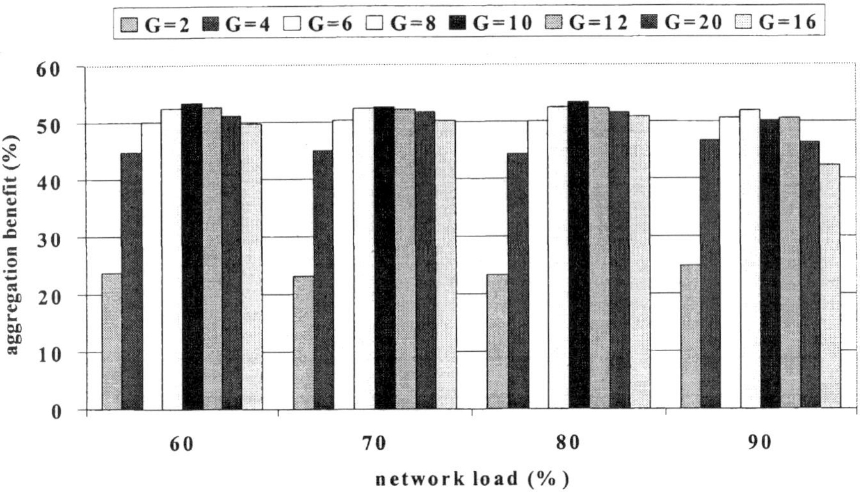


*Figure 2.8.* EON: offline algorithms on load.

## 2.4 Hierarchical Routing on Hybrid Cross-Connects: Non-Uniform Wavebands

### 2.4.1 Non-Uniform Wavebands

Aggregation of wavelengths into uniform wavebands (each comprised of $G$ wavelengths), introduces the aggregation overhead adversely affecting the hierarchical node's performance. Consider an optical switching node with $M$ output fibers and suppose that the input fiber carries $N$ wavelengths to be switched to any of $M$ outputs. Depending on the breakdown of $N$ input wavelengths among $M$ output ports (i.e., the set of the numbers of wavelengths switched to each output ports), the packing efficiency of their aggregation into wavebands may vary.

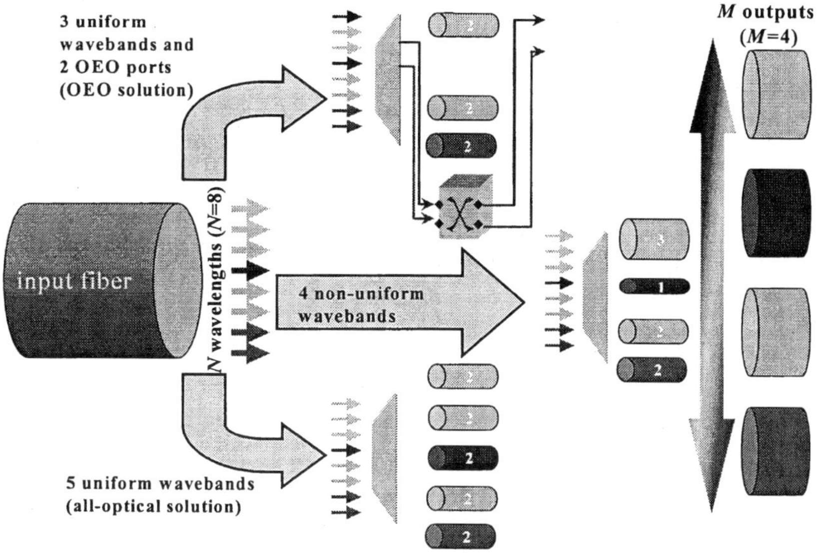


*Figure 2.9.* Example of uniform and non-uniform wavebands.

Consider the example in Figure 2.9. It shows an input fiber carrying $N$=8 wavelengths that have to be switched into $M$=4 output fibers (shown as four pipes in the right side of Figure 2.9). The numbers of wavelengths to be switched to the four output fibers are equal to (3,1,2,2). In *Figure*, the wavelengths to be switched to the same output fiber are painted in the same color as that of the output fiber; for example, the upper three "light" wavelengths are to be switched to the uppermost "light" output, one "dark" wavelength is to be switched to the "dark" output, etc. Figure 2.9 shows the uniform waveband granularity $G$=2: the wavelengths can be aggregated into pre-configured uniform wavebands of the size of two wavelengths each.

In this example, two switching solutions can be employed. In the first approach (*OEO solution*, shown in the upper part of Figure 2.9), two expensive OEO ports are used to switch two of the wavelengths in the OEO layer, while three wavebands are used to switch the remaining wavelengths. In the second approach (*all-optical solution*, shown in the lower part of Figure 2.9), five wavebands are used to switch all the traffic in the all-optical domain. However, the same wavelength demand (3,1,2,2) could have been switched optically if the wavebands had been pre-configured in the way shown as *non-uniform solution* (shown in the middle part of Figure 2.9): two wavebands containing two wavelengths each, one waveband containing three wavelengths and one waveband containing onc wavelength.

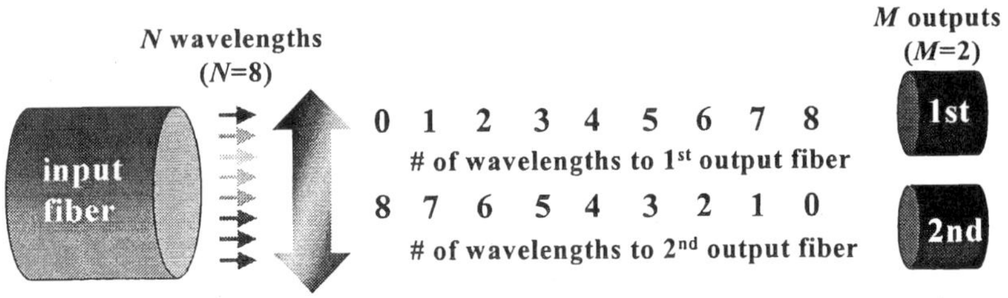


| wavelengths directed to 1st fiber | wavelengths directed to 2nd fiber | uniform wavebands $2\lambda+2\lambda+2\lambda+2\lambda$ | | All-optical cost | OEO cost | non-uniform wavebands $1\lambda+1\lambda+2\lambda+4\lambda$ | | cost |
|---|---|---|---|---|---|---|---|---|
| 0 | 8 | 0 | 2+2+2+2 | 4wb | 4wb | 0 | 1+1+2+4 | 4wb |
| 1 | 7 | 2 | 2+2+2+2 | 5wb | 3wb+2oeo | 1 | 1+2+4 | 4wb |
| 2 | 6 | 2 | 2+2+2 | 4wb | 4wb | 2 | 1+1+4 | 4wb |
| 3 | 5 | 2+2 | 2+2+2 | 5wb | 3wb+2oeo | 1+2 | 1+4 | 4wb |
| 4 | 4 | 2+2 | 2+2 | 4wb | 4wb | 4 | 1+1+2 | 4wb |
| 5 | 3 | 2+2+2 | 2+2 | 5wb | 3wb+2oeo | 1+4 | 1+2 | 4wb |
| 6 | 2 | 2+2+2 | 2 | 4wb | 4wb | 1+1+4 | 2 | 4wb |
| 7 | 1 | 2+2+2+2 | 2 | 5wb | 3wb+2oeo | 1+2+4 | 1 | 4wb |
| 8 | 0 | 2+2+2+2 | 0 | 4wb | 4wb | 1+1+2+4 | 0 | 4wb |

*Figure 2.10.* Switching costs of uniform and non-uniform wavebands.

Figure 2.10 illustrates all breakdowns of $N$=8 wavelengths between $M$=2 output fibers and the equipment costs required to carry those wavelengths (for uniform and non-uniform wavebands). Depending on the particular approach, uniform wavebands approach requires up to five wavebands (all-optical solution) or three wavebands and two OEO ports (OEO solution) to carry all possible traffic loads, whereas the approach based on non-uniform wavebands consistently requires only four wavebands for all traffic distributions. In general, depending on the cost assumptions, mechanisms of filling the wavebands, etc., the packing improvement provided by non-uniform wavebands can be anywhere between 20% and 40%.

The non-uniform wavebands framework is a unique optical solution that expands the flexibility of wavelength aggregation. While uniform wavebands can be compared to "one-size-fits-all fat pipes", non-uniform wavebands are in fact, "diverse pipes", which can be manipulated and assigned in a way that creates a system of appropriately sized all-optical shortcuts through the network with minimum aggregation overhead.

The example shown in Figure 2.10 gives rise to the following two issues. The first one, *the waveband selection problem*, is how to pre-configure a minimum set of wavebands that can be used to represent an arbitrary breakdown of input flow of $N$ wavelengths into $M$ output fibers. The second one, *the waveband assignment problem*, is how to assign these pre-configured wavebands for optical switching of $N$ wavelengths into $M$ output fibers. Both problems can be formally defined in the context of partition theory [11].

The optimal answer to the waveband selection problem is given the following algorithm where $B$ denotes the set of optimal waveband sizes.

*Waveband Cover Construction (WCC)*

1) Input the parameters $N$ and $M$.
2) Create the set $B=\{\varnothing\}$.
3) Assign $N^* = \lceil N/M \rceil$.
4) Add the element $N^*$ to the set $B$.
5) Assign $N = N - N^*$.
6) If $N = 0$, stop. Else go to step 3.

*Example.* Let $N$=6, $M$=2. Following WCC algorithm gives $B$={1,2,3}. The wavebands from $B$ can represent any arbitrary breakdown of $N$ wavelengths into $M$ fibers:

{0, 6} is covered as 6 = 3 + 2 + 1;
{1, 5} is covered as 1 = 1, 5 = 3+2;
{2, 4} is covered as 2 = 2, 4 = 3+1;
{3, 3} is covered as 3 = 2 + 1, 3 = 3.

The assignment problem is solved using the following algorithm. Let $V=\{v_1,v_2,\ldots,v_M\}$ denote an instance of a breakdown of $N$ wavelength into $M$ fibers. Let $B=\{b_1,b_2,\ldots,b_k\}$ denote the optimal waveband sizes. Both $V$ and $B$ are ordered sets with elements are given in descending order.

*Waveband Cover Assignment (WCA)*

The input $(N,M)$-partition $V=\{v_1,v_2,\ldots,v_M\}$ is stored as a heap. At the $j^{\text{th}}$ step of the algorithm, the following steps are taken.

1) The topmost element $V_t$ of the heap is deleted from the heap.
2) The element $b_j$ is assigned to $V_t$.
3) The element $V_t - b_j$, if it is non-zero, is inserted into the heap.

The $j^{\text{th}}$ step of the algorithm includes a deletion and an insertion to the heap; it requires $O(\log_2 M)$ time. Since these steps are carried $k$ times in order to assign all $b_j$ for $j = 1,\ldots,k$ the overall assignment can be completed in $O(k \log_2 M)$ time.

## 2.4.2. Performance of Switching Nodes

Consider now the switching throughput of non-uniform wavebands for the case of an optical switching node with $M$ output fibers with $N$ input wavelengths from a given fiber. Depending on the breakdown of wavelengths among the output fibers, it may or may not be possible to aggregate them into wavebands for optical switching.

Denote the number of input wavelengths to the hierarchical node by $N$. Also, denote the number of wavelengths that can be aggregated into wavebands and switched in OOO by $P$. The *switching throughput* $S$ is then defined as $P/N$. The switching throughput thus refers to the ratio of the wavelengths that can be transparently switched to the total number of input wavelengths. The switching throughput is the key component of the aggregation benefit: if more traffic can be switched transparently, the number of expensive OEO ports can be reduced.

Consider first the example in Figure 2.11. It shows an input fiber carrying $N$=8 wavelengths that have to be switched into $M$=4 output fibers. Let the number of wavelengths switched to four output fibers are (3,1,2,2) respectively. The case of using uniform wavebands with size two is shown in Figure 2.12. Since $P$=6 wavelengths can be aggregated into wavebands and switched optically, $S_U$=6/8. However, the same wavelength demand (3,1,2,2) could have been switched optically if the wavebands had been preconfigured in the way shown in Figure 2.12: two wavebands containing two wavelengths each, one waveband containing three wavelengths and one waveband containing one wavelength. In this case, $P$=8 results in $S_{NU}$=1.0.

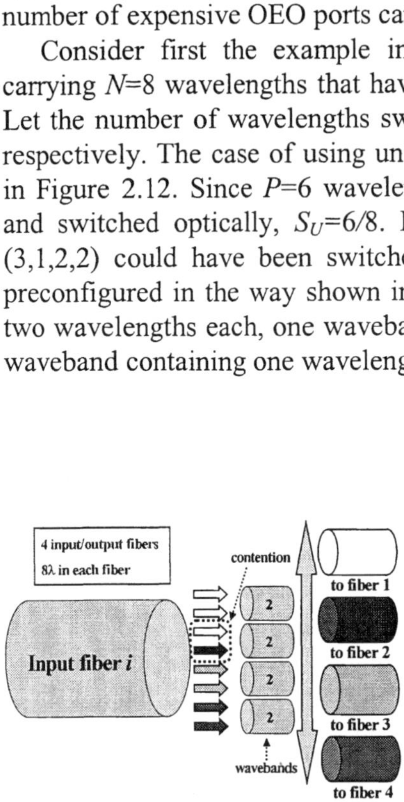


*Figure 2.11.* Example of uniform wavebands.

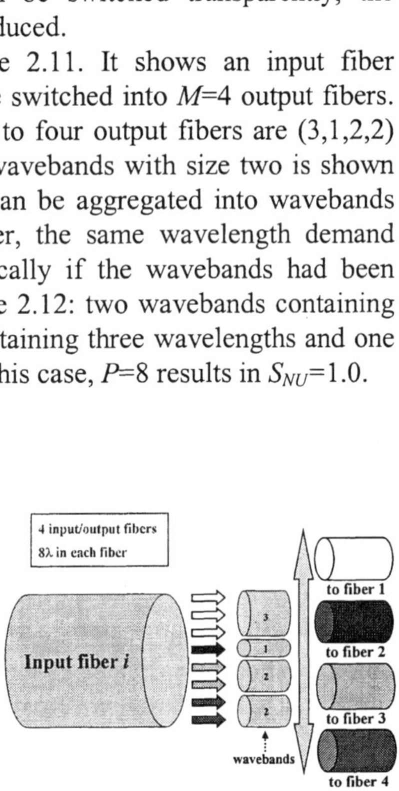


*Figure 2.12.* Example of non-uniform wavebands.

In order to compare the performance of non-uniform wavebands versus that of uniform wavebands, consider a switch with a single input port receiving $N$ wavelengths that are switched to $M$ output ports. In this model, input wavelengths that are aggregated into wavebands are switched optically in OOO. A valid aggregation of $L$ wavelengths into a waveband has to meet the following two conditions.

1) There exists an unused waveband of size $L$.

2) All $L$ wavelengths are switched to the same output ports.

Consider a single node with different numbers of output ports (4,6, and 8) serving an input fiber with 40 input wavelengths. The input wavelengths are randomly switched to the outputs ports. For fairness in comparison, consider uniform and non-uniform wavebands having the same number $K$ of bands.

The results for uniform traffic distribution are shown in Figure 2.13. They demonstrate that non-uniform wavebands consistently deliver superior switching throughput, which translates into a significant increase in aggregation benefits of the hierarchical optical switching. As the number of wavebands increases, the switching throughput for both uniform and non-uniform wavebands also increases. The difference between the respective throughputs is larger when higher number of *wavebands* switching wavebands throughput wavebands are used. For example, with 4 output ports, at $k = 4$, the difference between the throughputs is 5.15% whereas at $k = 10$, the difference is 15.7%.

| number of wavebands | *wavebands* | | | | | | | | | | | switching throughput |
|---|---|---|---|---|---|---|---|---|---|---|---|---|
| | **4 output ports** | | | | | | | | | | | |
| K=4 | NU | 13 | 9 | 9 | 9 | | | | | | | 60.90% |
| | U | 10 | 10 | 10 | 10 | | | | | | | 55.75% |
| K=5 | NU | 9 | 9 | 9 | 7 | 6 | | | | | | 81.10% |
| | U | 8 | 8 | 8 | 8 | 8 | | | | | | 67.60% |
| K=8 | NU | 10 | 7 | 7 | 5 | 4 | 3 | 2 | 2 | | | 95.70% |
| | U | 5 | 5 | 5 | 5 | 5 | 5 | 5 | 5 | | | 80.45% |
| K=10 | NU | 10 | 8 | 5 | 5 | 4 | 3 | 2 | 1 | 1 | 1 | 99.90% |
| | U | 4 | 4 | 4 | 4 | 4 | 4 | 4 | 4 | 4 | 4 | 84.20% |
| | **6 output ports** | | | | | | | | | | | |
| K=4 | NU | 15 | 9 | 8 | 8 | | | | | | | 36.98% |
| | U | 10 | 10 | 10 | 10 | | | | | | | 17.78% |
| K=5 | NU | 9 | 9 | 8 | 8 | 6 | | | | | | 59.93% |
| | U | 8 | 8 | 8 | 8 | 8 | | | | | | 42.28% |
| K=8 | NU | 7 | 7 | 6 | 6 | 5 | 4 | 3 | 2 | | | 90.00% |
| | U | 5 | 5 | 5 | 5 | 5 | 5 | 5 | 5 | | | 70.97% |
| K=10 | NU | 8 | 7 | 6 | 6 | 4 | 3 | 2 | 2 | 1 | 1 | 94.90% |
| | U | 4 | 4 | 4 | 4 | 4 | 4 | 4 | 4 | 4 | 4 | 77.60% |
| | **8 output ports** | | | | | | | | | | | |
| K=4 | NU | 14 | 14 | 6 | 6 | | | | | | | 30.00% |
| | U | 10 | 10 | 10 | 10 | | | | | | | 4.70% |
| K=5 | NU | 14 | 8 | 6 | 6 | 6 | | | | | | 42.00% |
| | U | 8 | 8 | 8 | 8 | 8 | | | | | | 18.07% |
| K=8 | NU | 7 | 7 | 6 | 5 | 5 | 4 | 3 | 3 | | | 83.70% |
| | U | 5 | 5 | 5 | 5 | 5 | 5 | 5 | 5 | | | 59.25% |
| K=10 | NU | 7 | 6 | 5 | 5 | 4 | 4 | 3 | 3 | 2 | 1 | 90.70% |
| | U | 4 | 4 | 4 | 4 | 4 | 4 | 4 | 4 | 4 | 4 | 69.93% |

*Figure 2.13.* Switching throughput for uniform traffic.

As the number of output ports increases, the switching throughput decreases since the routing of wavelengths becomes more diverse (wavelength divergence) and less suitable for aggregation. For example, consider $M = 4$ and $M = 8$ with $k = 10$. In this case, the difference between the throughputs is 15.7% for $M = 4$, while it is 20.77% for $M = 8$. This example illustrates that non-uniform wavebands can handle wavelength divergence in a more efficient manner than uniform wavebands.

### 2.4.3. Non-Uniform Wavebands in Ring Network

In this section, the Waveband Cover Assignment algorithm is analyzed in case of a single hub metropolitan ring topology. Ring topology was studied in the context of wavebands in [5]. Specifically, consider the ring network that is used as an access network, where all the traffic is from the access nodes in the ring to the hub and routed along shorted path. Our objective here is to aggregate the lightpaths from access nodes to the hub into wavebands, in a way that minimizes the overall port costs.

Consider a ring with $M$ access nodes and a single hub. For each access node, let $L_1$ and $L_2$ be the two incoming and outgoing links (all links are assumed to be able to accommodate $N = 160$ wavelengths), in the clockwise and counterclockwise directions, respectively (Figure 2.4). For a given access node $i$ denote by $T_1(i)$ the traffic load (number of wavelengths) from node $i$ to the hub using the link $L_1$; similarly, define $T_2(i)$ as shown in Figure 2.4. As a result, there are two sets (clockwise and counter-clockwise) of traffic flows denoted by $S_j = \{T_j(i) | i = 1,\ldots,M\}$, for $j$=1,2 (shown in Figure 2.4).

Consider now the waveband aggregation for each of the two sets $S_j$. Denote by $D$ the number of all wavelengths for a given set $S_j$; in other words, $D=T_j(1)+\ldots+ T_j(M)$. If $D=N$, the problem of aggregating $N$ wavelengths converging to a single hub from $M$ access points is similar to the problem of arbitrarily breaking down $N$ wavelengths into $M$ ports. Therefore, a pre-configured set of wavebands $B$ can be constructed (as discussed in Section 2.2) so that any breakdown of wavelengths is covered. Given the set of wavebands $B$, the wavebands are assigned to each traffic flow $T_j(i)$), for $i=1,\ldots,N$, using the WCA algorithm described in Section 2.2. The same method for aggregation is used in the case of the other set $S_j$. Because of the way of waveband construction and assignment, it is ensured that each traffic wavelength path is aggregated into a waveband that originates at the corresponding access node and extends to the hub node. As a result, the only OEO ports used are those at the source (access) nodes for add/drop purpose; the remaining paths are completely in the optical domain.

If $D<N$, the problem is slightly modified in the following way. Assign imaginary traffic load $R=N-D$ from the node closest to the hub. The problem then becomes the same as the one analyzed in the previous paragraph and can be solved in the same way. For computing the overall performance, the wavebands that are assigned to the imaginary traffic are excluded from cost computation.

The performance of the described approach was simulated on four rings of different sizes ($M$=10,20,30 and 40) using the same assumptions of OEO and OOO port costs a in Section 0. For simplicity, assume the traffic is routed along the shortest path to the hub node. For each of the four rings, multiple instances of traffic matrices were simulated by assuming that each node sends to the hub node a random number of wavelengths.

Simulation results are shown in Figure 2.14. This figure shows the cost benefit (relative cost reduction, as compared to the uniform wavebands) that can be achieved using non-uniform wavebands. For all simulated scenarios (ring sizes, loads and traffic distributions), the cost benefit provided by non-uniform wavebands is consistently within the range 25%-30%. This cost benefit is smaller than the average value observed in Section 2.3. The node-level benefit of non-uniform wavebands demonstrated in Section 2.3 becomes smaller in network-level scenarios due to relative aggregation efficiency of ring networks.

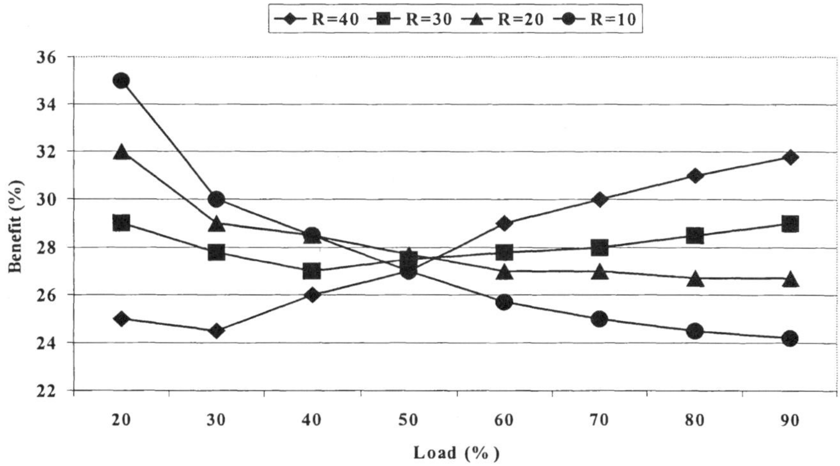


*Figure 2.14.* Ring networks: aggregation cost benefit.

## 2.4.5. Non-Uniform Wavebands in Mesh Network

In the previous section, non-uniform wavebands were used to aggregate wavelength flows from the access nodes to the hub. With a few modifications, the same approach can be applied to general mesh topology. Consider a specific node $i$ that receives non-zero wavelength flows from a number $M(i)$ of sources. The wavelength path for each flow connects the node $i$ through one of its incidence links. Denote by $N(i)$ the maximum number of wavelengths paths that use any of the incidence links of node $i$. In these definitions, $N_{max} = \max_i[N(i)]$, _I and $M_{max} = \max_i[M(i)]$. WCC algorithm with $N = N_{max}$ and $M = M_{max}$ can now be employed to find the optimal set of non-uniform wavebands to be used in the hybrid nodes of the mesh network.

Since the waveband cover construction is obtained for the worst case scenario, one or more wavebands can always be used to aggregate traffic from all sources to the single destination (by following the WCA algorithm as in the case of a ring network). This strategy will not work with multiple destinations for the following reason. Consider two nodes *i* and *j* that are receiving traffic flows from a source *s* and assume that the wavelength paths for these flows share a common link. There might arise a case where the same waveband needs to be assigned to traffic from *s* to *i* and from *s* to *j*, something that will make the waveband assignment infeasible.

The above case can be dealt if it is assumed that wavebands do not have to be completely filled: a waveband of size *G* does not necessarily have to have exactly *G* active wavelengths. The assumption of incomplete wavebands is a natural step in network evolution: by provisioning incomplete wavebands, the benefits of all-optical layer can be obtained from the start, while the wavebands can be filled as the traffic load increases. Under this assumption, WCA algorithm can be modified so that assigns the maximum size waveband that is available in the entire path from source to destination. The formal description is as follows.

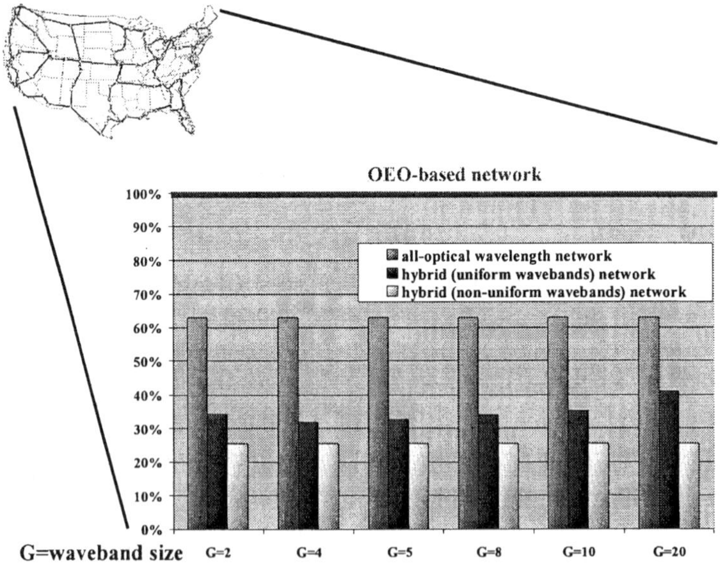


*Figure 2.15.* Mesh network: aggregation cost benefit.

First, WCC algorithm is used create the set of wavebands *B*. Next, for any undirected link *l* of the network, we denote by *A(l)* the set of available wavebands in the link. Initially, *A(l)* = *B* for all links in the network. Let *T*[*i*,

$j$] denote the total number of wavelengths used by the traffic flow from node $i$ to node $j$. Next, the following steps are performed for assignment.

*Waveband Assignment in Mesh Networks (WAM)*

1) Find $i_{max}$, $j_{max}$ for which $T[i, j]$ is the maximum.
2) If $T[i_{max}, j_{max}] \leq 0$ stop; else goto step 3.
3) Let $L$ be the set of links from $i_{max}$ to $j_{max}$.
4) Find the largest waveband size $B_{max}$ available in all links in $L$.
5) Assign $B$max to traffic from $i_{max}$ to $j_{max}$.
6) Update $T[i_{max}, j_{max}] = T[i_{max}, j_{max}] - B_{max}$.
7) Exclude the waveband $B_{max}$ from all links in $L$.
8) Goto step 1.

Simulation of the WAM algorithm is performed for a general US network (shown in Figure 2.15) with relative traffic distribution proportional to the population in the source nodes. Four scaled versions of the traffic distribution were simulated (matching the doubling of traffic volume every year). The port cost using wavebands is discussed in Section 0.

The results shown in Figure 2.15 compare the cost benefits that could be obtained by replacing OEO cross-connects with hybrid hierarchical nodes with both uniform and non-uniform wavebands. In consistency with the previous sections, the results illustrate the competitive cost reduction provided by non-uniform wavebands under different traffic load conditions. In particular, the advantage becomes greater as the traffic load increases.

## 2.5 Architecture and Technology

The preceding sections discussed various cost and performance characteristics of hybrid hierarchical cross-connects. In this section, general architecture is described in order to show how to implement such cross-connects using existing optical technology.

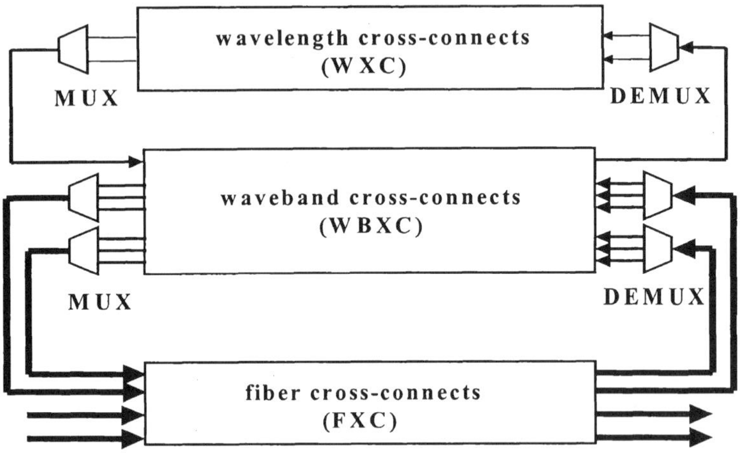


*Figure 2.16.* Hierarchical cross-connect node.

A general architecture for optical cross-connect with multi-granularity switching without using OEO can be found in [17] and is show in Figure 2.16. In this architecture, if the switching cannot be performed in the fiber layer at FXC, the fiber input is forwarded to next waveband layer WBXC where the fiber input is demultiplexed into wavebands. Similarly, if the wavebands cannot be switched at WBXC, they are forwarded to WXC or the wavelength layer for final switching.

An example [1] of hybrid hierarchical optical node is presented in Figure 2.17. It shows an OXC with $M$ input and $M$ output fibers, each carrying $N$ wavelengths $\lambda_1,\ldots,\lambda_N$. The wavelengths in each input fiber are deaggregated (using waveband deaggregators) in $K$ wavebands $B_1,\ldots,B_K$. The optical processing of wavebands is done at $K$ waveband switches forming the OOO part of the hybrid OXC. A waveband $B_i$ at any fiber is connected to a corresponding waveband switch $i$, which optically switches only wavebands with index $i$ from input to output fibers. Similar architecture was proposed in [17].

The optical processing of wavebands may not be possible. A contention for the same output fiber among different wavebands cannot be resolved in the waveband part of the device (in waveband switch). The waveband switch also cannot process a waveband if different wavelengths in it have to be switched into different output fibers. For these and other related tasks (such as adding a wavelength into a waveband), one or more wavebands have to be dropped to the OEO part of OXC.

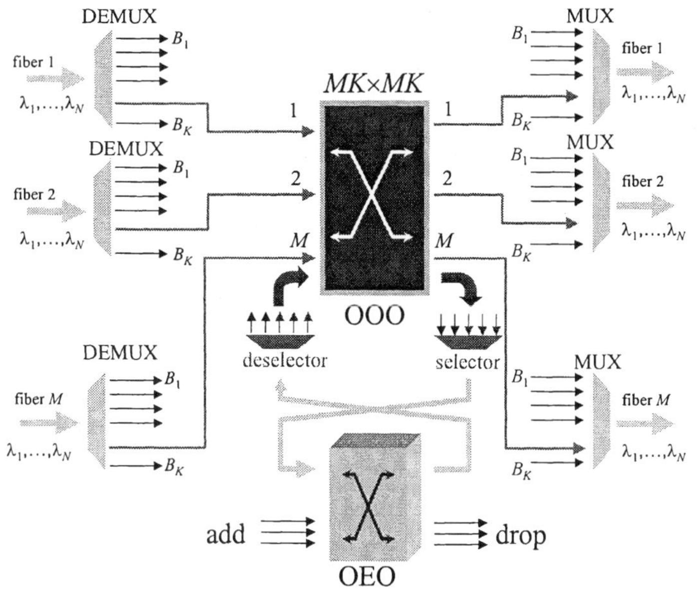


*Figure 2.17.* Hybrid hierarchical cross-connect node.

With properly designed routing and resource assignment algorithm, most of the incoming wavelengths are being switched by the all-optical (OOO) part of the hybrid, while the OEO part of the hybrid handles add/drop traffic along with contention resolution and signal regeneration.

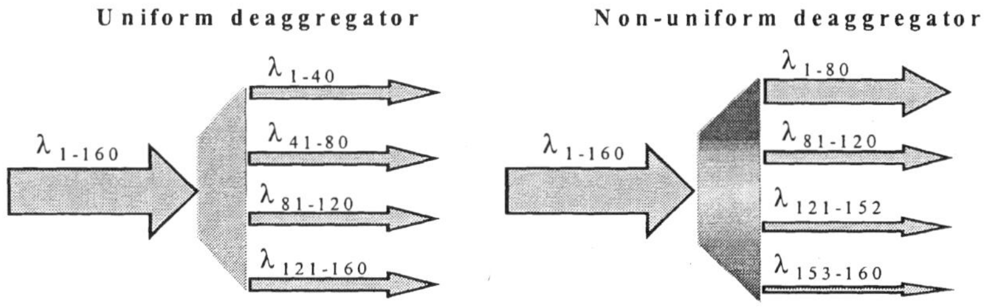


*Figure 2.18.* Waveband deaggregators.

As shown in [15], the performance of this architecture, under general assumptions of wavelength switching requirements, exhibits the characteristics similar to those discussed in Section 2.4: there exist a natural granularity of wavebands (around 4-6 wavelengths per waveband) delivering the best cost-performance ratio.

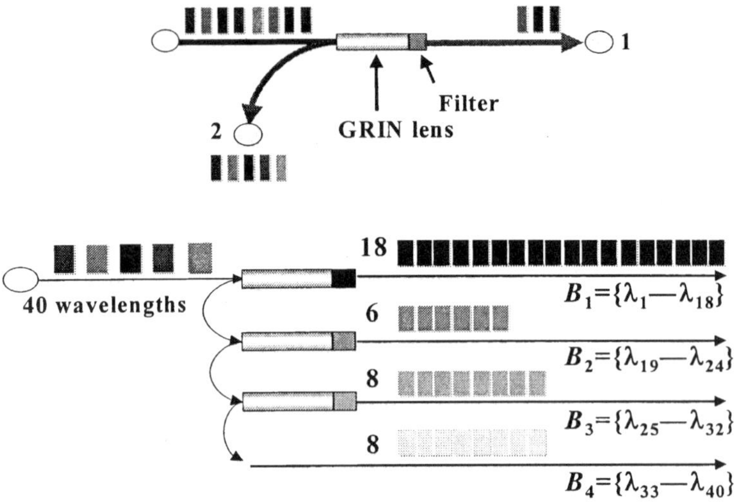


*Figure 2.19.* Three-port filters and waveband deaggregators.

A WDM DEMUX separates the set of wavelengths on an incoming fiber into a number of wavelength subsets. These wavelength subsets can be uniform or non-uniform fixed groups (wavebands), shown in Figure 2.18. A DEMUX subsystem may produce both fixed and arbitrary wavelength subsets. Non-uniform waveband deaggregators can be used instead of DEMUX in hybrid hierarchical optical cross-connects (shown in Figure 2.17), further reducing the cost of the optical cross-connect node by employing flexible set of wavebands containing different number of wavelengths.

In order to maintain the best optical performance, the preferred way to separate waveband and wavelengths is a three-port filter, which allows some of the wavelengths to pass through and reflects the rest. A three-port optical wavelength selective component is shown in the upper part of Figure 2.19. The component consists of three optical fibers, a self-focused GRIN lens, and a thin-film interference filter. The upper part of Figure 2.19 shows wavelength band separation, where a wideband DWDM filter passes a band of three wavelengths to output fiber 1 and reflects all others back into output fiber 2.

Three-port filters can be aggregated into a multi-stage system (waveband deaggregator), which delivers the desired waveband separation. The architecture of the waveband deaggregator is a cascade of (non-uniform) bandpass operations and recombinations, as illustrated in the lower part of Figure 2.19: the incoming set of 40 wavelengths is being separated into four fixed wavebands $B_1,\ldots,B_4$ containing 18, 6, 8 and 8 wavelengths respectively. The waveband sizes are determined by the bandpass operations.

## 2.6 Summary

In this chapter, we outlined the concept and potential advantages hybrid hierarchical approach for optical networking. By leveraging the strengths of opto-electronic and optical technology along with flexibility and cost-efficiency of employing wavebands, this approach carries the promise of becoming a part of the scalable and profitable solution for long haul and metro core networks.

Hybrid hierarchical optical nodes have a potential of providing various benefits to optical networks. The hybrid technology promises significant capital expenditure savings by replacing a large portion of expensive OEO part with optical one (along with reduction of power consumption and the footprint). The hierarchical technology further reduces capital expenditure since the same optical port can process multiple wavelengths simultaneously. The non-uniform wavebands reduce the capital expenditure even further by improving the optical throughput of the node and provide for diverse and flexible system of waveband paths. The all-optical part of the hybrid, being bit-rate free and protocol free, provides operational expenditure savings. Wavebands can be created and filled up with wavelengths in service, thus simplifying the network management, providing seamless upgrade and further reducing the operational expenditure. The all-optical part of the hybrid also provides necessary future proofing, while the electronic part delivering a seamless migration path to all-optical networks.

While promising a high switching and routing performance in a cost-efficient manner, the concept of non-uniform wavebands comes with new challenges. On the node level, new architectural solutions have to be explored for further improvement of switching performance. On the network level, new elements (uniform and non-uniform wavebands) are introduced into the traditional control and design problems. For hybrid hierarchical optical networks, the relationship between various components of optical control plane becomes tighter and requires intelligent routing and resource (wavelength and waveband) assignment for optimal performance.

## References

1. Araki, S. Ganguly, G., Izmailov, R., Maeno, Y., Suemura, Y., and Wang, T. (2002). Hybrid Cross-Connects and Hierarchical Optical Networks. *NEC R&D Journal*, April 2002.
2. Chlamtac, I, Ganz, A. and Karmi, G. (1992). Lightpath Communications: A Novel Approach to High Bandwidth Optical WAN-s, *IEEE Transactions on Communications*, Vol. 40, No. 7, July 1992.
3. Chlamtac, I, Farago, A and Zhang, T. (1996). Lightpath (Wavelength) Routing in Large WDM Networks, *IEEE Journal on Selected Areas in Communications jointly with IEEE/OSA Journal of Lightwave Technology*, Vol.14, No. 5, June 1996.
4. Ciaramello, E. (2000). Introducing Wavelength Granularity to Reduce the Complexity of Optical Cross Connects. *IEEE Photonic Technology Letters*, Vol. 12, No. 6, June 2000, pp. 699-701.
5. Faure, J., Noirie, L., Bisson, A., Sabouret, V., Leveau, G., Vigoureux, M., and Dotaro, E. (2001). A scalable transparent waveband-based optical metropolitan network. *In: ECOC 2001*, vol. 6, pp: 64-65.
6. Ganguly, S., Izmailov, R., Wang, T., Araki, S., and Maeno, Y. (2002). Non-uniform wavebands and hierarchical optical networks. *In: NFOEC 2002.*
7. Gerstel, O., Ramaswami, R., and Foster, S. (2002). Merits of hybrid optical networking. *In: OFC 2002.*
8. Harada, K., Shimizu, K., Kudou, T., and Ozeki, T. (1999). Hierarchical Optical Path Cross-Connect Systems for Large Scale WDM Networks. *In: OFC 1999.*
9. Ho, P. and Mouftah, H. (2001). Path Selection with Tunnel Allocation in the Optical Internet Based on Generalized MPLS architecture. *In: IEEE GLOBECOM 2001.*
10. Ho, P. and Mouftah, H. (2002). Routing and wavelength assignment with multigranularity traffic in optical networks. *Journal of Lightwave Technology*, vol. 20(8), Aug 2002, pp: 1292 –1303.
11. Izmailov, R., Ganguly, S., Kleptsyn, V., and Varsou, A. (2003). A. Non-Uniform Waveband Hierarchy in Hybrid Optical Networks. *In: IEEE INFOCOM 2003.*
12. Izmailov, R., Ganguly, S., Maeno, Y., Suemura, Y., Nishioka, I., and Araki, S. (2002). All-Optical Mesh Networks: Waveband Aggregation and Transmission Impairments Aware Routing. *In: COIN-PS 2002.*
13. Izmailov, R., Ganguly, Wang, T., Araki, S., and Maeno, Y. (2002). Switching performance of non-uniform wavebands in hierarchical optical networks. *In: OECC 2002.*
14. Izmailov, R., Ganguly, Wang, T., Suemura, Y., Maeno, Y., and Araki, S. (2002). Hybrid hierarchical optical networks. *IEEE Communications Magazine*, Volume 40 Issue 11, Nov 2002, pp 88 –94.
15. Izmailov, R., Kolarov, A., Fan, R., and Araki, S. (2002). Hierarchical optical switching: a node-level analysis. *In: IEEE Workshop on High Performance Switching and Routing 2002.*
16. Kolarov, A., and B.Sengupta, Waveband Routing and Wavelength Assignment in Hybrid Hierarchical Optical Networks. *In: ITC 2003.*
17. Lee, M., Yu, J., Kim, Y., Kang, C.H., and Park, J. (2002). Design of Hierarchical Crossconnect WDM Networks Employing a Two-Stage

Multiplexing Scheme of Waveband and Wavelength. *IEEE Journal on Selected Areas in Communications*, Vol. 20, pp. 166-171, No. 1, 2002.

18. Lingampalli, R., Vengalam, P. (2002). Effect of wavelength and waveband grooming on an all-optical networks with single layer photonic switching. *In: OFC 2002.*
19. Myungmoon, L., Jintae, Y., Yongbum, K., and Jinwoo, P. (2001). WDM network design with waveband and wavelength multiplexing scheme. *In: CLEO/Pacific Rim 2001*, vol. 2, pp: 568-569.
20. Noirie, L., Vigoureux, M., and Dotaro, E., (2001). Impact of Intermediate Traffic Grouping on the Dimensioning of Multi-Granularity Optical Networks. *In: OFC 2001.*
21. Ramaswami, R., and Sivarajan, K. (1998). *Optical Networks: A Practical Perspective.* Morgan Kaufmann Publishers, 1998.
22. Ramaswami, R., and Sivarajan, K. (1995). Routing and wavelength assignment in all optical networks, *IEEE Transaction on Networks*, Oct, 1995.
23. Suemura, Y., Nishioka, I., Maeno, Y., Araki, S., Izmailov, R., and Ganguly, S. (2002). Hierarchical routing in layered ring and mesh optical networks. *In: IEEE ICC 2002*, Volume 5, 2002, pp. 2727 –2733.
24. Suemura, Y., Nishioka, I., Maeno, Y. and Araki, S. (2001). Routing of hierarchical paths in an optical network. *In: APCC 2001.*
25. Varsou, A., Ganguly, S., and Izmailov, R. (2003). Waveband Protection Mechanisms in Hierarchical Optical Networks. *In: IEEE HPSR 2003.*
26. Xiaojun, C., Anand, V., Xiong, Y. and Qiao, C. (2003). Performance Evaluation of Wavelength Band Switching in Multi-fiber All-Optical Networks, *In: IEEE INFOCOM 2003*, Apr 2003.

Chapter 3

# ADVANCES IN PASSIVE OPTICAL NETWORKS (PONS)

Amitabha Banerjee[1], Glen Kramer[2], Yinghua Ye[3], Sudhir Dixit[3] and Biswanath Mukherjee[1]

[1]*Department of Computer Science, University of California, Davis, CA*
[2]*Teknovus Inc., Petaluma, CA*
[3]*Nokia Research Center, Burlington, MA*

Email: abanerjee@ucdavis.edu, glen.kramer@teknovus.com, yinghua.ye@nokia.com, sudhir.dixit@nokia.com, mukherje@cs.ucdavis.edu

**Abstract** This chapter describes the recent advances made in broadband access network architectures employing Passive Optical Networks (PONs). The potential of PONs to deliver high bandwidths to users in access networks and their advantages over current access technologies have been widely recognized. PONs have made strong progress in terms of standardization and deployment over the past few years. In this chapter, we first review the Ethernet PON (EPON) [3], which is currently being standardized by the IEEE 802.3ah task force. Next, we discuss the ATM PON (APON) and the Gigabit PON (GPON). We then review the technologies available for introducing wavelength-division multiplexing (WDM) in PONs, and the progress of research in this area. Finally, we examine the issues related to deploying PONs in access networks.

**Keywords:** Broadband Access, Passive Optical Networks, Ethernet PON, ATM PON, Gigabit PON.

## 3.1 Introduction

The access network, also known as the "*first mile*" network, connects the service provider central offices to businesses and residential subscribers. This network is also referred to in the literature as the *subscriber access network*, or the *local loop*. Residential subscribers demand first-mile access solutions that have high bandwidth, offer media-rich Internet services, and are comparable in price with existing networks. Similarly, corporate users demand broadband infrastructure through which they can connect their local-area networks to the Internet backbone.

### 3.1.1 Challenges in Access Networks

Much of the focus and emphasis over the years has been on developing high-capacity backbone networks. Backbone network operators currently provide high-capacity OC-192 (10 Gbps) links. However, current generation access-network technologies such as Digital Subscriber Loop (DSL) provide 1.5 Mbps of downstream bandwidth and 128 kbps of upstream bandwidth at best. The access network is, therefore, truly the bottleneck for providing broadband services such as video-on-demand, interactive games, and video conferencing to end users.

In addition, DSL has a limitation that the distance of any DSL subscriber to a central office must be less than 18,000 feet because of signal distortions. Typically, DSL providers do not provide services to distances more than 12,000 feet. Therefore, only an estimated 60% of the residential subscriber base can avail of DSL. Although variations of DSL such as very high bit-rate DSL (VDSL), which can support up to 50 Mbps of downstream bandwidth, are gradually emerging, these technologies have much more severe distance limitations. For example, the maximum distance which VDSL can be supported over, is limited to 1,500 feet.

The other alternative available for broadband access to end users is through Cable Television (CATV) networks. CATV networks provide Internet services by dedicating some Radio Frequency (RF) channels in co-axial cable for data. However, CATV networks are mainly built for delivering broadcast services, so they don't fit well for distributing access bandwidth. At high load, the network's performance is usually frustrating to end users.

Faster access-network technologies are clearly desired for next-generation broadband applications. The next wave of access networks promises to bring fiber closer to the home. The FTTx model – Fiber to the Home (FTTH), Fiber to the Curb (FTTC), Fiber to the Building (FTTB), etc. – offers the potential for unprecedented access bandwidth to end users. These technologies aim at providing fiber directly to the home, or very near the home, from where technologies such as VDSL can take over. FTTx solutions are mainly based on the Passive Optical Network (PON). In this chapter, we shall review major developments in PON in recent years – EPON, APON, GPON and the WDM PON. Finally, we shall review the issues related to deployment of PONs.

### 3.1.2 Passive Optical Network (PON) Architectures

A Passive Optical Network (PON) is a point-to-multipoint optical network. An Optical Line Terminal (OLT) at the central office is connected to many Optical Network Units (ONUs) at remote nodes through one or multiple 1:N optical splitters. The network between the OLT and the ONU is passive, meaning that it doesn't require any power supply. An example of a PON using a single optical splitter is shown in Figure 3.1. The presence of only passive elements in the network makes it relatively more fault tolerant, and decreases its operational and maintenance costs once the infrastructure has been laid down.

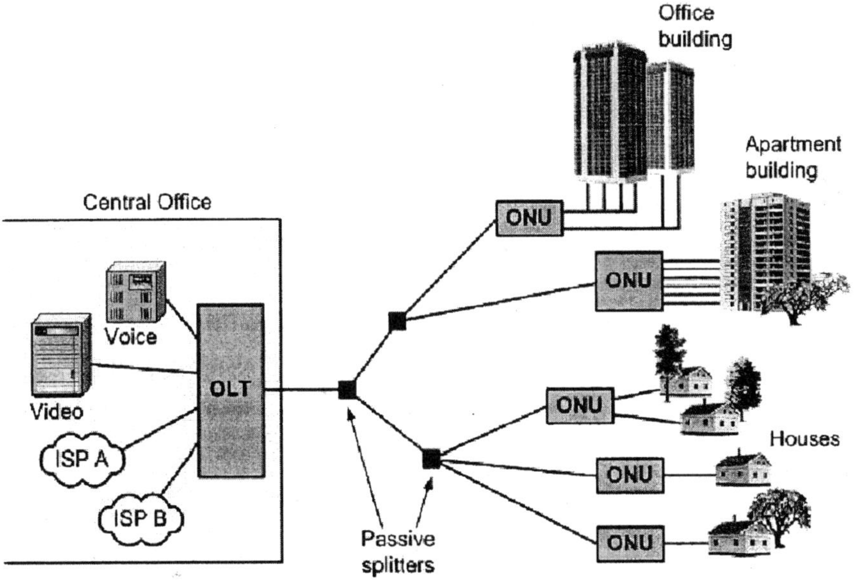


*Figure 3.1.* A Passive Optical Network (PON) connecting a central office to residential customers and business establishments.

Passive Optical Networks (PONs) have been considered for the access network for quite some time. A typical PON uses a single wavelength for all downstream transmissions (from OLT to ONUs), and another wavelength for all upstream transmissions (from ONUs to OLT), multiplexed on a single fiber through coarse wavelength-division multiplexing (CWDM).

## 3.2 Ethernet PON (EPON) Access Network

Ethernet PON (EPON) is a PON-based network that carries data traffic encapsulated in Ethernet frames (defined in the IEEE 802.3 standard). It uses a standard 8b/10b line coding (in which 8 user bits are encoded as 10 line bits), and it operates at standard Ethernet data rates.

### 3.2.1 Why Ethernet is Gaining Prominence?

The first-generation PON standardized by ITU–T G.983 employed ATM as the medium-access control (MAC) layer protocol. When its standardization effort was started in 1995, the telecom community believed that ATM would be the prevalent technology in backbone networks. ATM had the advantages of streamlining voice and data services while providing operational and performance guarantees. However, since then, Ethernet has grown vastly popular. Ethernet linecards are cheap, and they are widely deployed in LANs today. Since access networks are focused towards end users and LANs, ATM has turned out to be not the best choice to connect to Ethernet-based LANs.

In addition high-speed Gigabit Ethernet deployment is widely accelerating and 10 Gigabit Ethernet products are becoming available. Ethernet is a very efficient MAC protocol to use compared to ATM which imposes a considerable amount of overhead on variable-length Internet Protocol (IP) packets. Newly adopted quality-of-service (QoS) techniques have made Ethernet networks capable of efficiently supporting voice, data, and video. These techniques include full duplex transmission mode, prioritization (802.1p), and virtual LAN (VLAN) tagging (802.1Q). 802.1p is a specification which allows for prioritization of traffic into different priority classes. 802.1Q defines an architecture for VLANs. Although 802.1Q doesn't directly define any QoS support, it defines a frame-format extension allowing Ethernet frames to carry priority information. EPONs, therefore, have much more promise in future access networks compared to ATM PONs (APONs). The following subsections describe the operation of the EPON, as stated in the draft D2.0 of IEEE P802.3ah [1].

### 3.2.2 EPON Principle of Operation

In the downstream direction (OLT to ONUs), Ethernet frames transmitted by the OLT pass through a 1:N passive splitter and reach each ONU. Typical values of N are between 8 and 64. Packets are broadcast by the OLT and extracted by their destination ONU based on a Logical Link Identifier (LLID), which the ONU is assigned when it registers with the network. Figure 3.2 shows the downstream traffic in EPON.

In the upstream direction, data frames from any ONU will only reach the OLT and will not reach any other ONU due to the directional properties of a passive optical combiner. Therefore, in the upstream direction, the behavior of EPON is similar to that of a point-to-point architecture. However, unlike in a true point-to-point network, in EPON, data frames from different ONUs transmitted simultaneously may collide. Thus, in the upstream direction, the ONUs need to employ some arbitration mechanism to avoid data collisions and fairly share the channel capacity. A contention-based media-access mechanism (similar to Carrier Sense Multiple Access with Collision Detection (CSMA/CD))

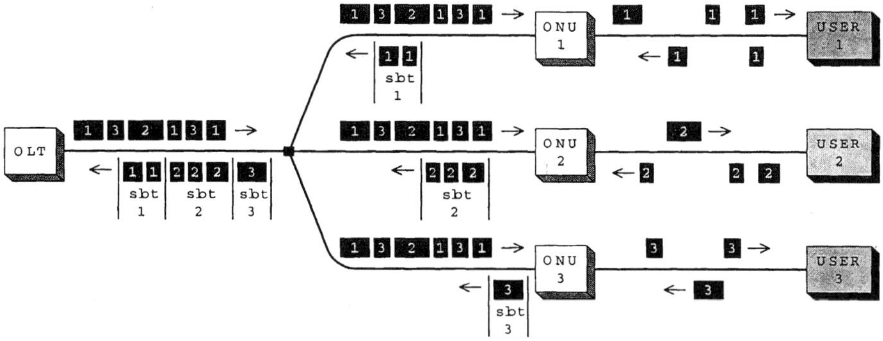


*Figure 3.2.* Downstream and Upstream operation in EPON.

is difficult to implement because ONUs cannot detect a collision in the fiber from the combiner to the OLT due to the directional properties of the combiner. The OLT could detect a collision and inform the ONUs by sending a jam signal; however, propagation delays in PON (the typical distance from the OLT to ONUs is 20 km), greatly reduces the efficiency of such a scheme. To introduce determinism in frame delivery in the upstream direction, different non-contention schemes have been proposed. Figure 3.2 illustrates an upstream, time-shared, data flow in an EPON.

All ONUs are synchronized to a common time reference and each ONU is allocated a timeslot in which to transmit. Each timeslot is capable of carrying several Ethernet frames. An ONU should buffer frames received from a subscriber until its timeslot arrives. When its timeslot arrives, the ONU would burst all stored frames at full channel speed. If there are no frames in the buffer to fill the entire timeslot, an idle pattern is transmitted.

Thus, timeslot assignment is a very crucial step. The possible timeslot allocation schemes could range from static allocation (fixed time-division multiple access (TDMA)) to a dynamically-adapting scheme based on instantaneous queue size in every ONU (a statistical multiplexing scheme). In the dynamically-adapting scheme, the OLT can play the role of collecting the queue sizes from the ONUs and then issuing timeslots. Although this approach may lead to higher signalling overhead between the OLT and the ONUs, the centralized intelligence may lead to more efficient use of bandwidth. More advanced bandwidth-allocation schemes are also possible, including schemes utilizing notions of traffic priority, Quality of Service (QoS), Service-Level Agreements (SLAs), over-subscription ratios, etc. [4].

### 3.2.3 Multi-Point Control Protocol (MPCP)

The Multi-Point Control Protocol (MPCP) is a supporting protocol to facilitate a dynamic timeslot-allocation scheme. It is being developed by the IEEE 802.3ah Ethernet in First Mile (EFM) task force. MPCP aims to define a signalling protocol between the OLT and the ONUs, and does not in any way define any bandwidth-provisioning scheme.

MPCP consists of three functions, as follows.

1 *Discovery Processing*: In this step, an ONU is discovered and registered in the network, and the round trip time (RTT) for a signal to travel from the OLT to the ONU and back is measured.

2 *REPORT Handling*: ONUs generate REPORT messages through which bandwidth requirements are transmitted to the OLT. The OLT needs to process the REPORT messages so that it can make bandwidth assignments.

3 *GATE Handling*: GATE messages are used by the OLT to grant a time slot at which the ONU can start transmitting data. Timeslots are computed at the OLT while it makes bandwidth allocations.

**Discovery Processing.** Discovery is the process in which newly-connected or uninitialized ONUs register in the network. The steps are shown in Figure 3.3, and they are elaborated below.

1 OLT: The OLT periodically makes available a discovery time window during which the offline ONUs are given the opportunity to register themselves with the OLT. A DISCOVERY GATE message, which mentions the starting and the ending time of the discovery window is broadcast to all ONUs.

2 ONU: Any offline ONU which wishes to register, waits for a random amount of time within the discovery window, and then transmits a REGISTER_REQ message that contains the MAC address of the ONU. The random wait is required to eliminate the possibility of REGISTER_REQ messages transmitted by multiple ONUs from colliding persistently.

3 OLT: The OLT, after receiving a valid REGISTER_REQ message, registers the ONU and allocates to it a Logical Link Identifier (LLID). The OLT now transmits a REGISTER message to the newly-discovered ONU which contains the ONU's LLID.

4 OLT: The OLT now transmits a standard unicast GATE message, indicating a timeslot to transmit data.

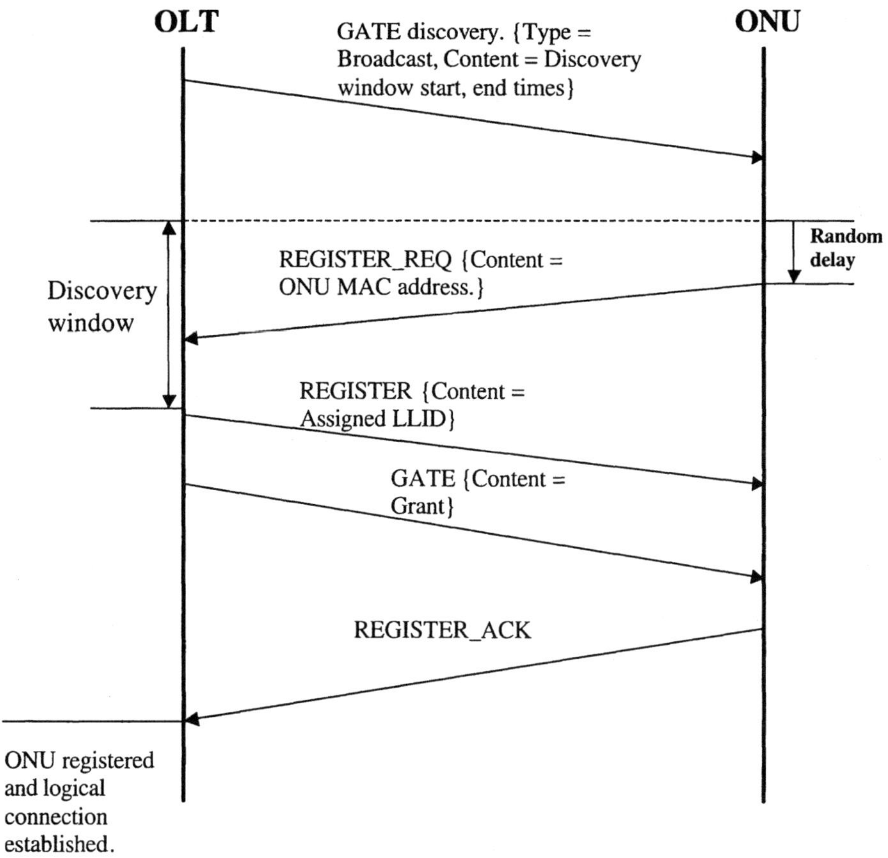


*Figure 3.3.* Discovery phase message exchange.

5 ONU: Upon receiving the GATE message, the ONU responds with a REGISTER_ACK message in the assigned timeslot. Upon receipt of this message, the discovery process is completed and, now, normal operation may start.

**REPORT Handling.** REPORT messages are sent by the ONUs in their assigned transmission windows along with data frames. A REPORT message is generated in the MAC control client layer and is timestamped in the MAC control unit of the ONU. Typically, REPORT would contain the desired size of the next timeslot, based on the ONU's queue size. REPORT messages are generated periodically, even when no request for bandwidth is being made. This prevents the OLT from de-registering an ONU. Thus, for proper operation of this mechanism, the OLT must grant the ONU a transmission window periodi-

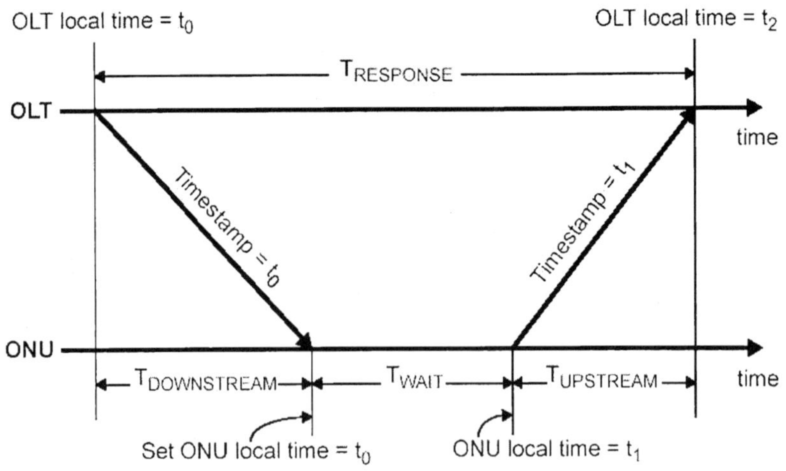


$T_{DOWNSTREAM}$ = downstream propagation delay
$T_{UPSTREAM}$ = upstream propagation delay
$T_{WAIT}$ = wait time at ONU = $t_1 - t_0$
$T_{RESPONSE}$ = response time at OLT = $t_2 - t_0$

$$RTT = T_{DOWNSTREAM} + T_{UPSTREAM} = T_{RESPONSE} - T_{WAIT} = (t_2 - t_0) - (t_1 - t_0) = t_2 - t_1$$

*Figure 3.4.* Calculation of round-trip time (RTT).

cally. At the OLT, the REPORT is processed and this data is used for the next round of bandwidth assignments.

**GATE Handling.** The transmission window of an ONU is indicated in the GATE message from the OLT. The *transmission start* time and *transmission length* are specified. Upon receiving a GATE message matching the ONU's LLID, the ONU will program its local registers with the *transmission start* time and *transmission length.* The ONU will also verify that the time (according to the local clock at the ONU) when the GATE message arrives is close to the timestamp value contained within the message. If the difference in values exceeds some predefined threshold, the ONU will assume that it has lost its synchronization and will switch itself into off-line mode. The ONU shall then attempt to register again during the discovery process.

When the time at the local clock of the ONU reaches the *transmission start* time, the ONU starts transmitting data.

**Clock Synchronization.** MPCP depends on clock synchronization between the OLT and the ONU, which compensates for the RTT. The RTT is expected to be different for each ONU as they may be located at different distances from the OLT. Clock synchronization compensating for RTT is important because the OLT does not have to keep track of the different RTTs of different ONUs, when it issues timeslots in GATE messages.

The above clock synchronization is achieved as follows. Whenever the ONU receives a MPCP message, it sets its local time from the timestamp of that message. When the OLT receives a MPCP message, it calculates the RTT as the difference between its local time and timestamp of the message (Figure 3.4). Any significant change in RTT implies that the OLT and the ONU clocks are not in synchrony any more, and the OLT now issues a de-register message for that particular ONU. The ONU will then attempt to register in the network again through the discovery process.

## 3.3 Other Types of PONs

Besides EPONs, other technologies have also been developed for PONs. Two such PONs – the APON and the GPON – will be reviewed in this section.

### 3.3.1 ATM PON (APON) Access Network

APON is based on Asynchronous Transfer Mode (ATM) as the MAC layer protocol. The downstream frame shown in Figure 3.5 consists of 56 ATM cells (53 bytes each) for the basic rate of 155 Mbps, scaling upto 224 ATM cells for 622 Mbps. There are two dedicated cells called Physical Layer Operation, Administration and Maintenance (PLOAM) cells; one at the beginning of the frame, and one in the middle. The remaining 54 cells are data ATM cells.

The upstream transmission (Figure 3.6) is in the form of bursts of ATM cells, with a 3-byte physical overhead appended to each 53-byte ATM cell to allow for burst-mode receivers. Burst-mode receivers are required at the OLT to synchronize to the different ONUs which may be located at different distances from the OLT. The ATM cell may be either an ATM data cell, or a PLOAM cell.

In the downstream direction, the PLOAM cells are used to carry grants from the OLT to the ONUs. Each grant is a one-time permission for an ONU to transmit payload data in an ATM cell. 53 grants for the 53 upstream frame cells are mapped into the PLOAM cells. The OLT sends a continuous stream of grants to all the ONUs in the PON. Thus, the OLT can moderate the portion of the upstream bandwidth assigned to each ONU. In the upstream direction, the PLOAM cells are used by the ONUs to transmit their queue sizes to the OLT. This information shall be used by the OLT for bandwidth allocation.

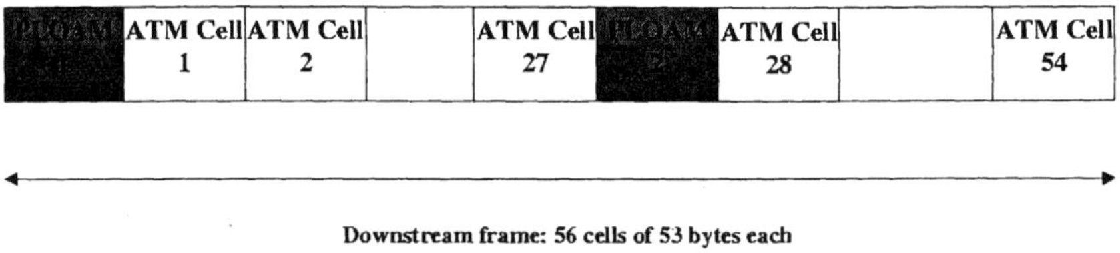


*Figure 3.5.* APON downstream frame format.

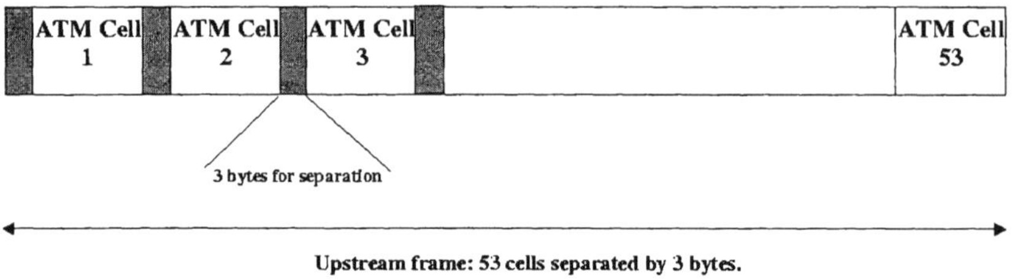


*Figure 3.6.* APON upstream frame format.

Initial work on ATM PONs was started in the mid 1990's by the Full Service Access Network (FSAN) [11] initiative which was driven by service providers. A further extension to APON called the Broadband PON (BPON) was developed. BPON supports upto 622 Mbps of downstream and 155 Mbps of upstream bandwidth. BPON has been standardized by the International Telecommunication Union (ITU) specification G.983.1. It provides overlay capabilities for services such as video and Ethernet traffic.

### 3.3.2 Gigabit PON (GPON) Access Network

The GPON is being standardized by the ITU in specification G.984.x. It proposes bit rates of upto 2.5 Gbps. It also aims towards providing higher efficiency while carrying multiple services over the PON. It proposes a protocol using Generic Framing Procedure (GFP) [18]. GFP provides a generic mechanism to adapt traffic from higher layers (Ethernet MAC/ IP), over a transport layer such as Synchronous Optical Network / Synchronous Digital Hierarchy (SONET/SDH). Other functionalities like dynamic bandwidth assignment, operation and maintenance etc. are borrowed from APONs.

As discussed in section 3.2.1, both APON and GPON have the disadvantages of complex protocols and implementations, because of which they have not gained much popularity among users and equipment vendors. Various service providers such as NTT, BellSouth etc., have tested initial deployments and testbeds of the BPON [12].

## 3.4 Wavelength-Division Multiplexed PON (WDM–PON) Access Network

This section reviews research contributions to WDM–PON access networks. We first emphasize the importance of WDM in PON. We then discuss various WDM–PON architectures which have been proposed in the research literature.

### 3.4.1 Need for WDM in PONs

Although the PON is a significant step towards providing broadband access to the end user, it is not very scalable. Since the basic form of PON employs only a single optical channel, the available bandwidth is limited to the maximum bit rate of an optical transceiver, which, under current technologies, is 1 - 2 Gbps. The attenuation due to splitting limits the maximum number of ONUs to 64. This limits the network's scalability. Since the deployment cost of laying fiber in the access network is high, it is important to consider technologies which may help scale the PON capacity in the future.

A lot of emphasis has been put on employing WDM in PONs in the recent literature. A WDM–PON is a point–to–point access network (as opposed to point–to–multipoint in PON), in which, typically, there exists a separate wavelength between the OLT and each ONU. In a WDM–PON, different ONUs can be supported at different bit rates, if necessary. Each ONU can operate at a rate up to the full bit rate of a wavelength channel; therefore, it does not have to share the available bandwidth with any other ONU in the network. Use of individual wavelengths for each ONU also facilitates privacy and reduces security concerns. Finally, WDM–PON architectures based on the Arrayed Waveguide Grating (AWG) (described in section 3.4.2) are highly scalable owing to the periodic routing pattern of an AWG (described in Section 3.4.4). Keeping in view such advantages of WDM–PONs, WDM has been recommended as an upgrade to the PON in the ITU-T G.983 [16].

### 3.4.2 Arrayed Waveguide Grating (AWG)

Wavelength routing in WDM–PONs may be implemented using an Arrayed Waveguide Grating (AWG). The AWG is a passive device with a fixed routing matrix, so it fits well with the PON philosophy. An AWG provides fixed routing of an optical signal from a given input port to a given output port, based on the wavelength of the signal. Signals of different wavelengths coming into an input port will each be routed to a different output port. Similarly, different

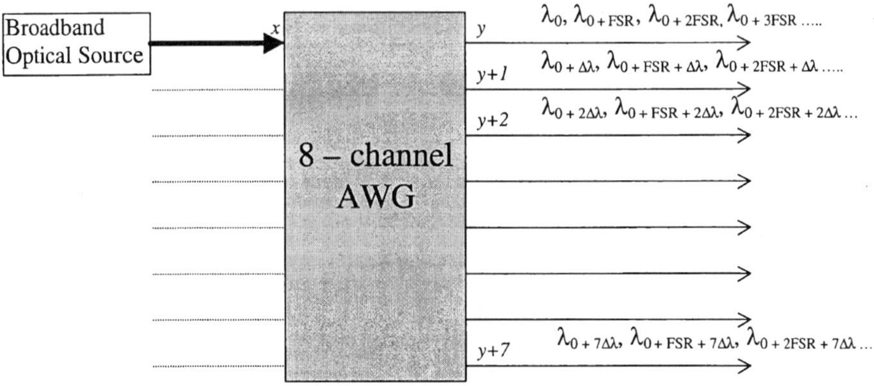


*Figure 3.7.* The periodic routing behavior of a AWG.

signals on the same wavelength may be directed from different input ports, and they shall be routed to different output ports.

One of the main advantages of the AWG is its periodic routing behavior shown in Figure 3.7. Consider a broad spectrum optical source entering input port *x*. For the optical signals entering port *x* and routed to a given output port *y*, the AWG routes wavelengths which are separated by a fixed wavelength interval called the free spectral range (FSR). Therefore, considering a base wavelength $\lambda_0$, the output wavelengths at port *y* are $\lambda_0$, $\lambda_{0+FSR}$, $\lambda_{0+2FSR}$, and so on. For output port *y* + *1*, the wavelengths routed from port *x* are shifted by a wavelength interval $\Delta\lambda$ compared to *y*. Thus, the output wavelengths at port *y+1* are $\lambda_{0+\Delta\lambda}$, $\lambda_{0+\Delta\lambda+FSR}$, $\lambda_{0+\Delta\lambda+2FSR}$, and so on. This periodic routing property of the AWG helps immensely in scaling the network as described in Section 3.4.4.

### 3.4.3 WDM-PON Architectures

All WDM–PON architectures typically employ a separate optical wavelength channel for each ONU in the downstream direction from OLT to the ONUs. However, the various proposed architectures differ in the amount of resources used in the upstream direction from the ONUs to OLT. Upstream communication differs from downstream communication due to two main reasons. ONU equipment (transmitters) must be inexpensive if ONUs are to deployed in a large scale. Moreover, it is preferable not to have wavelength-specific equipment at the ONU, because it is difficult to manage and maintain different inventories of end-user equipment.

One of the easiest approaches to employing WDM in downstream direction in the basic PON, is to employ WDM with an array of transmitters at the OLT and to employ a filter at each ONU for the corresponding wavelength. Each

ONU can thus be tuned to receive a particular wavelength. However, this architecture still has limitations in scalability owing to power splitting loses at the optical splitter.

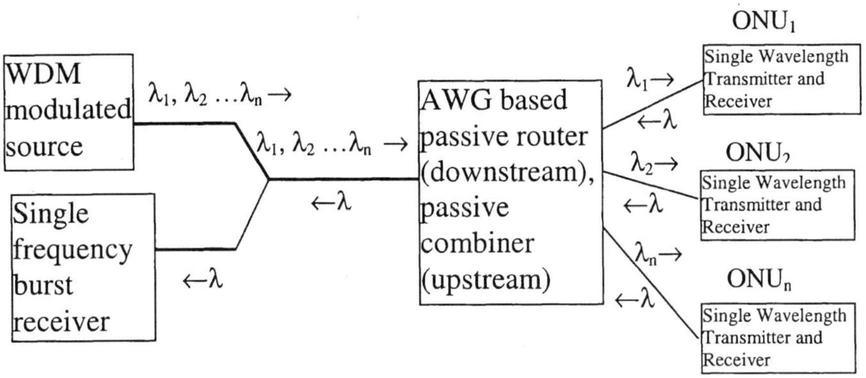


*Figure 3.8.* Composite PON (CPON).

To solve the limitations of scalability, WDM–PON architectures based on the AWG were proposed. One of the earliest WDM–PON architecture proposals based on this concept, employed WDM in the 1550 nm band in downstream and a single upstream wavelength in the 1300 nm band shared through time-division multiple access (TDMA) [8]. The upstream and downstream transmissions may be on a single fiber through coarse WDM (CWDM). This architecture has been referred to as Composite PON (CPON) in the literature [2]. A single-wavelength, burst-mode receiver is used at the OLT to receive the upstream signal. A burst-mode receiver is required to synchronize to the clock signals of different transmitting ONUs, which may be at different distances from the OLT. Figure 3.8 shows the layout of a CPON.

A limitation of the CPON architecture is that a single-frequency laser, such as a DFB laser, at the ONU may be economically prohibitive. Moreover, it may be difficult to control the wavelength changes that may arise due to temperature fluctuations at the remote (ONU) end.

The LARNET (Local Access Router Network) architecture [9] attempts to work around the above limitations by using a broad–spectrum source at the ONU such as an inexpensive edge-emitting LED, whose spectrum is sliced by the AWG-based router in the upstream direction. The edge-emitting LED emits a broad spectrum of wavelengths centered around a single wavelength, as compared to DFB lasers which emit only one wavelength of light. As illustrated in Figure 3.7, when a broad–spectrum source is directed into one input port of the AWG, the various constituent wavelengths are directed to different output ports. The LARNET architecture is shown in Figure 3.9. In LARNET,

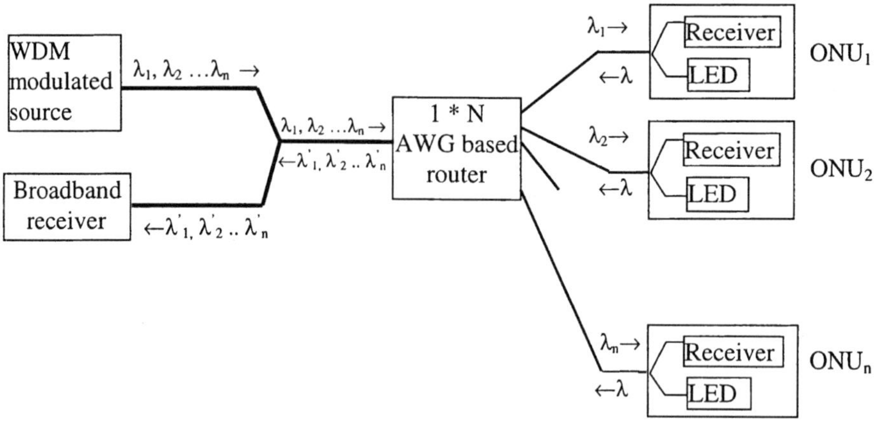


*Figure 3.9.* LARNET architecture.

one broad–spectrum source from each ONU is input to different ports of the AWG device. Depending on the input port, the wavelength component which is sliced at the output port is different. Therefore at the output port we observe many wavelength components, each corresponding to an input source from an ONU. The OLT employs a broadband burst-mode receiver (unlike a single-wavelength burst-mode receiver employed in CPON), which can receive any spectral component of the edge emitting LED. The OLT is thus able to receive from each ONU. Since there exists only one receiver, therefore the upstream channel must be shared by the ONUs using a scheme like TDMA or dynamic bandwidth allocation.

LARNET is attractive because the technology for edge-emitting LEDs is quite mature and such devices have been commercially available for some time. Edge-emitting LEDs are much cheaper compared to DFB lasers, so they help in the economics of the cost of the ONU equipment. The limitation is that spectrally slicing a broad–spectrum source by an AWG leads to very high power loss. Therefore, the distance from the OLT to the ONU is considerably reduced in LARNET.

Recently, variations to the LARNET architecture have been suggested in which the upstream signal from an ONU can be looped back downstream to all other ONUs from the AWG, through appropriate wiring at the AWG [5] based on the periodic routing property of the AWG. The AWG is typically deployed very close to the ONU. Because the propagation delay from the ONU to the AWG is very small, a CSMA/CD MAC protocol such as Ethernet can now be used for contention resolution of upstream traffic.

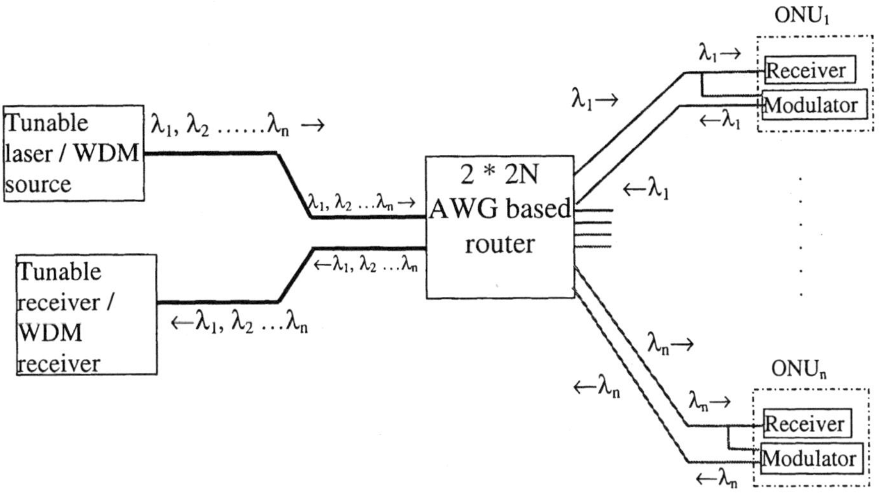


*Figure 3.10.* RITENET architecture.

The RITENET (Remote Interrogation of Terminal Network) architecture [6] aims to avoid the transmitter at the ONU by modulating the downstream signal from the OLT and sending it back in upstream direction. The signal from the OLT is shared for downstream and upstream through time-sharing. A frame is split into two parts: one used for downstream transmission, and the other for upstream transmission. A 2 * 2N AWG-based router is used to route the wavelengths.

In RITENET, it was proposed that a tunable laser be used, which can be tuned to the individual wavelengths. Thus, both the upstream and downstream channels have to be shared by the ONUs using TDMA or dynamic bandwidth allocation. Alternatively as has been suggested in some recent architectures [13], an array of transmitters and receivers may be employed at the OLT. In this case, channel sharing between ONUs is not required.

Since the same optical channel is used for both upstream and downstream, they must be separated on two different fibers. Figure 3.10 shows the architecture of RITENET.

While RITENET helps in reducing end-terminal costs at the ONU, the distance from the OLT to the ONU is much less as the signal at the OLT now has to travel double the distance. Also, since the signal is now shared between the two ends, the bit rate of the PON must be doubled. Moreover, the number of fibers employed is also doubled, which doubles the cost of deployment and maintenance. Employing either a tunable laser or an array of transmitters and receivers at the OLT, makes RITENET a more expensive architecture compared to LARNET and CPON. However, a significant advantage of

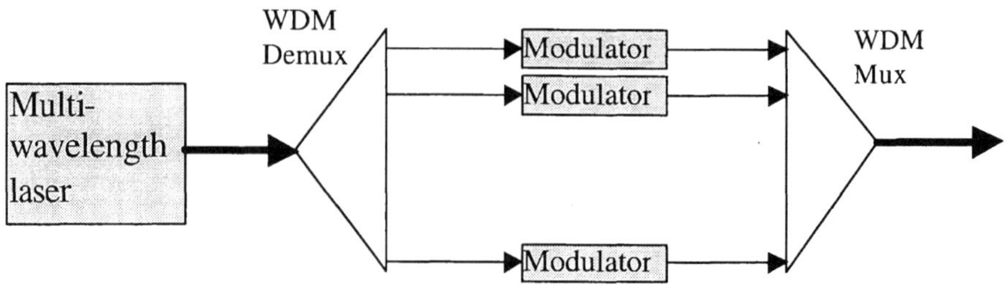


*Figure 3.11.* Modulating a multi-wavelength laser source.

the RITENET is the availability of symmetrical bandwidth in downstream and upstream directions, unlike LARNET and CPON. A number of WDM–PON architectures suggested in the recent literature, use variants of the RITENET architecture.

All of the above architectures employ a single multi-wavelength laser source at the OLT. Commercial products, which produce a multi-wavelength optical spectrum composed of many stable individual optical frequencies that can be locked into the standard ITU grid, are now available. The multi-wavelength source can then be modulated with independent modulators, as shown in Figure 3.11. A multi-wavelength laser source also implies greater wavelength stability in the network compared to using numerous DFB lasers in an array, because the single laser source can be very easily and efficiently controlled for temperature variations.

AWG-based routers are the building blocks of any WDM-PON architecture. Integrated optics technology has matured over the years and the number of channels supported has also scaled very well. 40-channel AWG devices are commercially available today. As more of these devices are deployed, their cost is expected to reduce. Since the AWG is an outdoor deployment in an access network, thermal stability is a very important issue. Outdoor temperatures may vary between -$40^0C$ to $85^0C$. Temperature variations cause the passbands at which the AWG operates to drift. Various measures have been suggested in the literature to improve the thermal performance of the AWG. Some proposed solutions such as keeping the AWG in a constant temperature environment is not suited for passive networks, as the router will then have to be supplied with power. Other solutions are based on drifting the input and output wavelengths as the temperature changes the refractive index, to accommodate for the passband drifts of the AWG.

Various field trials based on the above architectures have been reported in the literature recently. An experiment on an optical access network providing Gigabit Ethernet access to over 100 users was demonstrated in [13]. A variation of the RITENET architecture is used, a difference being that, instead of using one wavelength per ONU, two are used, one for upstream, the other for downstream. This eliminates the need for time-sharing. Thus, 256 wavelengths are used to support 128 users. An Optical Carrier Supply Module (OCSM) is used to generate 256 wavelengths with 25 GHz spacing.

A testbed using a variation of the LARNET architecture has been described in [14]. A variation of the CPON architecture, using optical amplifiers and called the SuperPON, has been described in [15].

### 3.4.4 Scalability of WDM–PON

Any network architecture must be easily scalable in order to be of value. For an access network, scalability is required both in terms of bandwidth and the number of end access points (ONUs) supported. Since the deployment cost of an optical network is high owing to the high cost of manual labor in most countries, it is important to ensure that scalability can easily be achieved without much deployment overhead. The costs of some of the optical devices such as the AWG are also quite high; therefore, it must be ensured that such devices can be reused and they do not have to be replaced when scalability is desired. Similarly, since a large number of ONUs are deployed and such devices are typically located in end-user homes, buildings, or communities, it is desired that replacements, if needed due to scalability, should be minimum. Since all end users may not wish to upgrade to higher bandwidths at the same time, it should be ensured that legacy users can be supported while scaling the network.

The combination of all the above needs make scalability in WDM-PON architectures a challenging issue. A novel solution in [7] proposes exploiting the periodic routing pattern of the AWGs by deploying AWGs in series. Figure 3.12 shows how additional AWGs may be deployed to scale from a 8-wavelength, 8-ONU WDM-PON architecture to a 32-wavelength, 32-ONU WDM-PON architecture. The subscript of the wavelength denotes the wavelength number while the superscript denotes the source (e.g., $\lambda_1^2$ denotes wavelength 1 from laser 2). This idea has several merits. The legacy ONUs – $ONU_1$ to $ONU_8$ – remain unaffected and continue to use wavelengths $\lambda_1^1$ to $\lambda_4^1$ and $\lambda_1^2$ to $\lambda_4^2$. The legacy 2 * 8 channel AWG is maintained and 8 new 1 * 4 channel AWGs are used to scale the network. This architecture also has the benefit of wavelength re-use. For example, in Figure 3.12, laser 1 and laser 2 share the same wavelength domain $\lambda_1$ to $\lambda_{16}$, but cater to different ONUs.

Figure 3.12 only shows how scalability can be achieved in the downstream direction. In upstream, scalability can be achieved by using the downstream wavelengths, by modulating them for upstream communication just as it is done in RITENET.

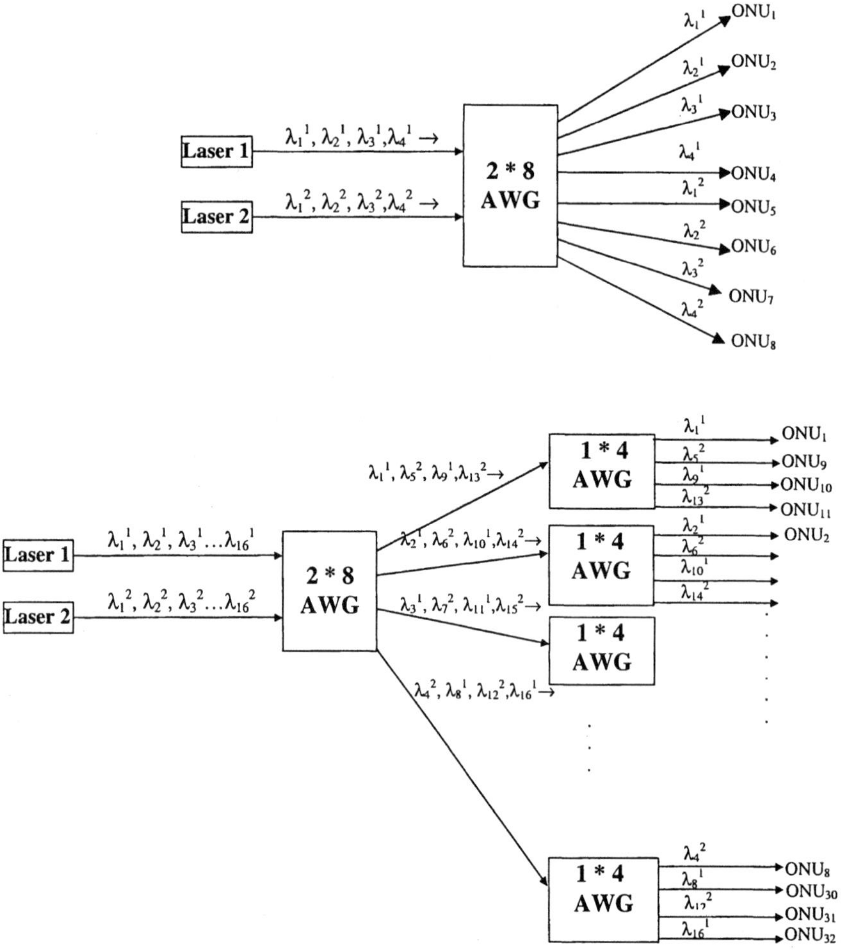


*Figure 3.12.* Scalability of a WDM-PON architecture. An 8-ONU WDM-PON architecture is scaled to 32 ONUs. The subscript of the wavelength denotes the wavelength number while the superscript denotes the source of the wavelength.

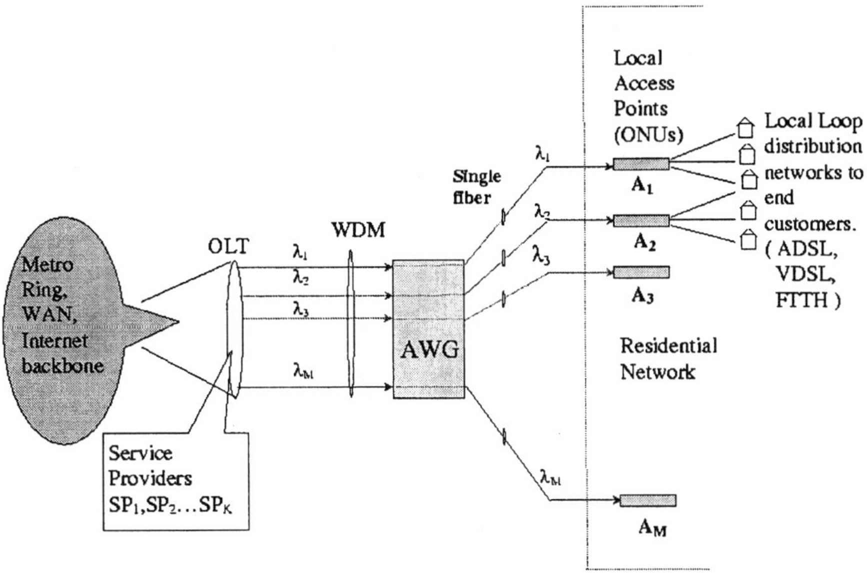


*Figure 3.13.* A WDM-PON-based FTTC network deployed as an open access network. Service providers use this infrastructure to serve end users by leasing bandwidth from the access-network operator (ANO).

## 3.5 Deployment of WDM-PONs

Various models have been proposed for the deployment of fiber in the access network. While Fiber To The Home (FTTH) is the ultimate objective, Fiber To The Curb (FTTC) and Fiber To The Building (FTTB) have been proposed as intermediary goals. In FTTC/FTTB, ONUs located at the curb or in a building serve as distribution points of bandwidth to end residential customers. End customers are provided broadband access through xDSL technologies over twisted-pair copper. The advantage of such a scheme is that the distance covered by the twisted-pair copper infrastructure is now much less, which makes technologies such as VDSL (which has a distance limitation of 1,500 feet) feasible.

Thus, a FTTC network would act as a single broadband access infrastructure through which different service providers can provide numerous application services to the end users. The access network needs to be shared because it is not feasible for every service provider to deploy its own access network because of deployment and operational costs and right-of-way issues. We call such an access network, an *open access network*, as shown in Figure 3.13. This access network could be deployed and maintained by an *access-network operator (ANO)*.

The OLT could be connected to a metro ring network, a wide-area LAN, or a long-haul optical network. Service providers (SPs) could lease bandwidth from the ANO and make the bandwidth available to the end users. It is expected that each SP will have a service-level agreement (SLA) with the ANO, and the SLA will specify the minimum bandwidth ($W_{MIN}$) that must be available at all times to the service provider, and the monthly service charge.

The ONUs are defined as Local Access Points (LAPs) which act as distribution centers for bandwidth to end users. It is anticipated that residential customers will be connected to the LAP using technologies such as ADSL, VDSL, or EPON. There are many advantages of such a model. First, it is much easier to provide optical fiber access to a residential curb than to every house in an already-established building or neighborhood. Second, not all users require high-bandwidth solutions such as EPON. Some users may be satisfied with incumbent technologies such as ADSL, others may prefer VDSL, whereas only a small subset might demand EPON solutions. Thus, the above network aggregates all incumbent and future access technologies on a common platform and provides a clear migration path from twisted-pair copper infrastructure to a fiber-based network.

End users subscribe to the SPs for different application services. A user may subscribe to different SPs for different services. For example, user $U_1$ might subscribe to $SP_a$ for VoIP, to $SP_b$ for interactive gaming, and to $SP_c$ for video-on-demand.

We consider the following model for bandwidth allocation. Since application requirements arise from end users, a REPORT message identifying bandwidth requirements from each SP originates from the end user. These REPORT messages are processed at the OLT, and bandwidth allocation is done. Thus, an efficient *admission control* and *bandwidth-allocation policy* must be implemented at the OLT.

It is desired that the *bandwidth-allocation policy* maintain the minimum-bandwidth requirement ($W_{MIN}$) guaranteed to the SP in the SLA. If the cumulative amount of bandwidth requested by all users from a SP exceeds $W_{MIN}$, the bandwidth in excess of $W_{MIN}$ can be allocated only if the $W_{MIN}$ of all the SPs has been met. Fairness is desired for allocating the excess bandwidth in the *bandwidth-allocation policy.* There may be numerous metrics of fairness, one example being the blocking probability (BP). It is also desired that fairness be achieved, not only to all the SPs, but also to all the end users.

The next issue is managing a wide range of application technologies which may be using the access network. Examples of broadband applications available today are video-on-demand, packetized voice, interactive video conferencing, interactive games, Internet browsing, etc. These different applications have different QoS requirements. Meeting diverse QoS requirements over the same optical channel is a difficult challenge.

A scheme for delivering multiple independent broadcast services over a WDM-PON network was shown in [17]. This approach uses the periodic routing property of the AWG mentioned in Section 3.4.2 to deliver different broadcast services on different wavelengths to the same end user. The broad-spectrum of a LED is adjusted to match the FSR of the AWG. Filters are deployed at the ONU to select the desired wavelength for the desired service. A different wavelength LED is used for each broadcast service.

The above implementation can be thought of as assigning different wavelengths to different applications. This approach has the merit of satisfying QoS with ease for each application, as the applications are on separate channels. However, applications in an access network are not limited to broadcast. Therefore, it may be inefficient in terms of bandwidth to reserve different wavelengths for different unicast applications. The presence of different service providers, which was not considered in the approach leads to further bandwidth inefficiencies. Moreover, in the above approach, each ONU must receive multiple wavelengths, which increases its deployment and operational costs.

*Open access networks* have been gaining a lot of ground in recent years. The issues of high deployment costs, long-term investment requirements, and right-of-way issues have prompted the debate for governments, municipalities, and utilities to operate such networks. Many municipalities have begun test trials among limited segments of customers to determine the feasibility of these networks [19]. Clearly, a lot of research on supporting protocols is desired before commercial deployment of such open access networks.

## 3.6 Summary

In this chapter, we first reviewed the Ethernet PON (EPON) which is currently being standardized by the IEEE 802.3ah EFM Task Force. Several commercial initiatives on E-PON are undergoing and products are expected. Then, we studied the ATM PON (APON) and the Gigabit PON (GPON). We discussed the WDM-PON, which is gaining attention as an attractive solution for solving our broadband access needs in the future. WDM-PONs are still in the process of test trials by numerous organizations and research communities. These technologies will clearly go a long way towards meeting future end-user broadband requirements. Finally, we discussed various aspects related to the deployment of PONs and the concept of *open access networks* which is gaining a lot of ground.

## References

[1] IEEE 802.3ah EFM EPON baseline technical proposals. http://grouper.ieee.org/groups/802/3/efm/baseline/p2mpbaseline.html.

[2] M. Zirngibl *et al.*, "An evaluation of Architectures Incorporating Wavelength Division Multiplexing for Broad-Band Fiber Access," IEEE/OSA Journal of Lightwave Technology, vol. 16, no. 9, pp. 1546-1558, September 1998.

[3] G. Kramer and G. Pesavento, "Ethernet Passive Optical Network (EPON): Building a Next Generation Optical Network," IEEE Communications Magazine, vol. 40, no. 2, pp. 66-73, February 2002.

[4] G. Kramer, B. Mukherjee, and G. Pesavento, "IPACT: A Dynamic Protocol for an Ethernet PON," IEEE Communications Magazine, vol. 40, no. 2, pp. 74-80, February 2002.

[5] B. N. Desai *et al.*, "An optical implementation of a packet-based (Ethernet) MAC in a WDM passive optical network overlay," Proceedings OFC 2001, Anaheim.

[6] N. J. Frigo *et al.*, "A wavelength-division multiplexed passive optical network with cost shared components," IEEE Photonic Technology Letters, vol. 6, no. 11, pp. 1365-1367, November 1994.

[7] G. Mayer, M. Martinelli, A. Pattavina, and Elio Salvadori, "Design and Cost performance of the Multistage WDM PON Access Networks," IEEE/OSA Journal of Lightwave Technology, vol. 18, no. 2, pp 121-142, February 2000.

[8] Y. K. Lin and D. R. Spears, "Passive optical subscriber loops with multi-access," IEEE/OSA Journal of Lightwave Technology, vol. 7, no. 11, pp 1769-1777, November 1989.

[9] M. Zirngibl *et al.*, "LARNET, a local access Router Network," IEEE Photonics Technologies Letters, vol. 7, no. 2, pp. 215-217, February 1995.

[10] P. Iannone *et al.*, "Simultaneous WDM and broadcast transmission using a single multi-wavelength waveguide-grating-router laser," IEEE Photonic Technology Letters, vol. 8, no. 10, pp. 1397-1399, October 1996.

[11] FSAN - Full Service Access Network.
http://fsan.mblast.com/default.asp.

[12] H. Ueda *et al.*, "Deployment Status and Common Technical Specifications for a B-PON system," IEEE Communications Magazine, vol. 39, no. 12, pp. 134-141, December 2001.

[13] J. Kani *et al.*, "A WDM based optical access network for wide-area Gigabit access services," IEEE Optical Communications Magazine, vol. 1, no. 1, pp. S43-S48, February 2003.

[14] M. Oksasen *et al.*, "Spectral Slicing Passive Optical Access Network Trial," Proceedings OFC 2002, Anaheim, CA.

[15] I. Van de Voorde *et al.*, "The SuperPON Demonstrator: An Exploration of Possible Evolution Paths for Optical Networks," IEEE Communications Magazine, vol. 40, no. 2, pp. 74-82, February 2002.

[16] F. J. Effenberger, H. Ichibangase, and H. Yamashita, "Advances in Broadband Passive Optical Networking Technologies," IEEE Communications Magazine, vol. 39, no. 12, pp. 118-122, December 2001.

[17] P. P. Iannone, K. C. Reicjmann, and N. J. Frigo, "High-Speed Point-to-Point and Multiple Broadcast Services Delivered over a WDM Passive Optical Network," IEEE Photonics Technology Letters, vol. 10, no. 9, September 1998.

[18] E. Hernandez-Valencia, M. Scholten, and Z. Zhu, "The Generic Framing Procedure (GFP): An Overview," IEEE Communications Magazine, vol. 40, no. 5, pp. 63-71, May 2002.

[19] "City of Palo Alto Utilities- FTTH Trial", http://inet-gw.city.palo-alto.ca.us/ftth/.

Chapter 4

# REGIONAL-METRO OPTICAL NETWORKS

Nasir Ghani

*Tennessee Technological University, Cookeville, TN 38505*

Email: nghani@tntech.edu

**Abstract** Regional and metropolitan networks are undergoing rapid transformations, propelled by shifting bandwidth and market paradigms. These evolutions have revealed some serious limitations with legacy SONET/SDH infrastructures and opened up many new avenues for the application of new, maturing optical technologies. This chapter presents a critical look at some of the major evolutions and challenges shaping this vital networking space.

**Keywords:** Metro area networks, regional networks, metro optical networks, metro transport, metro edge, optical DWDM rings, metro DWDM, coarse WDM (CWDM), next-generation SONET / SDH, Ethernet resilient packet ring (RPR), IEEE 802.17, network migration, network evolution, sub-rate grooming.

## 4.1 Introduction

Modern transport networking hierarchies are broadly segmented along geographic lines, i.e., long-haul, regional, metropolitan, and access. Long-haul optical backbones support larger tributaries over inter-regional distances (1,000 km or more) and are optimized for distance and speed transmission. Hence related issues focus on amplifier and regeneration concerns [1]. Meanwhile, access networks implement "last-mile" business and residential connectivity, i.e., 5-20 miles, and reflect a diverse mix of incumbent, intermediate, and emergent technologies – dial-up, digital subscriber line (DSL), digital cable, air-fiber, etc. In between lie the complex metropolitan and regional domains, also termed as "metro", with coverages ranging anywhere from 20 km (suburban loops) up to 500 km (regional rings). Today this segment is undergoing rapid evolutions owing to key developments in the associated market and technology sectors.

The primary driver in the "metro" market has been the continued growth of end-user bandwidth demands. Foremost, the expansion of the Internet coupled with improving "last-mile" technologies is yielding ever-increasing data volumes. In fact, IP/Ethernet traffic is the dominant type across most network domains. In addition, specialized *storage area network* (SAN) technologies have been evolved to address critical business continuity needs. Moreover, the demand for legacy leased line services has continued to grow at a steady pace. Concurrently, the broader market has seen extensive deregulation—and subsequent consolidation—yielding a very diverse array of "metro" operators, e.g., incumbent local and inter-exchange carriers, competitive local exchange carriers, cable multi-service organizations, even utilities. Inevitably, this intensified competition has increased margin pressures and is forcing operators to field a diverse array of service types.

Traditionally, "metro" networks have been built using a two-level hierarchy comprising *metro/regional cores* and *metro edge* domains [2]. In both cases, *synchronous optical network* (SONET)/*synchronous digital hierarchy* (SDH) has been the technology of choice, as related demands largely comprised of voice or leased line services. Over the years, these *time division multiplexing* (TDM) setups have also been adapted to carry data traffic using specialized intermediate protocol layers/overlays, e.g., *asynchronous transfer mode* (ATM) and *frame relay* (FR). However, as is well-known, such "legacy" setups pose acute scalability, cost, and operational complexities for emerging "non-TDM" demands [3],[4]. Hence new solutions are required in order to address critical scalability and multi-service requirements and streamline network cost structures.

Optical *dense wavelength division multiplexing* (DWDM) [5] technology addresses many operator concerns and is gradually gaining traction in broader regional/metro cores [2],[7],[8]. Foremost, DWDM provides capacity and distance scalability along with protocol transparency. Despite these saliencies, the adoption of this technology has been complicated by various technical and business factors. Hence, several renditions of "metro DWDM" have emerged to facilitate a more cost-sensitive migration path. Meanwhile, increased service diversity presents its own complications at the metro edge. In particular, there is critical need to economically field a wide array of client interfaces, many of which operate at "sub-wavelength" speeds. As a result, metro edge networks are trending towards new *opto-electronic* multi-protocol aggregation/grooming setups, with the major focus being on increased data efficiency, e.g., *"next-generation" SONET* [3] and Ethernet *resilient packet ring* (RPR) [9].

This chapter studies the major ongoing evolutions that are shaping the critical regional/metro networking space. First, a brief review of existing SONET/SDH architectures is given in Section 4.2 along with a summary of pertinent market and technology developments, Section 4.3. Subsequently, Section 4.4 looks at various regional/metro core solutions, detailing specialized DWDM adaptations and related migration strategies. Finally, Section 4.5 addresses the more diversified metro edge, presenting new developments in SONET/SDH, Ethernet, and "cost-optimized" optics. Concluding thoughts and future directions are presented in Section 4.6.

## 4.2 Legacy Infrastructures: An Overview

As is well-known, SONET/SDH is heavily-entrenched across metro-edge and metro/regional core domains. Briefly, SONET/SDH is TDM technology that uses fixed 125 μs frames and defines a multiplexing hierarchy to aggregate traffic, i.e., OC-n/STM-n. Additional networking functionalities (such as transport, multiplexing, add/drop, cross-connection/switching, regeneration, and protection) are also provided [10]. SONET/SDH delivers excellent resiliency and supports various configurations—hub, linear chain, ring, even mesh. In regional and metro domains, however, *hierarchical* ring architectures are by far the most common, using smaller edge rings to aggregate traffic onto faster metro/regional core *inter-office* (IOF) rings, i.e., between *central office* (CO) sites, Figure 4.1. Undoubtedly, these legacy networks have had a huge impact on the evolution of this space.

### 4.2.1 Metro Edge Networks: Services Aggregation

Metro edge rings vary from 20-65 km and typically run at OC-3/STM-1 (155 Mbps) or OC-12/STM-4 (622 Mbps). The main building-block here is the *add/drop multiplexer* (ADM) node, which aggregates traffic from low-speed interfaces, e.g., DS1, DS3, T1, etc. Metro edge ADM devices are usually linked to *customer premise* (CP) networks that groom client traffic onto TDM tributaries. Examples include *digital loop carrier* (DLC) setups, enterprise networks (T1), telephony *public branch exchanges* (PBX), etc. These rings essentially handoff to larger metro core rings at CO hubs (Figure 4.1) using either manual interconnection (patch-panels) or via *wide-band digital cross-connect* (WB-DCS) nodes. In the latter case, WB-DCS nodes can also perform bandwidth management, providing tributary termination, segregation, and grooming (i.e., multiplex/de-multiplex) at various levels such as VT1.5 (1.544 Mbps) or STS-1 (51.84 Mbps). This "centralized" DCS-based setup is commonly referred to as "back-hauling" [10],[11]. Note that many newer ADM designs provide more capable edge aggregation interfaces, helping reduce "back-hauling" port counts (Section 4.1).

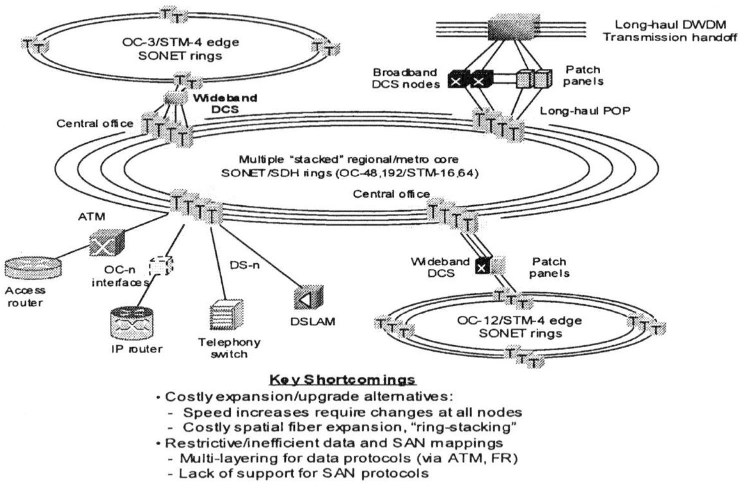


*Figure 4.1.* Legacy hierarchical SONET/SDH ring infrastructures.

Now "hubbed" traffic profiles are most prevalent in edge rings since demands are largely outbound from local regions. This essentially "fixes" routes between access sites and central hubs. Hence SONET/SDH *uni-directional path-switched rings* (UPSR) [11] are very efficient, reducing the need for more complex (expensive) shared protection schemes. Moreover, since legacy TDM uses "multi-layering" to map data traffic onto leased lines, operators have to maintain added "adaptation" platforms (ATM, FR nodes) to field data clients. Albeit acceptable for moderate demands, such "box-stacking" is very costly and induces multiple levels of bandwidth/port overheads. In particular, the latter comprises recurring charges such as footprint space, power consumption, maintenance, etc. To alay these concerns, many routers are being equipped with direct *packet over SONET* (POS) ports, e.g., 2.5 Gbps OC-48c/STM-16c. Nevertheless, overheads are still significant due to the incongruency between the Ethernet and SONET/SDH hierarchies. Hence, some proprietary mapping/interleaving solutions have emerged to increase efficiency, see [3] for more discussions.

## 4.2.2 Regional and Metro Cores: High-Speed Transport

Regional and/or metro core rings interconnect major *central office* (CO) locations at larger hand-off sites, Figure 4.1. These infrastructures are also termed as *inter-office fiber* (IOF) or collector rings [7],[11] and comprise dense core rings interconnected via larger DCS nodes. This achieves both increased scalability and increased coverage. Naturally, regional/metro core domains represent a higher level of aggregation, and hence must operate at higher tributary speeds. Today, OC-48/STM-16 (2.5 Gbps) and OC-

192/STM-64 rates are most common and rings sizes vary anywhere from 50-500 km (average span 40-80 km [1]). Note *per-hop* SONET/SDH regeneration precludes impairment concerns for most metro/regional cores.

Due to increased geographic spread, regional/metro core traffic profiles are typically more "meshed" in nature, i.e., "any-to-any". Hence, operators have typically relied upon more complex *bi-directional line-switched ring* (BLSR) designs [11], as they yield improved bandwidth efficiency (time-slot re-use). These rings can implement loop-back span protection switching for both logical (two-fiber) and physical (four-fiber) rings. In addition, ring interconnection is another key feature here, as most large core domains can easily comprise well over 10 fiber rings. Due to increased traffic scales herein, many operators use robust *dual ring interworking* (DRI)/*drop-and-continue* schemes, i.e., *dual homing*, to prevent against catastrophic single-hub/interface and even dual failures [11].

Clearly, metro core interconnection costs and complexities are much higher than those in metro edge domains. In some instances where nodal fiber connectivity levels are significantly higher (i.e., beyond degree two), operators may choose larger *broad-band DCS* (BB-DCS) nodes to provide path-level restoration/protection over *mesh* topologies, i.e., *logical layer* recovery. Although these schemes can be much more efficient in terms of spare capacity utilization (over-subscription), most solutions are vendor-proprietary and require complex software and pre-planning [11]. More importantly, mesh recovery timescales are usually much longer (seconds to minutes), and hence high-end services will still utilize ring-based protection.

## 4.3 Key Developments

Today's metro market is being driven by notable shifts in technology and business paradigms. These include advances in optical components, new access technologies, burgeoning end-user demands, and broad industry deregulation. In order to qualify further discussions, this section briefly reviews these developments and highlights the associated challenges facing legacy networks. Interested readers are referred to [1],[7] for more details.

### 4.3.1 Optical Components

Advances in DWDM components and high-speed integrated circuit design have laid the foundation for modern optical networking. In particular, there have been notable developments in lasers, amplifiers, filters, switching devices, and fibers. Consider a brief summary:

- Lasers: New integration techniques have enabled narrow line-width ITU-T grid lasers with very good thermal stability. In particular, directly-modulated designs, e.g., *distributed feedback lasers* (DFB), can deliver 2.5 Gbps speeds across most metro domains (100 km). Meanwhile, more powerful (costly) *externally-modulated* variants can overcome dispersion issues at 10 Gbps speeds [5]. Tunable lasers are also maturing, and will offer many gains in wavelength provisioning flexibility and reduced sparing costs.

- Amplifiers: Multi-channel amplification is a key DWDM advantage as it eliminates costly *per-channel* electronic regeneration. Today, *erbium-doped fiber amplifier* (EDFA) [6] devices are most common, yielding good noise/gain flatness across the C and L-bands and increased 200-600 km reach. Lately, vendors are also offering "sub-band" EDFA devices (i.e., "amplets") coupled with integrated *variable optical attenuator* (VOA) devices for gain balancing. Advances in compact *erbium-doped waveguide amplifier* (EDWA) devices and broadband Raman amplification are also noteworthy.

- Filters: Optical filtering devices offer wavelength channel/band management (i.e., mux, de-mux, bypass) and currently three types are widely used—thin-film, planar waveguide, and fiber-gratings [12]. Thin-film filters are ideal for wider channel spacing (200 Ghz) and deliver good temperature stability and passband isolation. Meanwhile, increased C and L-band densities (40 channels at 100 Ghz/0.8 nm, or 80 channels at 50 Ghz/0.4 nm) can be achieved using planar waveguides, e.g., *arrayed waveguide gratings* (AWG), and/or fiber-grating types (although temperature stability and insertion loss issues can arise). Emerging developments in *tunable* filter technologies are also showing much promise.

- Switching/add-drop: Switching elements are critical for "dynamic" optical functionality, and currently various technologies have evolved, e.g., lithium niobate, *semiconductor optical amplifier* (SOA) gate, beam-steering, liquid crystal, etc, see [14]. Most notably two-dimensional *micro electro-mechanical system* (MEMS) designs have gained the most attention, yielding sub-millisecond switching times and low crosstalk levels. Overall, many advances are expected in this field over the coming years.

- Fiber cable: Regional/metro domains largely comprise *single mode fiber* (SMF, G.652), which is ideal for single channel (1310 nm) transmission and has relatively low attenuation in the 1550 nm DWDM band (0.2-0.3 db/km). Still, SMF poses serious chromatic dispersion challenges for bit-rates over 10 Gbps and requires compensation for spans over 60 km. Hence, newer *non-zero dispersion shifted fiber* (NZDSF) and *negative dispersion fiber* (NDF) types have emerged, delivering extended uncompensated span reach (over 200 km). Also, other "metro-optimized" fibers deliver added capacity by removing the SMF 1350-1450 nm "water-peak", i.e., *low water-peak fiber* (LWPF) [13].

The above "building blocks" facilitate many advanced *networking* capabilities at the optical circuit layer, e.g., add/drop, bypass, switching, protection. Moreover, intense component vendor competition has yielded steady price declines. Clearly, future advances will help further improve component price-performance envelopes.

### 4.3.2 Networking Standards

Standards are crucial for the successful adoption of optical networking technology. Today, the core components for an interoperable optical layer have emerged [14], and although a detailed review is out of scope, a very brief summary is given. At the physical and link layers, many interfaces are already well-defined, e.g., such as SONET/SDH framing/control, 1.0/10 Gbps Ethernet, ITU-T digital wrappers (G.709), ITU-T wavelength channel grids, etc. Moreover, new developments in "dynamic" SONET/SDH framing and Ethernet packet rings are also being finalized and are proving particularly amenable for broader metro-area applicability (Section 4.5).

Meanwhile, higher-layer *network* control architectures are also starting to mature, Figure 4.2 [33]. For example, the ITU-T *optical transport network* (OTN) architecture defines a three-layer transport hierarchy comprising optical channel, multiplex, and transport sections. Also, the IETF and OIF have specified detailed signaling/control protocols for dynamic optical layers. Perhaps the most notable evolution is the *generalized multi-protocol label switching* (GMPLS) framework [14] which abstracts the label concept to a much broader set of entities—TDM timeslots, wavelengths, bands, fibers (the ITU-T is also defining a related *automatic switched transport network* (ASTN) framework [3]). GMPLS/ASON effectively increase horizontal control plane integration (data-optical) by eliminating overlapping features in traditional multi-layered setups, e.g., addressing, signaling, protection, etc. Furthermore, the OIF optical *user network interface* (UNI) provides critical "optical dial-tone" capabilities for clients to request/release capacity, and further *network node interface* (NNI) efforts [14] are expanding into inter-domain issues. Collectively, these developments will facilitate a wide range of automated *network-level* provisioning features.

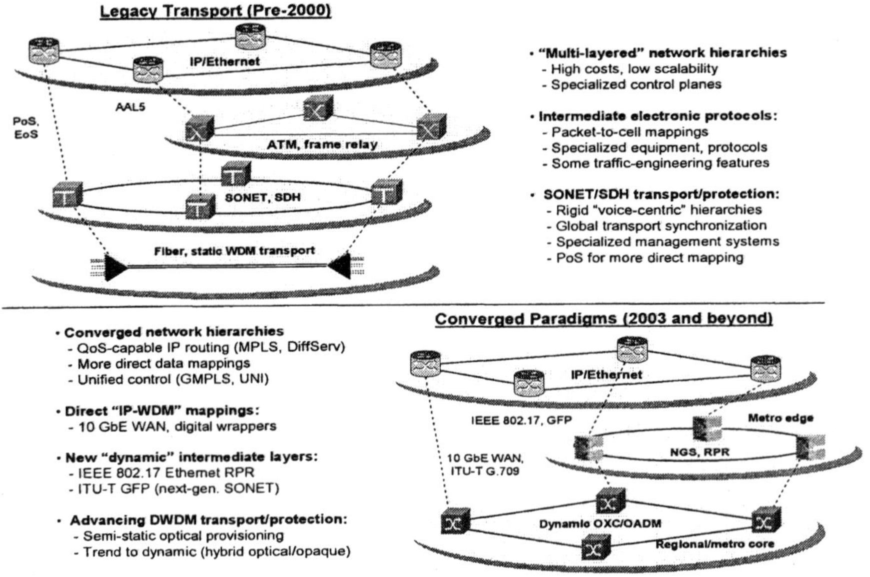


*Figure 4.2.* Network hierarchy transformation (from [33]).

## 4.3.3 Evolving End-User Domains

Concurrent to technology advances, fundamental changes have occurred across client access domains in recent years. Foremost, bandwidth demands have grown sizably, both in terms of scale and diversity. This growth has been accelerated by the deployment of improved *broadband* access technologies that largely surpass basic dial-up capacities. Collectively, the effects on the metro space have been profound, particularly due to increased end-user proximities. Briefly, consider some of the main developments:

- <u>*Broadband Access:*</u> Numerous "last-mile" access solutions are being deployed and evaluated today. For example, digital cable dominates North American residential broadband and the latest "data-over-cable" standards are using advanced silicon/signal processing to deliver multi-megabit bi-directional speeds (e.g., 30 Mbps downstream, DOCSIS 2.0). Meanwhile, global DSL growth has also been solid, and the latest offerings are achieving over 5 Mbps (1 Mbps) download (upload) speeds over copper. Concurrently, a host of high-speed wireless technologies are maturing to offer megabit speeds, e.g., 3/3.5G wireless, wireless LAN (IEEE 802.11), wireless MAN (IEEE 802.16). Furthermore, select businesses located in dense, fiber-constrained settings are even using *free-space optics*

(FSO) to achieve localized "wire-like" transmission, e.g., 155 Mbps up to 10 km. Finally, there are now definitive trends towards fiber in the last-mile, e.g., *fiber-to-the-home* (FTTH), *fiber-to-the-curb* (FTTC). Particularly, advances in Ethernet *passive optical networks* (EPON, IEEE 802.3ah) are promising in excess of 20 Mbps per user along with advanced service guarantees.

- Residential Growth: Internet growth has been the primary driver for residential bandwidth demands. Today, end-users have adopted a very broad range of "content-rich" applications (web-browsing, email, instant messaging, etc), and many new "real-time" offerings are on the horizon, e.g., packet telephony/video, video-conferencing. Hence many operators—particularly those with broadband access infrastructures—are now offering "bundled" service packages that furnish a full range of services, e.g., voice, high-speed Internet (megabits/sec), video-on-demand, etc. Overall, these changes have propelled packet-data volumes well beyond legacy voice levels, and are yielding much higher levels of traffic asymmetry and unpredictability. More importantly, studies indicate that resultant complexities are shifting to the metro/edge/access domains, as extensive long-haul core builds are largely complete [15].

- Business/Enterprise Expansion: Businesses have made extensive use of voice trunking and leased line services (e.g., DS0, DS3, T1, T3). Conversely, smaller enterprises have relied upon Ethernet-based offerings for smaller campus settings. However, given ubiquity of Internet communications, many organizations are now looking to adopt more formalized, robust "packet-based" frameworks. In particular, there is a strong push to deploy gigabit-speed Ethernet connectivity to host a wide range of applications, e.g., LAN extension, *virtual private networks* (VPN) services, and even "packetized" telephony/video. Furthermore, many corporations have deployed specialized SAN protocol technologies to meet their burgeoning archival needs and achieve robust disaster recovery across metro/regional zones [16]. Examples include Fiber Channel, ESCON, FICON, and "IP-based" *Internet SCSI* (iSCSI).

Given the above developments, metro traffic patterns are quickly moving away from earlier, more predictable legacy profiles, e.g., voice and leased line growth averaging 7-20% per year [13],[15]. By contrast, even long-haul cores exhibit more gradual demand variations, largely due to the higher levels of service aggregation. In all, these trends clearly underscore the need for improved, scalable regional/metro solutions.

### 4.3.4 Challenges and Requirements

With changing service paradigms, legacy metro infrastructures pose many limitations. Foremost, demand growth has already caused notable capacity exhaust on metro core rings. Here, traditional *spatial* TDM expansion is proving to be very unscalable and costly, i.e., increasing ring speeds or "stacking" multiple rings. The former mandates equipment upgrades at all ring nodes (plus ring-interconnection costs), whereas the latter implies added conduit pulls or altogether new roll-outs. Such expansions are only feasible in "fiber-rich" settings as rights-of-way and construction overheads can easily complicate and delay expansion otherwise [17]. Overall, these issues present a huge "bottleneck" in tapping abundant long-haul capacity.

Furthermore, legacy "multi-layering" imposes notable penalties for burgeoning data-traffic profiles, e.g., wasted bandwidth, increased port counts, intermediate protocol layers (Section 4.2.1). In particular, bandwidth inefficiency actually exacerbates capacity exhaust and leads to higher "cost-per-bit." Finally, SAN tributary support is also very difficult due to the lack of standard mappings, i.e., for 200 Mbps ESCON, 1.0 Gbps Fiber Channel. Hence, many operators must resort to costly "dark fiber" provisioning. In all, much-improved offerings are needed to address a host of concerns:

- Network Scalability: A paramount operator requirement is abundant capacity scalability over a full range of metro/regional distances. This is particularly important given the increasing speed of client interfaces and unpredictable growth levels.

- Multi-Protocol Support: As metro protocols diversify, operators are looking to achieve multi-service support over a common "de-layered" infrastructure, i.e., *transparency*. Another related need is "backwards compatibility" for maximizing return on existing investments and facilitating smoother, cost-sensitized migrations.

- Provisioning Flexibility/Efficiency: New metro platforms must also deliver significant gains in terms of provisioning capabilities. In particular, operators require increased service velocity (termed "on-demand"), efficient resource utilization, and selective/differentiated service offerings, e.g., priority, resiliency, etc [14].

- Service Survivability: Carriers are very familiar with SONET/SDH protection and newer technologies must match this capability (50 ms bound). Beyond this, multi-level "tiered" protection capabilities are also needed given the widely differing stringencies of client services, e.g., unprotected, moderate protection, full protection [20].

- *Improved Cost:* Cost is an overriding concern since metro-area user bases are relatively smaller and much more price-conscious (versus long-haul client bases). Moreover, network economics are continually dictating lower "cost-per-bit" even as tributary speeds increase. Hence new offerings must offer operational savings (power consumption, footprint) and also be amenable to staged, modular deployments, i.e., "pay-as-you-grow" [1].

Overall, a full breadth of technologies are being evolved to address these concerns. Interestingly, many of these solutions are still classified along traditional metro/regional core and metro edge taxonomies. This is in part due to entrenched operational delineations between these two domains and also due to the higher levels of service aggregation in larger metro cores.

## 4.4 Regional and IOF Core Domains

Regional and metro core networks implement scalable interconnection across large inter-city zones, ranging anywhere from 100-1000 km. Here, associated traffic flows are usually multiplexed into larger gigabit speed tributaries—OC-48/STM-16, OC-192/STM-64, 10 Gbps Ethernet—either via larger client switching/routing platforms or metro edge networks (Section 4.5). Although this loosely parallels long-haul setups, regional/metro cores feature many more add-drop points and generally higher levels of traffic variability. Hence, related design concerns shift more towards network element design as opposed to basic line transmission [7].

Overall, DWDM technology provides many advantages for regional/ metro cores, e.g., terabit scalability, multi-channel amplification, service transparency, reduced footprint/operations costs, etc. Advances in "soft-optics" are also furthering new automated service capabilities. Nevertheless, the induction of DWDM within this space has been relatively cautious, owing to notable market and technical factors. Hence, several renditions have emerged in order to facilitate more cost-sensitized, staged transitions, Figure 4.3. Indeed, the choice of a given solution is very operator-specific and depends upon a many variables, e.g., demands/service projections, cost, etc.

### 4.4.1 Early Renditions: Point-To-Point Fiber Relief

Early regional/metro DWDM deployments have comprised point-to-point transmission systems for fiber-relief on heavily-loaded IOF spans, [17]. Termed as "first-generation" DWDM [1], Figure 4.3, these solutions evolved from long-haul designs and can pack up to 100 channels/fiber at 10 Gbps per channel—a huge increase in scalability. A sample design is shown in Figure 4.4, consisting of short-reach client interfaces, wavelength transponders interfaces, amplifiers, and wavelength mux/de-mux filters. Here, transponder units perform optical modulation for client signals received from "legacy" 850 or 1310 nm onto ITU-T 1550 nm wavelengths. Note that many current SONET/SDH and Ethernet/IP platforms are already equipped with 1550 nm lasers for direct interconnection with DWDM transport. More recently, the emergence of newer compact pluggable interface transceivers is furthering support for a wide-range of client-side gears.

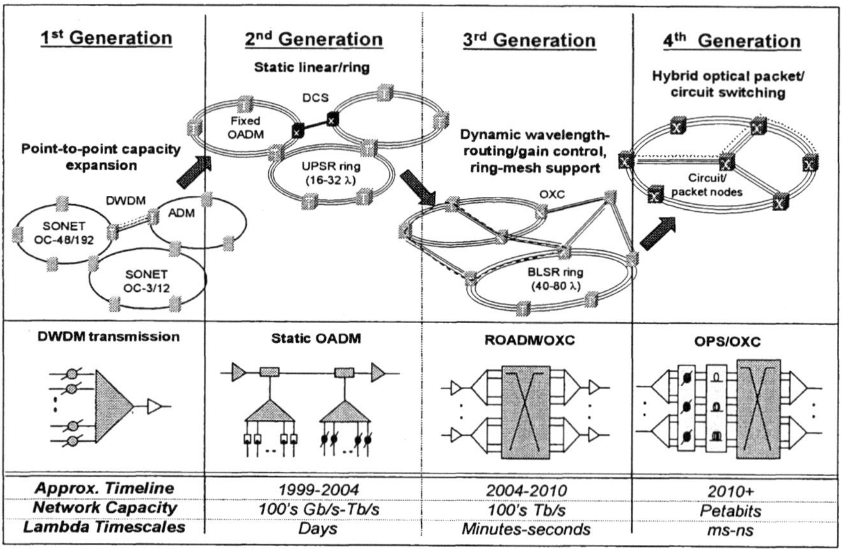


*Figure 4.3.* High-level view of DWDM migrations in the regional/metro core.

Metro DWDM systems feature many design innovations to help reduce per-channel costs. Most notably, *modularized* passive filter setups are used, exploiting channel banding to facilitate incremental "pay-as-you-grow" capacity expansions. Examples here include serial cascades, parallel fan-outs, and hybrid variants, Figure 4.4 [21]. Generally, serial cascades require added stages and yield higher losses (about 2-3 dB per stage). Additionally, revertive protection switching is also necessary in order to avoid service disruption during upgrades (more complex). Now thin-film filters are most commonly used for band filtering smaller channel groups. However, if drop counts increase, deploying fewer high-channel-count filtering devices is much more cost-effective [12], e.g., wavelength gratings or AWG type devices. Also note that band filters have also been coupled with interleaving devices to increase channel densities with less costly wider channel filters.

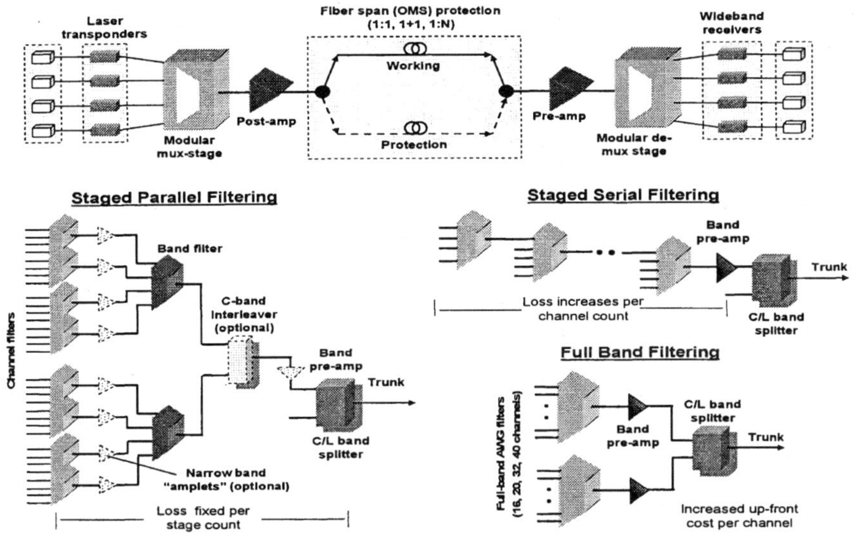


*Figure 4.4.* Point-to-point DWDM transmission and channel banding/interleaving.

Depending upon span length and insertion losses (filters, interfaces), "first-generation" DWDM systems may require various laser/amplifier provisions. For example, short SMF spans (under 50 km) can use basic DFB lasers, see [1] for sample references. Meanwhile at increased 10 Gbps speeds, *polarized mode dispersion* (PMD) effects further reduce SMF span lengths to about 60 km [22]. Here, designers have to use more powerful externally-modulated lasers, EDFA devices and possibly even *dispersion compensation fiber* (DCF) modules. Note that *forward error correction* (FEC) at transponder interfaces can also give notable gains, e.g., ITU-T digital wrappers (G.709) payload-independent FEC delivers 2-3 dB gain for about 6% bandwidth overhead. Inevitably, FEC imposes notable cost/ complexity for 10 Gbps speeds (OC-192/STM-64, 10 Gbps Ethernet).

Given the massive scalability of DWDM, service protection is a critical issue. For point-to-point systems, various protection schemes are possible, including both *optical multiplex section* (fiber span) or *optical channel* (OCh) level setups [20]. For example, *dedicated* 1+1 protection (non-signaled) can be done at either the wavelength/band/fiber levels via passive splitters to bridge/switch all protection entities onto two working/protection fibers. Alternatively, "non-dedicated" 1:1 or 1:N OMS protection (Figure 4.4) can also be done using active switching devices and fast protection signaling. Although these alternatives pose increased cost/complexity, resource efficiency is much higher as multiple working fibers and/or lower priority traffic can share idle capacity, see [1]. Moreover, optical-layer protection is of particular benefit for SAN applications, as many related protocols lack protection-switching features [28]. Overall, DWDM protection can significantly lower higher-layer (client) electronic protection costs, e.g., OMS protection with 1:N SONET/SDH protection, see [17].

### 4.4.2 Intermediate Transport: Static Add/Drop Rings

As point-to-point systems proliferate, the next logical step is the extension of wavelength channels/bands across *multi-hop* metro ring fiber-plants. A key objective here is to maintain optical transparency as much as possible in order to eliminate costly *service-specific* electronic bit-processing at intermediate sites. Considering that add-drop ratios in TDM core rings are typically about 25%, such "transponder-less bypass" achieves sizeable equipment savings, particularly at 10 Gbps OC-192/STM-64 speeds. As a result, designers have also evolved static *optical add-drop multiplexer* (OADM) rings, i.e., "second-generation" DWDM [1], Figure 4.3.

A static/fixed OADM is basically a "back-to-back" interconnection of transmission systems, see Figure 4.5 [8]. Here modular banding filter setups (Section 4.4.1) are augmented with wavelength/band-level *bypass-and-add-drop* filters—either thin-film, fiber Bragg grating, or circulators—to implement fixed routing relations. DWDM bypass lowers through-channel insertion losses considerably, to about 1-2 dB per node [18], and yields commensurate increases in link budgets/ring diameters. Moreover, bypass filtering can also prevent against passband-narrowing effects arising from multi-hop filter concatenations [19]. Carefully note, however, that modular serial filtering setups (Section 4.4.1, Figure 4.4) complicate ring expansion and will require forced protection switching to avoid service disruptions.

A central issue in static ring designs is amplification. Obviously, larger spans (over 80 km) require commensurate EDFA provisions. However, the increased number of metro add-drop hub sites usually mandates amplification in smaller rings to handle *nodal* losses on bypass channels. Hence most static OADM designs incorporate pre- and post-amplifiers, the former to overcome transmission losses and the latter to overcome node bypass losses (Figure 4.5). Note that *variable optical attenuators* (VOA) devices are essential here to equalize gain between bypass and add-drop channels, i.e., amplifier cascades. Moreover, VOA devices are also required to counteract gain variations arising from external factors such as temperature, aging, and polarization. Hence many filter modules directly incorporate manual VOA elements (Figure 4.5).

Now from an architectural viewpoint, the optical *dedicated protection ring* (OCh-DPRING) [1],[2],[20] architecture is the most common static ring, Figure 4.6. Also termed as *uni-directional path-switched ring* (UPSR), this scheme offers simplified, low-cost operation and extremely fast/non-signaled protection. OCh-DPRING draws strongly from related SONET concepts and uses two counter-propagating uni-directional fibers. Protection is done in a dedicated 1+1 manner using head-end bridging and receive-end switching, usually based upon power levels. This avoids complex protection signaling and performance monitoring *inside* the ring. However, associated per-channel hardware complexities limit scalability for fiber cut events—which can average about one per 10-20 km/year in metro cores.

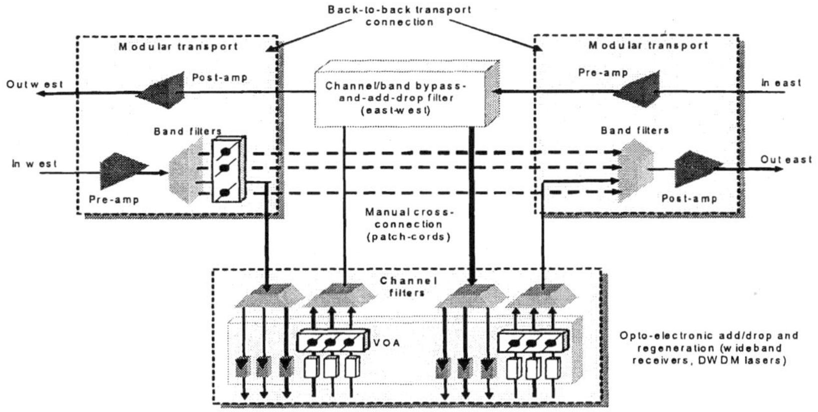


*Figure 4.5.* Fixed/static wavelength OADM design (two-fiber).

Overall, static rings are most amenable to long-standing demands, e.g., weeks/months. Hence related provisioning concerns center around wavelength (band) pre-planning and routing, i.e., static ring *routing and wavelength assignment* (RWA) [5]. Now it is well-known that UPSR designs are very efficient for hubbed demands, as routes are essentially pre-determined, i.e., linear relation between node and wavelength counts [32]. Hence, wavelength blocking can be avoided by simply allocating a unique/fixed set of wavelengths (bands) to each node, as per projected demand [8] (more robust dual-hub rings have also been studied [1]). Overall, static rings mandate careful demand projection, as inaccurate estimates can result in significant stranded capacity. Moreover, static UPSR is very inefficient for meshed or overly dynamic profiles. As such, their applicability in larger "meshed" regional/core settings may be limited.

Finally, static rings require careful link-budget analysis and amplifier placement to ensure adequate *signal-to-noise* (SNR) ratios. However, since amplifiers represent a sizeable cost, deploying "full-band" EDFA devices at all nodes is overly costly and can degrade performance due to excessive noise amplification. Designers have proposed several solutions here. Some have used careful link budget analysis to optimize amplifier placement on selected spans, e.g., per distance, fiber type, demands, see [23]. Meanwhile, others have optimized demand routing over selected bands and used smaller, more cost-effective "amplets" (Section 4.4.1) to boost bypass groups. Note that link budget planning requires gain equalization, adding notable manual operational overheads, e.g., continual fine tuning of VOA settings.

### 4.4.3 Evolving Schemes: Reconfigurable Add/Drop Rings

As traffic dynamics increase, static rings become less efficient due to excessive pre-planning and manual provisioning overheads. Moreover, IOF projections are touting much higher scalability, particularly for "meshed" demand profiles, i.e., quadratic wavelength/node-count relation [24]. Hence, operators are looking at *reconfigurable* OADM rings to implement expedited "on-demand" services provisioning with much-lower costs. These trends have led to the emergence of "third-generation" DWDM [1], Figure 4.3, essentially blending flexible optics with intelligent software control.

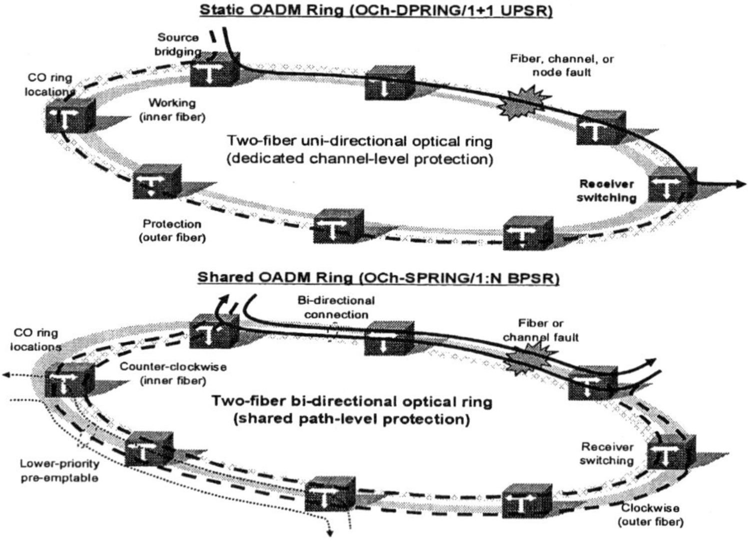


*Figure 4.6.* Static (UPSR) and dynamic (SPRING) ring protection.

Dynamic rings apply *shared protection ring* (SPRING) paradigms [20], extending upon earlier SONET *bi-directional line-switched ring* (BLSR) [11] concepts. Namely, two-way wavelength plans are defined in order to achieve spatial re-use and deliver increased efficiency and *multi-level* protection. SPRING routes bi-directional demands along the same set of ring nodes and allows working/protection traffic to travel in both directions. Note that the DPRING designs (Section 4.4.2) cannot achieve such *spatial re-use* [20], since connections traversing different ring segments are unable to use the same wavelength. Now *both multiplex-section* (OMS-SPRING) and *channel-level* (OCh-SPRING) variants are possible:

- OMS-SPRING: Both two and four-fiber multiplex-section (fiber) protection schemes are possible. The former partitions *intra-fiber* wavelengths into working and protection groups and performs loop-back switching for node/fiber faults [1]. Here, all failed channels are re-routed onto a protection fiber (opposite direction) between failure-adjacent nodes. Meanwhile, the four-fiber variant uses two fiber *pairs* to carry counter-propagating working/protection traffic and supports span and loop-back switching. Note that loop-back is very disruptive and yields longer channel distances (worst-case nearly double). Conversely, span switching simply routes failed channels onto a protection fiber but cannot overcome node faults.

- OCh-SPRING: Although OMS-SPRING is very efficient, it lacks protection selectivity. Hence OCh-SPRING schemes, i.e., *bi-directional path-switched ring* (BPSR) [25], have been developed to tailor protection per individual demands. OCh-SPRING uses end-to-end switching between channel end-points and thus only requires *edge* fault detection. More importantly, OCh-SPRING can achieve wavelength sharing (backup multiplexing) between multiple working paths and/or lower-priority traffic (Figure 4.6). Hence operators can differentiate services to meet a broader set of customer needs, e.g., dedicated (platinum), shared (gold), un-protected (silver), and pre-emptable (bronze) [1]. Note OMS-SPRING can also support pre-emptable traffic on idle protection channels/spans.

Overall, SPRING delivers high efficiency for "meshed" demands, and four-fiber schemes are most scalable, see [1],[20] for details. Nevertheless, these setups require rapid, distributed protection signaling to match 50-ms TDM recovery times, i.e., "optical" APS. To date, such protocols have not been standardized and related work is still ongoing in the ITU-T. Note that optical protection also requires careful coordination with higher-layer recovery since timescales can be similar (escalation strategies). For example, many operators use DWDM to host SONET/SDH rings, e.g., "ADM-in-lambda". Here, results have shown that OMS-SPRING can actually induce *cascaded* APS behaviors between the two layers, prolonging recovery times well beyond 50 ms [25]. Conversely, simpler OCh-DPRRING avoids these deleterious effects since recovery is usually much faster (under 5 ms). Hence the simplest strategy may be to disable protection at a given layer, e.g., SONET APS over unprotected wavelengths.

Now the main SPRING building block is the *reconfigurable OADM* (ROADM) node [18]. This device loosely resembles its static counterpart in that it comprises transport, amplification, and add-drop stages. However, the main difference lies in its dynamic operability and choice of implementation technologies. Namely, ROADM nodes can either be opaque (wavelength convertible) or transparent (wavelength inconvertible), and each type imposes its own saliencies/limitations [6]. Consider the former. Opaque elements perform *opto-electronic* conversion of all incoming channels and use standardized framing formats for regeneration, e.g., SONET/SDH, digital wrappers. Meanwhile, dynamic add/drop is done using electronic switching, either via DCS fabrics or *electronic cross-point switches* (EXC). For example, EXC chips can deliver high-port counts (tens-hundreds) and support speeds up to 3.5 Ghz, whereas DCS fabrics offer further regeneration/sub-rate grooming capabilities. Note that DCS fabrics can also perform *local* client-side tributary interconnection, i.e., "hair-pinning" [8].

For provisioning, ROADM rings must implement intelligent "on-demand" RWA, as requests can occur over many timescales. Now in opaque settings, ring RWA is notably simpler since ROADM nodes support opto-electronic *wavelength conversion.* For example, existing SONET/SDH circuit provisioning algorithms can easily be re-applied here. Nevertheless, despite these benefits, wavelength conversion achieves only modest efficiency gains in optical rings, e.g., under 5% [26]. Generally, results indicate much better (wavelength conversion) blocking probabilities in mesh topologies, where increased connectivity—beyond degree two—can reduce link load correlations [5]. Instead, optical rings benefit more from increased wavelength counts, especially for increased demand peakedness.

Overall, opaque (translucent) transmission offers some notable benefits. Foremost, full regeneration mitigates analog impairments—such as loss, dispersion, cross-talk—on most spans under 100 km. Another by-product is the availability of *inband* control and detailed performance monitoring information *inside* the ring. These provisions are crucial for fault isolation and protection signaling in OMS-SPRING operation, e.g., both SONET/SDH and digital wrappers define APS bytes. Nevertheless, opaque rings pose excessive cost and scalability limitations. For example, each ROADM node requires extensive transponder arrays for each fiber along with high-density DCS fabrics. These overheads are further exacerbated for high channel counts (over 32) and 10 Gbps speeds, i.e., externally-modulated transponders, thousands of STM-1 ports.

## 4.4.4. Transparency Considerations

Transparent ROADM platforms have also been considered to address the limitations/unscalabilities of opaque designs. These nodes use "soft-optics"—such as tunable filters, tunable lasers, optical switches, software-controlled VOA—to selectively route and add-drop wavelengths. A generic design is shown in Figure 4.7 (optional DCS grooming), and a variety of renditions are possible. For example, static channel/band filtering (Section 4.4.1) can be coupled with optical switching to achieve selective channel add/drop. Here MEMS/thermo-optic switching fabrics have shown switching times under 5 ms [6]. Alternatively, per-channel tunable filters can be placed on input trunks to drop selected channels, e.g., fiber Bragg gratings [8]. Overall, multi-hop "all-optical" transmission is quite challenging and requires many specialized provisions. Some of these are now discussed, and interested readers are referred to [1] for more details.

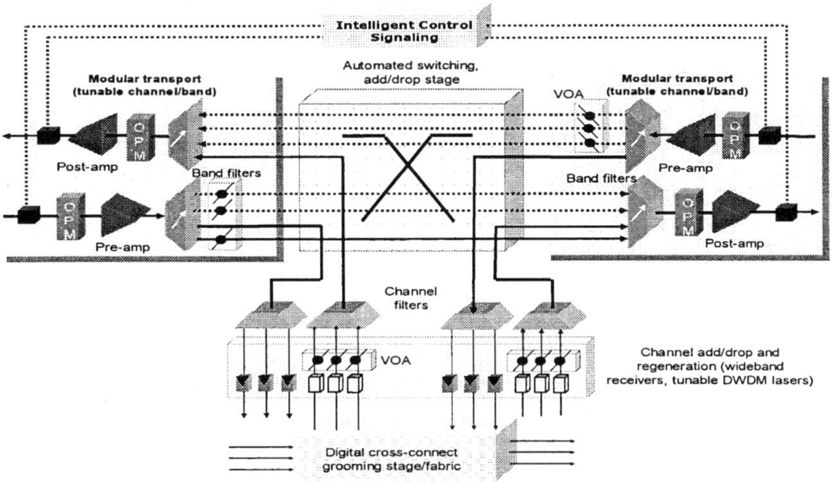


*Figure 4.7.* Integrated OADM/DCS design (two-fiber, out-band OSC control).

A major challenge for transparent rings is effective performance monitoring and fault localization. Currently, several non-intrusive schemes are available to monitor various parameters, e.g., fiber/wavelength powers, *optical* SNR, and other specialized metrics. However, it is well-understood that these offerings lack the bit-level resolution of *opto-electronic* overheads and are less effective in detecting degenerative conditions. Regardless, these features may still suffice for SPRING. For example, trunk power monitoring can rapidly isolate hard failures (fiber cuts, node faults), as needed for OMS-SPRING. However, since loop-back generally yields longer protection paths (increased signal degradation), OCh-SPRING schemes are more applicable.

These schemes preclude detailed *intra-ring* fault localization as error detection/recovery coordination is only done at the edge (e.g., even via electronic monitoring). Carefully note that transparent rings require on-overlapping wavelength plans in order to ensure switchovers [20].

Unlike translucent rings, transparent rings also require fast amplifier/ attenuation control for gain flatness over wide dynamic input ranges. The reasons here are several-fold. Foremost, dynamic add-drop (setup/takedown, protection) yields sizeable transients in cascaded amplifier settings [18]. Depending upon the number of channels, these shifts can severely disrupt active bypass channels and even damage components. Moreover, external/ageing factors can also affect component gain (Section 4.4.1). Hence *automatic gain control* (AGC) and software-controlled VOA devices (Figure 4.7) must be used to stabilize outputs. Here, sample feedback control schemes have shown impressive results, e.g., millisecond nodal settling times for surge levels under 2 dB and sub-second *network* stabilization, see references in [1]. Nevertheless, it is still difficult for these schemes to achieve 50 ms "SONET-like" operation [28]. Moreover, per-channel monitoring entails significant cost and packaging overheads with discrete components. Here, advances in optical integrated circuits will be of particular relevance here, as multiple functionalities can be coalesced at module level, e.g., filters/VOA/switches, switches/photo-detectors, etc. Such integration will also reduce insertion losses considerably, further scaling ring diameters.

Transparent ROADM rings also require more specialized *outband* control setups with physically separated data and control planes. In most cases this is done using a dedicated *optical supervisory channel* (OSC, 1510 nm) along with *per-fiber* add/drop filtering and transponders (Figure 4.7). Herein, some optimized designs implement OSC extraction/insertion directly within coarser band-separation modules. Now a key concern with outband OSC control is the lack standardized signaling mappings, e.g., protection signaling, routing updates, network management signaling, etc. As a result, some designers have chosen Ethernet OSC framing to carry packet-based signaling protocols such as GMPLS. Others have used more traditional SONET/SDH framing on slower OC-3/12 links to lower costs. Inevitably, these variations will complicate multi-vendor interoperability.

In terms of provisioning, transparent RWA is more involved due to ROADM wavelength selectivity. Namely, RWA computation now entails two steps—route resolution and wavelength selection—and numerous schemes have been studied, e.g., including shortest/longest/least-loaded path selection, and random/first-available/least-used wavelength selection, etc [7],[8]. Overall, efficient RWA yields good results for transparent rings, e.g., 90% utilization of static "a-priori" RWA and only 5% below full wavelength conversion in [26]. Note that dynamic RWA requires accurate resource information (e.g., wavelength usages, connection maps, protection routes), and this can be extracted via GMPLS resource updates [14]. Further details and theoretical treatment of optical rings can be found in [1],[32].

Given the many optical impairments in transparent settings, related provisioning algorithms must ensure adequate lightpath quality. This is particularly true for 10 Gbps speeds, where PMD effects alone can limit ring diameters to under 80 km. Now a simple solution can be to use geographic bounds to ensure adequate SNR levels, i.e., *enforced homogeneity* [22]. However, with expanding infrastructures and increasing speeds, this is an overly rigid and inefficient approach. Alternatively, some have studied more advanced schemes that explicitly incorporate impairments into the "logical" RWA process, e.g., loss, dispersion, see references in [31]. Nevertheless, related computational overheads are quite prohibitive, and it is still difficult to accurately model all effects. Consequently, many designers have to resort to explicit *hardware-based* considerations, especially for PMD effects. For example, fixed DCF coils can be placed along selective spans using static pre-planning. In the future, advances in *active* dispersion compensation can also be of benefit herein, and this application requires further attention.

Given the above concerns, *hybrid* architectures have also been tabled as a compromise between transparent and opaque setups. For example, the "optical islands" concept [22] proposes limited "all-optical" domains with bordering opaque elements. Alternatively, *translucent* [1] setups inter-mix transparent and opaque capabilities within a single domain, e.g., by augmenting ROADM nodes with opto-electronic transponders and DCS fabrics, Figure 4.7. Note that these schemes require that RWA algorithms with *partial* wavelength conversion. Here, some have studied two-step algorithms in which transparent RWA is attempted first and regeneration is only used if degradation on a searched route is excessive, see [31] for details. Generally, results for hybrid ring RWA indicate that selective wavelength conversion delivers blocking performance on par with full wavelength-conversion, e.g., only 10-20% nodes with full wavelength conversion [5],[6]. This is a growing area which will likely require more future research.

### 4.4.5 Network Migration Strategies

Despite its saliencies, the induction of regional/metro core DWDM has been complicated by some notable factors, e.g., entrenched legacy TDM, maturity concerns, challenging markets, etc [33]. Hence network migration is a key issue as operators seek to maximize existing infrastructures and evolve in a staged, cost-effective manner. Here, DWDM solutions must offer low "first-cost" and sizeable reduction in operational expenditures [16].

Today, most metro networks comprise "first-generation" setups for fiber-relief on congested spans. Commonly, these "virtual-fiber" systems are paired with flexible client-side transponders to field "non-TDM" services. To lower initial cost, modular filtering/interleaving schemes (Section 4.4.1) have been used extensively. Additionally, legacy/DWDM band-partitioning has also been done to alleviate migration concerns. Namely, inexpensive 1310/1550 nm filters are used to selectively induct DWDM (per-hop basis) over legacy rings, [27]. With increasing point-to-point deployments, some operators are even graduating to static OADM setups, effectively migrating SONET/SDH out of the core (ideal for long-standing demands).

The further migration to "third-generation" ROADM rings is a crucial issue for metro/regional carriers. In cases with relatively modest demands, carriers will likely retain static dedicated rings setups. Alternatively, shared ROADM designs are under serious consideration in larger settings in order to improve fiber utilization for "gigabit-level" demands. However, the widescale deployment of such shared paradigms is still lagging, as host of economic and technical issues remain [33]. In the case of opaque ROADM rings, transponders and DCS fabrics present a significant barrier to increased channel-counts and speeds. Conversely, transparent ROADM designs lack maturity and operators are very concerned about performance monitoring and impairment handing provisions. Given these realities, initial ROADM deployments will likely comprise smaller *opaque* setups [28].

Note that ROADM deployments raise the issue of multi-ring *interconnection* at large CO sites. Conceptually, such interconnection can be done by re-using SONET/SDH concepts, e.g., signaled/non-signaled, single dual homing [6]. However, with increasing nodal/spatial degrees and traffic dynamics, *optical switching* platforms are very much required (akin to WB-DCS in legacy cores, Section 4.4.2). Now recently, some have proposed *optical cross-connects* (OXC) using "all-optical" switch fabrics (Section 4.3.1). These systems are agnostic to bit-rate and can support wavelength/band cross-connection—very scalable. However, operators remain concerned about the maturity of these new technologies along with the lack of detailed "trail trace" features (a vital necessity for multi-operator handoffs). Hence for the foreseeable future, ROADM interconnection will use DCS platforms with full monitoring/ regeneration. Albeit less scalable, these offerings also enable wavelength conversion, which has been shown to be most effective at interconnection points [1]. Moreover, it is likely that future *hybrids* will incorporate combined DCS/OXC fabrics, Figure 4.7.

Further evolutions toward partial/full *mesh* topologies in regional/metro cores are more debatable. Still, it is conceivable that specialized scenarios will justify such undertakings. For example, operators may deploy new fiber routes to "break" ring topologies and relieve congestion at "hot-spot" locations. Alternatively, rights-of-way restrictions may mandate "non-ring" topologies to be deployed in new "greenfield" builds. The resultant networks can either be provisioned as multiple "virtual" rings or generic meshes, both of which have been well-studied [1]. The latter, however, can increase complexity as OXC/DCS have to implement more specialized OADM functionality. Clearly, more defining studies are required here.

## 4.5 Metro Edge Domains

Metro edge domains range from 20-50 km and form a merging point between metro cores and last-mile access. These networks present largely different contingencies owing to increased client proximity levels. As such operators must handle much higher protocol heterogeneity and provision slower-speed "sub-gigabit" interfaces, e.g., DS3, T1, T3, OC-3/STM-1, Fast Ethernet, ESCON. Furthermore, economic sensitivities are much higher as costs have to be amortized over smaller populations. Clearly, the direct application of large-tributary DWDM systems is not very feasible.

In response, advanced *opto-electronic* grooming solutions have been evolved to address metro edge needs. Some of these improve upon existing SONET/SDH and Ethernet technologies to better address multi-service needs. Alternatively, "cost-optimized" optical transport has also emerged. Ultimately, the choice of solution will depend upon a large range of factors, e.g., existing setups, economic sensitivities, client demands/projections, etc. Given that metro edge infrastructures largely out-number regional/metro core rings, indeed this is a huge market segment. In fact, some research even indicates that revenue opportunities are migrating to the network edge as long-haul/metro core capacities become more commoditized [15].

### 4.5.1 "Next-Generation" SONET/SDH and MSPP

Entrenched TDM plays a vital role in data-optical convergence. For example, legacy leased line services (DS-1/3, T1/3) still dominate demand profiles and are widely used for data POS interfaces. Hence, designers have evolved SONET/SDH to better address emerging needs, i.e., *"next-generation" SONET*/SDH (NGS) or *multi-service provisioning* (MSPP) [3]. These solutions deliver several major enhancements in terms of dynamic circuit allocation, improved mappings, and layer two/three aggregation (Figure 4.8). A major advantage is that NGS re-uses existing TDM provisions for performance monitoring, protection switching, management, etc. This ensures much-needed "backward compatibility" and eases migrations.

A major shortcoming of legacy TDM is its inefficient "tributary-to-tributary" mapping, e.g., Gigabit Ethernet requires full 2.5 Gbps OC-48/STM-16. To resolve this, NGS applies inverse multiplexing using *virtual concatenation* [3], i.e., combining multiple smaller tributaries into a *VC group* (VCG) to better match "non-TDM" demands (ITU-T G.707). Today, DCS switch fabrics permit VC at both STM-1 and even VT1.5 increments (e.g., Gigabit Ethernet via seven STS-3c, Figure 4.9). Overhead performance monitoring/protection features can also be extended for VCG entities. Furthermore, the related *link capacity adjustment scheme* (LCAS) protocol (ITU-T G.7042) enables "hitless" re-adjustment of the number of assigned VC trails—ideal for fielding dynamic, varying demands. Moreover, each trail can be independently routed (even over legacy domains) to improve resiliency and efficiency, and connection asymmetry is also possible [4].

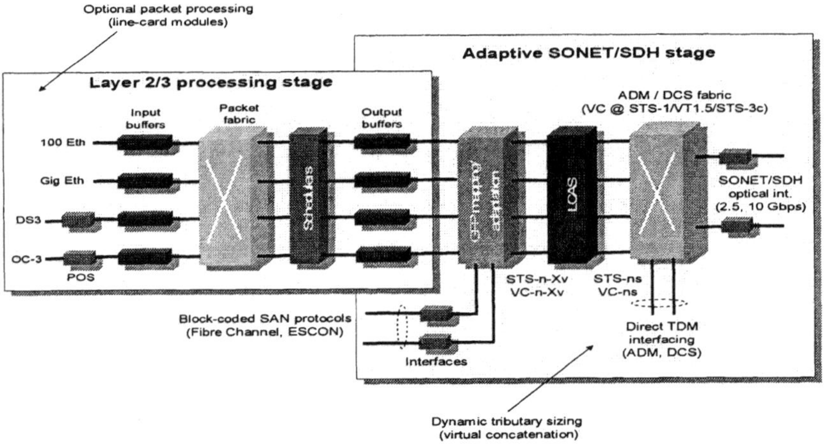


*Figure 4.8.* Next-generation SONET/SDH (MSPP) node architecture.

Another vital NGS addition is the *generic framing procedure* (GFP, ITU-T G.7041) [3] that maps diverse protocols onto byte-synchronous TDM channels. This mechanism uses robust error-controlled frame delineation (like ATM) and supports two payload mappings, frame and transparent. The former is geared for generic layer-two packets, e.g., Ethernet MAC or IP packets, and yields deterministic bandwidth overheads (versus POS). Meanwhile the latter transparently maps generic 8B/10B encoded payload types (Gigabit Ethernet, Fiber Channel, ESCON, FICON) and minimizes any packetization/ buffering delays. This is ideal for high-speed block-coded interfaces [4]. Note GFP also defines linear and ring extension header frames for multiplexing multiple client streams (ports) into a payload, i.e., akin to a "virtual" leased line. In all, coupling GFP adaptation with dynamic

VC/LCAS capabilities allows NGS to effectively "right-size" a full range of data/SAN service tributaries onto SONET/SDH pipes. Perhaps most important, "direct" bit-level mappings effectively collapse legacy multiplexing hierarchies and yield much lower operational overheads.

Finally, some MSPP designs even support "high-layer" statistical multiplexing features, i.e., "data-aware SONET", Figure 4.8. Examples include data packet aggregation (packet policing, discarding, class-based-queueing), LAN switching, and ATM switching [4]. Overall, coupling these features with VC/LCAS achieves very high time-slot efficiency and can lower port counts considerably. For example, three 100 Mbps Ethernet streams averaging 10% utilization can be policed/shaped into a single VCG of size 20 VT1.5 circuits (versus three OC-1 legacy interfaces)—a capacity savings of nearly 94%. In general, edge-aggregation provides significant improvement in overall ring efficiency versus more costly centralized back-hauling [10]. Architecturally, MSPP nodes can also leverage the GMPLS framework for equivalence mappings between circuit and packet labels and to interwork "hitless" LCAS control, see [3] for details.

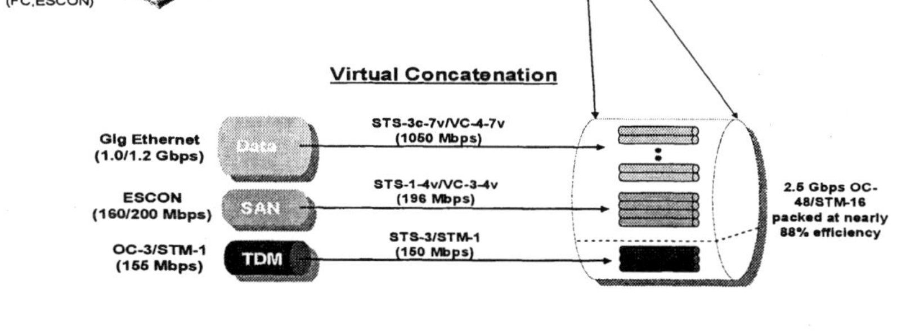


*Figure 4.9.* Sample of channelized SONET/SDH (virtual concatenation).

From a broader optimization perspective, NGS "sub-rate" grooming onto larger wavelengths is a key research area. Earlier work had looked at grooming "virtual" SONET/SDH rings over underlying DWDM networks, see [1],[29] and related references. More recently, however, the focus has shifted toward mesh-type topologies, with the goal of optimizing various metrics such as resource utilization/efficiency, port counts, wavelengths, specialized cost functions, etc. For example, some have used *integer linear programming* (ILP) to groom a-priori demands whereas others have studied more dynamic shortest-path heuristics, see [1],[29]. In particular, [29] develops a novel augmented graph model to incorporate multiple grooming granularities. Finally, sub-rate grooming of multicast connections has also been studied [35]. Overall, many of these results can be used in operational NGS settings and/or to help size sub-tending DCS stages on DWDM OXC or OADM nodes. Further studies in this broader area need to consider grooming protection and inverse multiplexing requirements. In particular, the operation of *multi-domain* transport networks comprising multiple switching granularities (layers) is an area of growing interest [34], i.e., inter-domain routing (topology/resource updates), signaling protocols, and protection/restoration schemes.

Overall, NGS has garnered the most interest out of all the metro edge technologies, largely due to its inherent synergies with legacy TDM. NGS allows incumbent carriers to re-coup stranded ring capacity and resolve deficiencies in data/SAN services—a key advantage considering that incumbents largely own the business loop. Herein, only *selected* premise equipment needs to be upgraded without costly changes to legacy core/management systems. Moreover, since most metro edge rings currently operate at slower TDM speeds (155 Mbps OC-3), sizeable gains are possible by simply upgrading to larger SONET/SDH tributaries. This is especially true in fiber-rich settings with moderate demands (DS3, OC-3, ESCON).

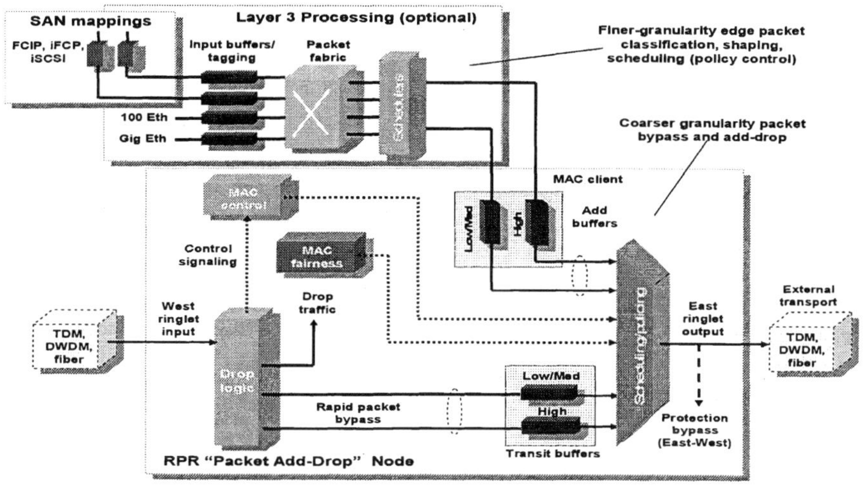


*Figure 4.10.* Ethernet RPR packet ADM node.

## 4.5.2 Resilient Packet Ring: The "Packet ADM" Solution

Despite its benefits, NGS is still a "data-over-circuit" approach that uses rigid underlying framing. For "data-centric" carriers lacking legacy TDM setups, these necessities incur notable cost/complexity. Hence the "packet rings" concept has been evolved to combine the saliencies of SONET/SDH (simplified connectivity, resiliency) with those of ubiquitous Ethernet (low cost, statistical multiplexing, etc). Termed *resilient packet ring* (RPR) [9], this solution focuses on per-flow fairness and ring capacity re-use.

RPR defines a modified Ethernet *media access control* (MAC) protocol running over dual counter-rotating "ringlets" comprising multiple "packet ADM" nodes, Figure 4.10. This protocol uses in-band (in-stream) signaling and is agnostic ("media-independent") to underlying transport layers, e.g., SONET/SDH, DWDM, CWDM, dark fiber, see Figure 4.11. RPR also provides an *automatic protection switching* (APS) protocol for rapid "50-ms" path recovery using both steering and wrapping mechanisms, akin to SONET K1-K2 byte APS. Additionally, new IP-based *quality/class of service* (QoS/CoS) frameworks, such as *multi-protocol label switching* (MPLS) and *Differentiated Services* (DiffServ), are also supported. Namely, RPR has full provision for features such as multi-level traffic policing, class-based queueing, and scheduling, Figure 4.10. Essentially, this allows RPR nodes to interface directly with DWDM cores at full wavelength line-rates, and at the same time, provide tailored "soft circuit" capabilities.

RPR uses various methods to increase bandwidth "re-use" across a ring, i.e., *bandwidth multiplication* [9]. Foremost, sink nodes perform *destination stripping* by removing unicast packets to improve "spatial re-use" (long-side). Additionally, both ringlets can simultaneously carry working traffic and only need to activate protection capacity during failure events. Finally, RPR provides advanced fairness mechanisms that allows nodes to share capacity in a fair, efficient manner. Namely, this is done via feedback flow-control setups, loosely similar to early ATM-based offerings. Also note that RPR is inherently optimized for packet multicasting since multiple ring stations can easily share a packet, i.e., drop-and-continue. Now a key issue is the design of effective distributed RPR fairness/re-use algorithms (left to vendor discretion). For example, [30] prototypes a novel solution which achieves good fairness with rapid convergence (under two ring times).

RPR standardization has attracted much focus in the recent years. Although the initial work originated in the vendor community, as interest grew a new IEEE 802.17 working group was created (2000). A key milestone was the successful harmonization of two competing proposals, i.e., Gandalf and Aladdin, into the new Darwin proposal (2002). Subsequently, the initial RPR framework draft was released, which included an RPR MAC specification [9]. Today, the IEEE 802.17 team continues to refine the standard, adding new physical layer interfaces, management enhancements, and improved bridging support, etc. The group is also developing simulation models to benchmark RPR performance.

In all, RPR is very cost-effective for packet transport and may become the solution of choice for "data-centric" operators. For example, business loops can deploy small packet ADM nodes to deliver a full range of converged services, e.g., data (Internet, LAN extension, storage extension), carrier-grade voice, video-conferencing, virtual leased line. Service flexibility is also very high as allocations can be re-optimized per demand variations—a vast improvement over cumbersome "telecom-adapter" solutions. Also, packet ADM nodes can be used for low-cost data aggregation in DSL hubs. Meanwhile, larger RPR systems can serve as *point-of-presence* (POP) aggregation devices at larger metro core handoff sites. Moreover, cable operators are also considering RPR for bundled services in the business and residential loops. In particular, the inherent multicasting capabilities of RPR are very cost-effective for packet video broadcasting.

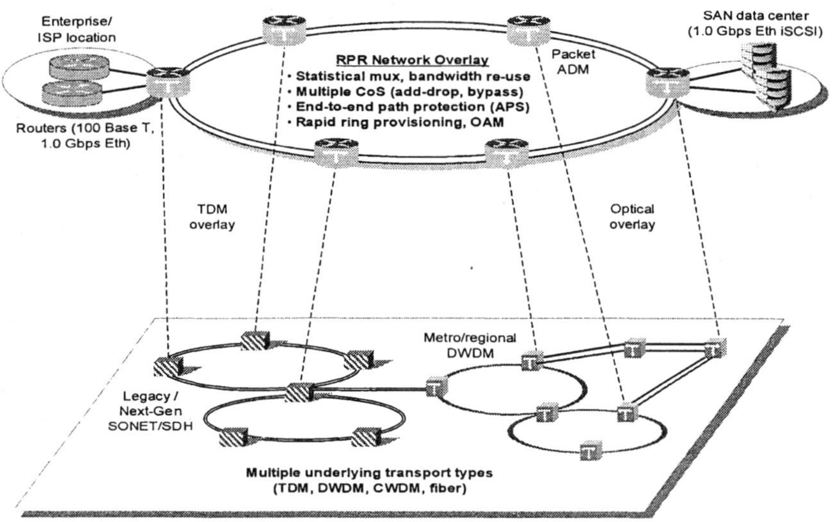


*Figure 4.11.* Ethernet RPR topology overlay (SONET/SDH, DWDM).

In incumbent "telco" domains RPR will also see targeted application as a "packet overlay" solution for underlying TDM rings, Figure 4.11. Already, RPR platforms come with SONET/SDH interfaces—OC-48/STM-16 or OC-192/STM-256—and the latest offerings can even support virtually concatenated SONET/SDH tributaries mappings, [3]. Note that related RPR-TDM control provisioning and resource allocation issues need further investigation, and this can draw from work in both sub-rate grooming and ring overlays (Section 4.4.1). However, RPR lacks standardized support for legacy TDM voice/leased line. Instead, only vendor-proprietary offerings that "packetize" TDM frames or use non-standard MAC implementations are available. Although generic "TDM-over-data" proposals are being tabled, their wider adoption in legacy settings is still uncertain.

## 4.5.3 Cost-Optimized Optical Transport

As DWDM prices decline, the feasibility of static "hubbed" metro/edge transport rings is also improving. In particular, reduced scales herein—both geographic and traffic—enable notable "optics-related" cost reductions. Foremost, shorter spans permit *passive* un-amplified transport, eliminating EDFA devices and permitting simpler DFB lasers (e.g., 2.5 Gbps up to 100km). Moreover, since moderate wavelength counts can deliver ample capacity, low-cost coarse filters can be used, e.g., 8-16 channel DWDM at 200 GHz. Finally, since *static* setups are quite efficient for "hubbed" edge demands, wavelength provisioning is much simpler (Section 4.4.2).

Recently, un-amplified *coarse WDM* (CWDM) has also gained favor in the metro edge. Interestingly, this technology was commercialized *after* DWDM in order to lower filter and transponder costs. CWDM uses a new wavelength grid (ITU-T G.694.2) with 20 nm channels from 1270 nm (O-band) to 1610 nm (L-band), i.e., well beyond DWDM bands. Since wider channel-widths can easily absorb center-frequency drifts (6 nm for 0-70°C), designers can use low-power *un-cooled* lasers without costly temperature control. Similarly, commensurate thin-film devices require much "fewer" layers, lowering filter costs considerably. Overall, CWDM can support about 8 channels over SMF and up to 18 channels over newer LWPF. Moreover, footprint and power consumption costs are much lower (nearly 50%), crucial for constrained premises. Nevertheless, transceiver link budgets limit CWDM ring circumferences to under 80 km and channel bit-rates are usually below 2.5 Gbps. Hence, some are considering more advanced CWDM capabilities, such as wide-band amplifiers and 10 Gbps transceivers. However, given the strong overlap of these provisions with more advanced DWDM designs, overall traction remains uncertain.

Although DWCM/CWDM edge nodes deliver ample capacities, they require careful *service mappings* to help mitigate the large "granularity gap" caused by gigabit wavelength capacities. For example, numerous studies on sub-rate SONET/SDH aggregation have indicated a rough "break-even" between DWDM rings and TDM "ring-stacking" at OC-12/STM-4 speeds [1] (likely lower for CWDM). Hence larger demands types, e.g., Gigabit Ethernet or Fiber Channel services, are best served by *direct* wavelength mappings, e.g., via "front-ending" with new pluggable interfaces. Moreover, programmable SONET/SDH transceivers provide even more flexibility as line rates can be adjusted per demand, eliminating the need for constant module upgrades. The further integration of SONET/SDH line termination functionality with CWDM/DWDM transport can also yield significant savings in electronic protection overheads.

The predominance of smaller edge tributaries still mandates *aggregation* to reduce wavelength counts and distribute costs over clients. Although standalone TDM or Ethernet devices can be used (higher footprint, complexity), many edge DWDM/CWDM systems now provide integrated "thin-mux" modules, either *synchronous* or *asynchronous* [1], Figure 4.13. Synchronous modules output SONET/SDH formats (2.5 Gbps OC-48/STM-16) and are ideal for grouping many hubbed legacy demands, e.g., 16:1 OC-3/STM-1. In fact, some modules even support client-side add-drop ("ADM-on-a-lambda") or combine data/SAN tributaries, e.g., 12:1 ESCON onto OC-48 (via GFP chipsets, Section 4.4.2). Overall, synchronous aggregation eases interoperability with extensive legacy cores and retains overhead monitoring for multiplexed aggregates—a key advantage. Meanwhile, *asynchronous* modules use packet-based designs and are geared for data port aggregation, e.g., 10:1 Fast Ethernet switch. Here, newer variants are even offering QoS/CoS features via edge packet classification/policing. Inevitably, these advanced "thin-mux" features will start to blur the boundaries between optical transport and more service-specific domains. For many operators such functional replication may not be desirable. Instead, low-cost optical transport may be used in a strictly "sub-systems" role to boost tributary counts and transparently host various infrastructures—legacy TDM, NGS, RPR (Figure 4.13). As such, edge CWDM/DWDM transport is very complimentary and will help accelerate infrastructure consolidation.

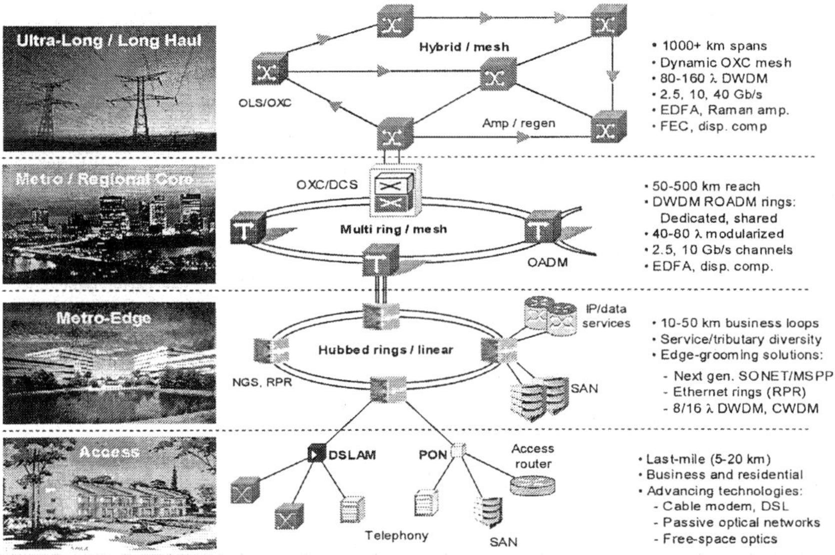


*Figure 4.12.* Emerging taxonomies: long-haul, regional, metropolitan, access.

## 4.6 Conclusions and Future Directions

Metro and regional network infrastructures occupy a strategic position in the network hierarchy. Although, legacy SONET/SDH has traditionally dominated this space, progressive shifts in customer demands have given rise to serious service provisioning and cost concerns. Hence, there is an urgent need for new evolutions that provide improved scalability, multi-service bandwidth delivery, and high service flexibility and cost-effectiveness. It is here that advances in optical networking technologies have proven particular beneficial and have essentially transformed the transport networking space (Figure 4.12 details the emerging taxonomy).

Within larger metro/regional cores, DWDM transport is fast emerging as the technology of choice. This solution offers many compelling saliencies—scalability, flexibility, lower cost—and has been heavily tailored for deployment in entrenched, cost-sensitive settings. These innovations range from simpler, static "first/second-generation" transmission systems to reconfigurable "third-generation" wavelength-routing transport. However, a host of challenges still remain to be addressed for this market to fully develop. Most notably, these include provisions for dynamic (multi-node) power balancing, dispersion compensation, rapid protection switching protocols, multi-ring interconnection, etc. Meanwhile, metro edge domains are migrating towards intelligent opto-electronic "grooming" setups with high service/interface diversity. Here, many different solutions have been developed, including "data-aware" next-generation SONET/SDH, carrier-grade Ethernet transport, and cost-optimized CWDM/DWDM transport. Herein, key open issues still remain to be investigated, such as sub-rate grooming/aggregation algorithms and multi-layer/domain (NGS/RPR-DWDM) internetworking and survivability.

Finally, given the continued evolution of end-user demands, it is important to also consider further evolutions, e.g., ten years and beyond. In particular, given the strong trends towards packet-based communications, many researchers are studying *optical packet switching* (OPS) [36] to overcome the inherent inefficiencies of circuit-switched optical networks, i.e., "fourth-generation" DWDM (Figure 4.3). The key objective of OPS is to perform as much of the packet routing operation—switching, buffering, even header look-up/processing—in the optical domain. Although this technology is currently in its infancy, continued component and architectural advances are showing much promise. As these paradigms mature, it is likely that OPS will gain increasing favor within regional/metro cores, where transmission distances are amenable to all-optical transmission (versus long-haul). For more details please refer to [1],[36].

## Acknowledgements

The author is highly indebted to the editors, Dr. Krishna Sivalingam and Dr. Suresh Subramanian, for their encouragement and overall patience in the preparation of this manuscript. The author is also grateful to Mr. Alexander Greene (Kluwer) for his support and friendship over the years.

## References

[1]N. Ghani, J. Pan, X. Cheng, "Metropolitan Optical Networks," *Optical Fiber Telecommunications (OFT) IV*, Academic Press, March 2002, pp. 329-403.
[2]A. Saleh, J. Simmons, "Architectural Principles of Optical Regional and Metropolitan Access Networks," *IEEE Journal of Lightwave Technology*, Vol. 17, No. 12, December 1999, pp. 2431-2448.
[3]P. Bonenfant, A. Moral, "Generic Framing Procedure (GFP): The Catalyst for Efficient Data Over Transport," *IEEE Communications Magazine*, Vol. 40, No. 5, May 2002, pp. 72-79.
[4]E. Hernandez-Valencia, "Hybrid Transport Solutions for TDM/Data Networking Solutions," *IEEE Communications Magazine*, Vol. 40, No. 5, May 2002, pp. 104-112.
[5]B. Mukherjee, *Optical Communications Networks*, McGraw Hill, New York, NY, 1997.
[6]T. E Stern, K. Bala, *Multiwavelength Optical Networks: A Layered Approach*, Addison-Wesley, Reading, MA, 1999.
[7]Y. Chen, *et al*, "Metro Optical Networking," *Bell Labs Technical Journal*, Vol. 4, No. 1, January/March 1999, pp. 163-186.
[8]D. Stoll, *et al*, "Metropolitan DWDM: A Dynamically Configurable Ring for the KomNet Field Trial in Berlin," *IEEE Communications Magazine*, Vol. 39, No. 2, February 2001, pp. 106-113.
[9]*IEEE Draft P802.17/D0.2*, "Part 17: Resilient Packet Ring Access Method & Physical Layer Specifications—Media Access Control (MAC) Parameters, Physical Layer Interface, and Management Parameter," LAN MAN Standards Committee, March 2002.
[10]M. Chow, *Understanding SONET/SDH: Standards and Applications*, Andan Publisher, New Jersey, 1995.
[11]T. H. Wu, *Fiber Network Service Survivability*, Artech House, Boston, 1992.
[12]W. Tomlinson, "Comparison of Approaches and Technologies for Wavelength Add/Drop Network Elements," *NFOEC 1998*, Orlando, FL, September 1998.
[13]R. Boncek, *et al*, "Fiber and Systems for Metropolitan Optical Networks," *NFOEC 1998*, Orlando, FL, September 1998.
[14]G. Bernstein, B. Rajagopalan, D. Saha, *Optical Network Control: Architectures, Protocols, and Standards*, Addison-Wesley Company, Boston, MA, July 2003.
[15]A. Odlyzko, "Internet Traffic Growth," *Optical Transmission Systems and Equipment for WDM Networking II, Proceedings of SPIE,* Vol. 5247, 2003, pp. 1-15.
[16]N. Ghani, "Storage Networking Extension Technologies," *National Fiber Optics Engineers Conference (NFOEC 2002)*, September 2002, Dallas, TX.
[17]P. Hatton, F. Cheston, "WDM Deployment in the Local Exchange Network," *IEEE Communications Magazine*, Vol. 36, No. 2, February 1998, pp. 56-61.
[18]T. Strasser, *et al*, "Optical Layer Innovation to Support Operationally Scalable, Regional Optical Networks," *NFOEC 2002*, Dallas, TX, September 2002.
[19]N. Antoniades, "Engineering the Performance of DWDM Metro Networks," *NFOEC 2000*, Denver, CO, August 2000.
[20]D. Zhou, S. Subramaniam, "Survivability in Optical Networks," *IEEE Network Magazine*, Vol. 14, No. 6, November/December 2000, Vo. 14, No. 6, pp. 16-23.
[21] J. Bautista, "Multiplexers Bring DWDM to the Metro/Access Markets," *WDM Solutions*, February 2000, pp. 11-14.

[22]A. Chiu, J. Strand, "Control Plane Considerations for All-Optical and Multi-Domain Optical Networks and Their Status in the OIF and IETF," *Optical Networks*, January/February 2003, Vol. 4, No. 1, pp. 26-35.

[23]M. Daoust, B. Lavallee, "Designing and Planning Metropolitan DWDM Solutions," *NFOEC 2000*, Denver, CO, August 2000.

[24]S. Johansson, *et al*, "A Cost Effective Approach to Introduce an Optical WDM Network in the Metropolitan Environment," *IEEE Journal on Selected Areas in Communications*, Vol. 16, No. 7, September 1998, pp. 1109-1122.

[25]R. Iraschko, *et al*, "An Optical 4-Fiber Bi-Directional Line-Switched Ring," *OFC 1999*, San Diego, CA, February 1999.

[26]D. Marcenac, "Benefits of Wavelength Conversion in Optical Ring-Based Networks," *Optical Networks Magazine*, Vol. 1, No. 2, April 2000, pp. 23-29.

[27]O. Gerstel, R. Ramaswami, "Upgrading SONET Rings with WDM Instead of TDM: An Economic Analysis," *OFC 1999*, San Diego, CA, February 1999.

[28]O. Gerstel, R. Ramaswami, "Optical Layer Survivability: A Post-Bubble Perspective," *IEEE Communications Magazine*, Vol. 41, No. 9, September 2003, pp. 51-53.

[29]K. Zhu, *et al*, "Traffic Engineering in Multigranularity Heterogenous Optical WDM Mesh Networks Through Dynamic Traffic Grooming," *IEEE Network Magazine*, Vol.17, No. 2, March/April 2003, pp. 8-15

[30]V. Gambiroza, *et al*, "Design, Analysis, and Implementation of DVSR: A Fair, High Performance Protocol for Packet Rings", *IEEE/ACM Transactions on Networking*, February 2004, Vol. 12, No. 1, pp. 85-102.

[31] B. Ramamurthy, "Transparent Versus Opaque Versus Translucent Wavelength-Routed Optical Networks," *Optical Fiber Telecommunications Conference (OFC) 1999*, San Diego, CA, February 1999.

[32]O. Gerstel, R. Ramaswami, G. Sasaki, "Fault-Tolerant Multi-Wavelength Optical Rings with Limited Wavelength Conversion," *IEEE Journal on Selected Areas in Communications*, September 1998, Vol. 16, No. 7, pp. 1166-1178.

[33]N. Ghani, S. Dixit, T. Wang, "On IP-WDM Integration: A Retrospective," *IEEE Communications Magazine*, Vol. 41, No. 9, September 2003.

[34]W. Alanqar, A. Jukan, "Extending End-to-End Optical Service Provisioning and Restoration in Carrier Networks: Opportunities, Issues, and Challenges," *IEEE Communciations Magazine*, Vol. 42, No. 1, January 2004, pp. 52-60.

[35]A. Kamal, R. Ul-Mustafa, "Multicast Traffic Grooming in WDM Networks," *SPIE OptiComm 2003*, Dallas, TX, October 2003.

[36]S. Yao, *et al*, "All-Optical Packet Switching for Metropolitan Area Networks: Opportunities and Challenges,' *IEEE Communications Magazine*, March 2001, Vol. 39, No. 3, pp. 142-148.

# II

# SWITCHING

# Chapter 5

# OPTICAL PACKET SWITCHING

George N. Rouskas and Lisong Xu
*Department of Computer Science, North Carolina State University, Raleigh, NC 27695*
Email: { rouskas,lxu2 } @csc.ncsu.edu

**Abstract** The concept of optical packet switching (OPS) is emerging as an alternative to coarser-grained switching in the optical domain. Despite the significant technological challenges it faces, OPS holds the promise of a highly reconfigurable, bandwidth-efficient, and flexible optical layer. In this chapter we study some of the architectural and design issues for OPS networks, we examine a number of enabling technologies, and we discuss some of the ongoing research and experimental efforts.

**Keywords:** Optical packet switching, wavelength division multiplexing (WDM), switch architectures, contention resolution techniques

## 5.1 Introduction

Optical transmission and switching technologies based on wavelength division multiplexing (WDM) have been increasingly deployed in the Internet infrastructure over the last decade in order to meet the ever-increasing demand for bandwidth. Given that point-to-point WDM transmission technology is quite mature today, while optical switching technologies continue to evolve at a rapid pace, the result has been the creation of *opaque* optical networks in which the optical signal undergoes *optical-to-electrical-to-optical* (OEO) conversion or regeneration at each intermediate node in the network. More recently, two trends have emerged in the design and deployment of WDM networks. The first is towards increasing *transparency* in the network so as to eliminate electronic bottlenecks and enable the handling of a broad range of heterogeneous signals regardless of protocol formats, bit rates, or modulation. The second trend is towards *reconfigurability* in optical networks, such that bandwidth can be created in real time between end-users to accommodate dynamically changing traffic demands. These trends reflect the vision of a future network in which optical switching technology plays a central role and bandwidth is relatively abundant, inexpensive, and readily available to end-users.

The migration of switching functions from electronics to optics will be gradual, and will take place in several phases. Already, the first phase is underway in the form of wavelength routed networks which offer circuit switching services at the granularity of a wavelength. Due to their circuit-switched nature, wavelength routed networks can be built with commercially available optical switch technologies, such as MEMS cross-connects [10], which are still relatively slow with switch configuration times in the order of milliseconds. While wavelength routing represents a significant step in the direction of transparent and configurable optical networking, optical circuits tend to be inefficient for traffic that has not been groomed or statistically multiplexed; moreover, the circuit-switching model does not fit well within the Internet philosophy of packet switching.

The next phase in the switching evolution is likely to involve the more recent optical burst switching (OBS) paradigm [27, 2]. Because it attempts to minimize the need for header parsing and buffering at intermediate network nodes, OBS is widely viewed as a promising technology for supporting finer switching granularity in the optical domain. Since the unit of transmission and switching is a burst, which is the aggregation of a flow of data packets, OBS is more efficient than circuit switching when the sustained traffic volume does not consume a full wavelength. OBS technology is still in the stage of research and experimentation, but at least one proof-of-concept testbed has been operational for the last year [1]; as optical switching speeds improve to microseconds or less, OBS networks are expected to become a reality within the next few years.

In the longer term, optical packet switching (OPS) promises almost arbitrarily fine transmission and switching granularity, evoking the vision of a bandwidth-efficient, flexible, data-centric all-optical Internet [3, 24, 15, 35, 33]. The realization of this vision, however, faces significant challenges in that OPS requires practical, cost-effective, and scalable implementations of optical buffering and packet-level parsing. We also note that each of the three optical switching technologies (wavelength routing, OBS, and OPS) have important application domains; hence, rather than each technology replacing the previous one, it is highly likely that all three will coexist in the optical network of the future.

In this chapter, we discuss some of the critical issues in designing and implementing OPS networks. In Section 5.2, we describe the architecture of an OPS node, and we take a close look at the building blocks, and the corresponding optical technologies, for realizing such a node. In Section 5.3, we discuss switch fabric architectures and contention resolution schemes. In Section 5.4,

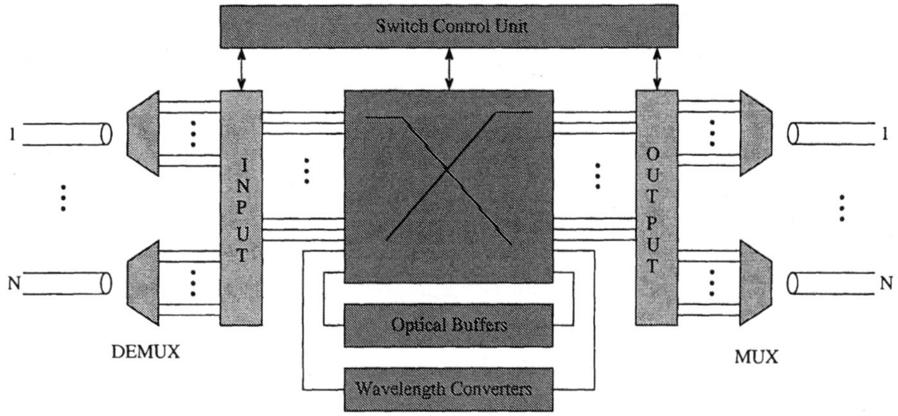


*Figure 5.1.* OPS node architecture.

we describe experimental efforts and testbeds, and we conclude the chapter in Section 5.5.

## 5.2 OPS Node Architecture

In Figure 5.1, we show the functional block diagram of a generic OPS node architecture. The architecture consists of a set of multiplexers and demultiplexers, an input interface, a space switch fabric with associated optical buffers (i.e., fiber delay lines) and wavelength converters, an output interface, and a switch control unit. Packets arriving on an input fiber are first de-multiplexed into individual wavelengths and are then sent to the input interface. Each packet consists of the payload and an *optical header* which is used for routing in the optical domain; note that any network layer header (e.g., IP header) is considered part of the payload for optical routing purposes. Among other functions, the input interface is responsible for extracting the optical packet header and forwarding it to the switch control unit for processing. The switch control unit processes the header information, determines an appropriate output port and wavelength for the packet, and instructs the switch fabric to route the packet accordingly. In routing the packet, the switch may need to buffer it and/or convert it to a new wavelength. The switch controller also determines a new header for the packet, and forwards it to the output interface. When the packet arrives at the output interface, the new header is attached, and the packet is forwarded on the outgoing fiber link to the next node in its path.

OPS networks can be classified along several dimensions depending on how the above packet switching and header processing functions are implemented.

*Synchronous vs. asynchronous switch operation.*
In a synchronous OPS network [18], time is slotted, and the switch fabric at

each individual node can only be reconfigured at the beginning of a slot. All packets in a synchronous network have the same size, and the duration of slot is equal to the sum of the packet size and the optical header length (plus appropriate guard bands). Note that, due to variable link propagation delays, packets arriving at a node over different interfaces may not be aligned with the local clock. Therefore, it is the responsibility of the input interface (refer to Figure 5.1) to synchronize arriving packets and align them with switching time slots. Synchronous optical switching fabrics, much like their electronic counterparts, are easier to build and operate, hence synchronous OPS networks have received more attention from the research community.

In an unslotted network [31], packets are of variable size, switch operations may take place at any point in time, and there is no need to align arriving packets at the switch input. Unslotted OPS networks are more flexible and robust than slotted ones, and they do not require segmentation or reassembly at the edges of the network.

*Electronic vs. optical control.*
The optical packet header contains information for routing the packet in the OPS network. Today, the lack of fast, scalable, and robust optical bit-level processing technologies means that electronic processing of the header is the only practical approach. However, all-optical header processing is an active research area, and considerable progress has been made in some critical functions, such as all-optical label swapping (AOLS) [4]. AOLS is a promising approach for routing packets in optics, in which packets are encapsulated with a small optical label as they enter the optical network. The packets are routed based on the information carried in the optical label, which is erased and rewritten at each OPS node, while the electronic packet header remains intact with the payload throughout the OPS network.

*Optical header format.*
There are two main approaches to optical header formatting: bit serial and subcarrier multiplexed; these are discussed in more detail in the next subsection.

*Switch fabric architectures.*
A wide variety of switch fabric architectures has been proposed for OPS networks, both for fixed-size and variable-size packets; due to their central role in the overall operation of an OPS node, we study a number of switch architecture designs in the next section.

*Contention resolution strategies.*
When two packets from different input port/wavelength pairs must be switched to the same output port/wavelength pair at the same time, contention arises. In this case, the switch controller and the switch fabric must employ some strategy to resolve the contention. Output port contention can be resolved in three dimensions: wavelength (using converters), time (using fiber delay lines), or

space (using deflection routing); strategies that combine more than one dimensions are also possible. We discuss and compare various contention resolution strategies in Section 5.3.1.

We note that, in addition to the data plane, contention is also possible in the control plane. Contention in the switch control unit may result in header loss or a significant delay such that the packet payload may precede the header; in either case, the packet has to be discarded. Therefore, proper buffer size dimensioning and efficient buffer management are of utmost importance; since, however, established techniques can be used for this purpose [34, 26], we will not consider contention in the control plane any further.

## 5.2.1 Enabling Technologies

Optical packet switching is still in its infancy compared to its electronic counterpart. We now discuss five functions of critical importance to the realization of practical OPS nodes.

**Optical switch fabrics.** The switch fabric at an OPS node must be capable of rapid reconfiguration on a packet-by-packet basis. At data rates of 40 Gbps and beyond, this requirement implies that switching times have to be on the order of a few nanoseconds. Other critical requirements include scalability of the technology to high port counts, low loss and crosstalk, and uniform operation across all signals independent of the path from input to output port; moreover, issues such as reliability, energy usage, and temperature independent operation are also important.

Most current optical switch fabrics, including those based on opto-mechanical, thermo-optic, or acousto-optic technologies, are limited to switching speeds in the millisecond or microsecond range. Two promising technologies include semiconductor optical amplifier (SOA) switches and electro-optic lithium niobate ($LiNbO_3$) switches, both capable of switching speeds in the nanosecond rage. However, both technologies have limitations that must be overcome before it becomes possible to build high-performance, reliable, and cost-effective optical packet switches. For a recent comprehensive survey of optical switch fabric technologies, the reader is referred to [26].

**Optical buffering.** The lack of an efficient way to store information in the optical domain represents a major difficulty in the design of OPS nodes. Research has focused on ways of emulating electronic RAM capabilities through the use of fiber delay lines (FDLs) to delay optical signals [19, 31, 20, 9]. FDLs are fibers of fixed length, and can hold a packet for an amount of time determined by the speed of light and the length of the FDL. Hence, unlike electronic RAM, FDLs cannot store a packet indefinitely, and, once a packet has entered an FDL, it cannot be retrieved until it emerges on the other side. Furthermore,

FDLs can be bulky and expensive, and introduce quality degradation to optical signals. As a result, the design of optical buffers that mitigate the effects of these limitations has emerged as an important research area for OPS. Among the important issues to be considered in designing FDL architectures include packet loss, cost, control complexity, packet reordering, and signal loss along the FDL. We discuss FDL buffer architectures in the next section.

**Wavelength conversion.** Wavelength conversion [28, 16] is the ability to convert an optical signal on a given input wavelength to some other output wavelength. One of the main applications of wavelength conversion is as a mechanism for contention resolution that can dramatically improve the utilization of resources in an optical network, especially in highly dynamic traffic environments such as OPS. Consequently, wavelength converters have become integral to the design of optical buffer and switch architectures for OPS networks. Wavelength translation can be achieved by OEO conversion; however, all-optical wavelength conversion is desirable for OPS. Important features of all-optical converters include large wavelength span, fast setup time, high signal-to-noise ratio for cascadeability, and bit-rate transparency. All-optical converter approaches include the use of cross-gain modulation (XGM) or cross-phase modulation (XPM) in SOAs, and wave mixing techniques. Unfortunately, none of the existing techniques exhibits all the desired properties listed above; for a more detailed discussion and comparison of wavelength converter technologies, see [16].

**Packet delineation and synchronization.** Packet delineation is required for both synchronous and asynchronous networks, and its purpose is to determine the beginning and end of the arriving packet. Current approaches perform delineation electronically as follows: a splitter taps a small amount of power from incoming packets and passes it to a bit-level synchronization circuit which locks the incoming bits in phase with the local clock in order to read the header information. Since this operation must be performed for each incoming packet, the circuit must be able to synchronize the header with its clock within a few bit times.

In addition to bit-level synchronization, OPS nodes in slotted networks must also synchronize incoming fixed-size packets to the local switching slots. This slot-level synchronization is accomplished by passing each incoming packet through a cascade of fiber delay lines and optical switches, in order to delay the packet by a sufficient amount of time for it to align with the beginning of a slot. This scheme introduces losses and crosstalk, resulting in a significant power penalty over long paths. A different strategy takes advantage of the fact that the propagation delay in a highly dispersive fiber depends on the signal wavelength. Each incoming packet is therefore passed through such a fiber, after its wavelength is first converted to achieve the desired delay.

**Optical header format and processing.** There are two main approaches to formatting the optical header associated with a packet [4]. In the *bit-serial* approach, the header is transmitted serially on the same wavelength; a guard band is placed between the header and payload to allow for the removal and reinsertion of the header at intermediate OPS nodes. The second method uses *subcarrier multiplexing*, in which the header is situated slightly higher in the spectrum than the payload bandwidth, and is subcarrier multiplexed with the baseband payload. Both approaches have relative advantages and disadvantages, and both are being pursued in the lab.

As we mentioned earlier, currently, the processing of the header is performed electronically [18, 21]. All-optical header processing [14] is an area of research that has received considerable attention, but the technology is still in the very early stages. In order to optically process headers, two functions have to be developed in optics: optical correlators to read a header, and all-optical flip-flop memory to store the header information. Currently, these functions have been demonstrated for headers containing only a few bits worth of information, limiting the switch size to only $1 \times 2$; for a review of optical header process techniques, see [14]. One area in which significant progress has been made is in all-optical label swapping, which refers to techniques used to extract and replace the optical header without the need for OEO conversion of the payload. The interested reader is referred to [4] for a description of all-optical label swapping technologies for both bit-serial and subcarrier multiplexed headers.

## 5.3 Optical Packet Switch Architectures

### 5.3.1 Contention Resolution Schemes

As we discussed in the previous section, contention in the data path of an OPS node can be resolved using one of three methods or combination thereof: optical buffering, wavelength conversion, or deflection routing.

**Optical buffering.** The most straightforward method for resolving output port contention is to exploit the time dimension. Specifically, one of the contending packets (i.e., those arriving on the same wavelength at the same time and requesting the same output port) is routed through the switch fabric, while the rest are sent to an FDL. When the stored packet(s) emerge from the FDL, the whole process is repeated.

Similar to their electronic counterpart, optical buffers may be placed at the input, output, or both, of a packet switch. However, to compensate for the lack of a true "random access" property, a number of optical buffer arrangements have been proposed, such as single- or multi-stage FDLs, feed-forward or feed-backward connections, etc. Each of these arrangements can be used to implement a variety of packet switch architectures, and some representative examples are discussed in the next subsection.

**Wavelength conversion.** With this method, when two or more packets contend for the same output port and wavelength, the wavelength of all but one of the packets is converted to another wavelength, thus resolving the contention. If such a capability is available, then only when all wavelengths of an output port are busy does it become necessary to buffer contending packets. As a contention resolution method, wavelength conversion has some highly desirable properties in that it does not introduce delays in the data path and it does not cause packet resequencing.

Converters may be fixed or tunable, and they may be placed at the input and/or output of a packet switch; moreover, each port of the switch may be equipped with its own dedicated converter, or the converters may be shared by all ports. Consequently, a variety of switch architectures are possible depending on the availability and placement of converters.

**Deflection routing.** This method exploits the space dimension to resolve contention. Specifically, packets that lose the contention are sent to a different output port than the one requested, and hence may take a longer route to their destination. Deflection routing introduces delays in the data path and may cause packets to arrive out of order. However, it does not require additional hardware (e.g., FDLs or converters), unlike the previous two methods. On the other hand, the effectiveness of deflection routing as a contention resolution scheme depends on the traffic pattern and the density of the network topology.

The above three contention resolution schemes may be used in pure form, or they may be combined to implement more sophisticated strategies. For instance, optical buffering may be used along with either conversion or deflection routing to allow for more flexibility in resolving contention. The three pure schemes, along with the various combinations, make possible a wide spectrum of contention resolution methods that offer various tradeoffs of performance versus hardware cost and complexity. A comprehensive performance study of contention resolution methods can be found in [34]. The main finding of the study was that wavelength conversion offers the most performance benefits, and that the most efficient strategy is to combine conversion with limited buffering and selective deflection.

## 5.3.2 Switch Fabric Architectures

A wide variety of switch fabric architectures have been proposed for OPS. In general, we can classify the switch architectures as follows.

- *Single-stage vs. multi-stage switches.*
  Switches may consist of a single stage, or they may be built by appropriately cascading a set of smaller, single-stage switches [19, 26]. Single-stage switches usually have a small number of input and output ports and small buffer capacity, and they are easy to implement and con-

trol. Due to cost (e.g., in terms of the amount of optical components required) or performance (e.g., in terms of power loss) considerations, switches with high port counts and/or large buffer capacity are usually implemented using multiple stages. Some considerations in building multi-stage switches include the number of smaller switches required, the blocking characteristics of the architecture, and the degree of loss uniformity along the various paths from input to output.

- *Space vs. wavelength-routing vs. broadcast-and-select switches.*
Space switch architectures are based on a non-blocking switch fabric, such as a crossbar, which is usually implemented using SOAs. A wavelength routing switch [8] is usually based on arrayed waveguide gratings (AWGs) [23], devices which implement a static permutation from input to output ports. A broadcast-and-select switch [22] is usually based on a WDM passive star coupler. AWG-based switches require fewer optical components (especially SOAs) than either space or broadcast-and-select switches. On the other hand, it is straightforward to implement broadcast or multicast with a space or broadcast-and-select switch but not with a wavelength-routing switch. However, due to splitting losses, neither space nor broadcast-and-select switches may scale to large numbers of ports. We also note that, in a large, multi-stage switch, multiple technologies may be used simultaneously.

- *Feed-forward vs. feed-backward buffers (FDLs).*
In a switch with feed-forward FDLs [19], a packet may be buffered only once: when such a packet emerges from the FDL after the specified delay, it is switched to an output port, and then leaves the switch. However, in a switch with feed-backward FDLs [19], a packet emerging from the FDL may be buffered multiple times by sending it (feeding it back) to the FDL; this situation may arise if the packet experiences contention again after emerging from the FDL. One advantage of a switch with feed-backward FDLs is that it can support priority scheduling of optical packets. That is, after leaving the FDL, an optical packet may be buffered again because of preemption by a later-arriving but higher-priority optical packet.

Due to the large number of different switch architectures that have been proposed for OPS, it is impossible to cover all of them in this chapter. In the following, we discuss some representative architectures to illustrate some of the possibilities in OPS switch design.

**Single-stage space switch with feed-forward FDLs.** In Figurc 5.2, wc present a single-stage space switch architecture with $N$ ports, $W$ wavelengths, and $D$ FDLs per output port, similar to the one in [11]; we can think of this as an output-queue architecture. Each incoming optical signal is first demultiplexed into the $W$ wavelengths, and each wavelength is then converted to a wavelength that is free at the destination optical output buffer. The space switch fabric consists of splitters, optical gates, and combiners. The optical signal of each packet is split into $ND$ identical signals, where $N$ is the number of output ports and $D$ is the number of FDLs per port. Once it is determined how long the packet has to be delayed in order to avoid output port collision, the packet is switched to the desired output port and corresponding FDL by closing the appropriate optical gate. It was shown in [12] that with at least $W = 11$ channels per fiber, a low packet loss rate of $10^{-10}$ can be achieved even without the optical buffers. A more cost-efficient variant of this architecture was studied in [17]; in the new architecture, the optical buffers (FDLs) were eliminated and the tunable wavelength converters were shared among all incoming wavelengths. Obviously, the scalability of both switch variants is limited by the loss incurred by splitting each signal.

**Single-stage broadcast-and-select switch with feed-forward FDLs.** Figure 5.3 shows the architecture of a broadcast-and-select switch proposed as part of the European ACTS KEOPS project [18, 29]. The switch has $N$ input and output ports, and it is equipped with $D$ FDLs such that a packet can be delayed for an integer multiple of the slot time $T$, up to $DT$. The architecture in Figure 5.3 assumes that each input fiber carries only one wavelength that is different than the wavelengths carried by the other input fibers; hence the total number of wavelengths is $N$. The switch may be modified to handle multiple wavelengths per input fiber, by introducing an additional stage to demultiplex the input signal into individual wavelengths, and replicating the architecture shown in Figure 5.3. However, because of the power loss due to splitting, the product of the number of wavelengths times the number of ports cannot be high, limiting the scalability of the switch.

The switch operates as follows. First, the packets from all input ports are combined and distributed through a WDM passive star coupler to all $D$ FDLs; note that in Figure 5.3, the multiplexer and splitter at the input play the role of the passive star coupler. At the output of the FDLs, optical gates are used to select the packets that have undergone an appropriate delay, of which only one packet is then selected and transmitted to the output port by yet another set of optical gates at the output ports. Note that performing broadcast or multicast is straightforward: all that is needed is for multiple output ports to select the same packet.

**Single-stage wavelength routing switch with feed-backward FDLs.** Figure 5.4 shows the wavelength routing switch architecture proposed as part of the WASPNET project [21, 25]. The AWG is used to switch packets either to the correct output port, or to the appropriate FDLs in case of packet contention. In this switch, each input fiber carries only one wavelength. The switch can

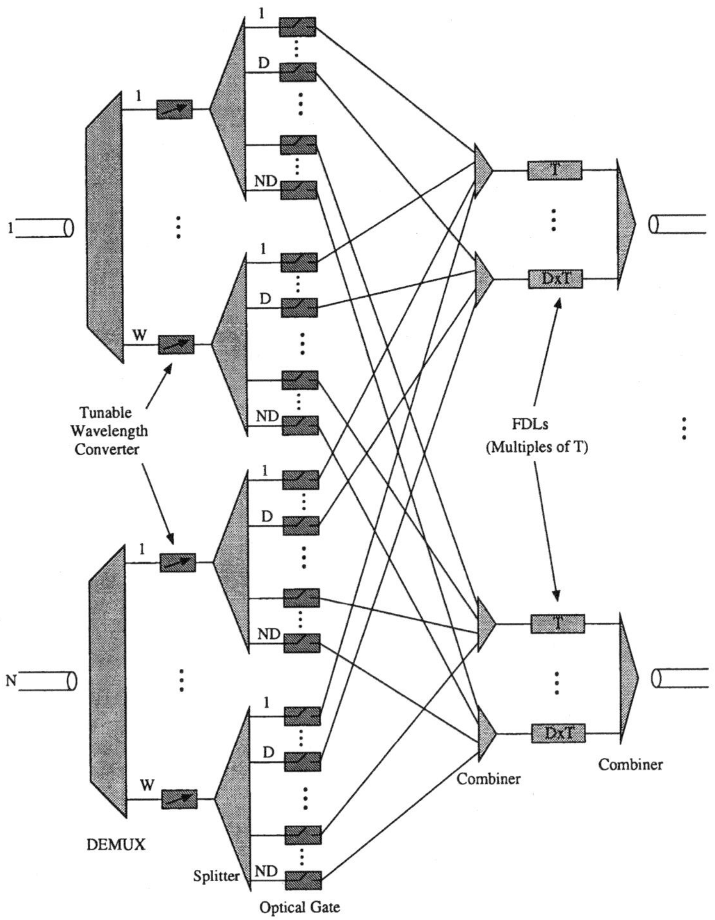


*Figure 5.2.* A single-stage space switch with $N$ ports, $W$ wavelengths, and $D$ FDLs.

be extended to handle multiple wavelengths as follows. First, the optical signals are demultiplexed into individual wavelengths, and then they are fed to multiple planes, one for each wavelength. Each plane has the same switch architecture as in Figure 5.4. The space switch is then used to prevent wavelength contention when combining packets from different planes.

**Multi-stage switch with feed-forward FDLs.** If the packet size is variable, the performance of FDL buffers is poor [30], and a low packet loss rate can be achieved only with a large buffer [6]. Figure 5.5 shows one stage of the multi-stage switch proposed in [7] to address this issue. The switch consists of three parts: an input part, a multi-stage FDL buffer, and an output part; the figure shows only stage $i$ of the multi-stage buffer. At the input part, the set of signals from all input ports are converted to the set of $W$ wavelengths used

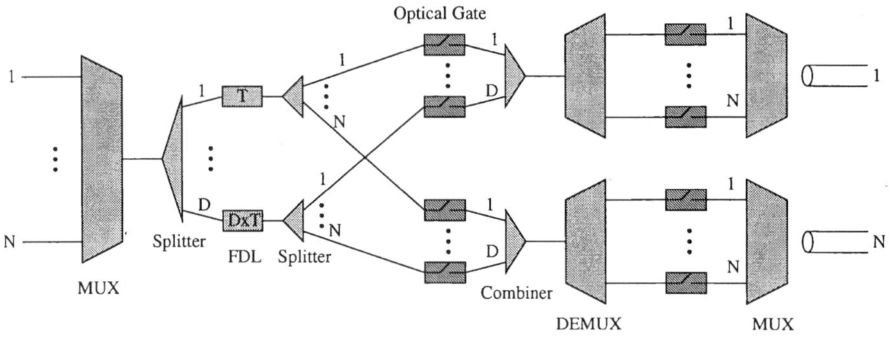


*Figure 5.3.* Broadcast-and-select switch with $N$ ports, $N$ wavelengths, and $D$ FDLs.

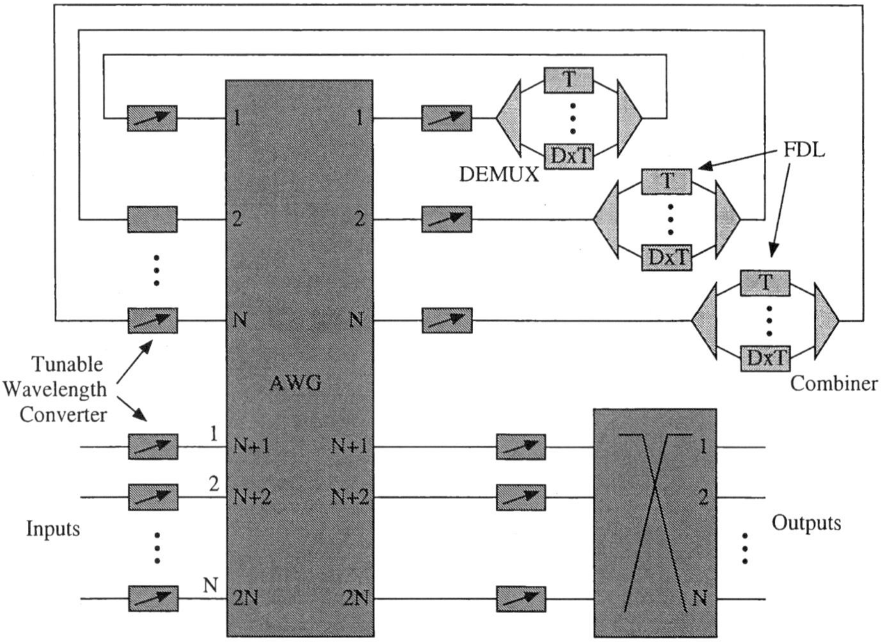


*Figure 5.4.* Wavelength routing switch with $N$ ports.

within the switch fabric. There is no wavelength conversion within the switch, so that a packet assigned a particular wavelength at the input will emerge on the same wavelength at the output. The large buffer is implemented as multiple stages of FDLs. Let $k$ represent the number of FDL stages, and let $D$ denote the number of FDLs per stage. Then, for a given $k$ and $D$, the FDLs are organized such that at stage $i, i = 1, \cdots, k$, the $D$ FDLs produce delays equal to: $0, D^{k-i}T, 2D^{k-i}T, \cdots, (D-1)D^{k-i}T$, where $T$ is the delay granularity. This arrangement makes it easy to find the indices of the FDLs to which a

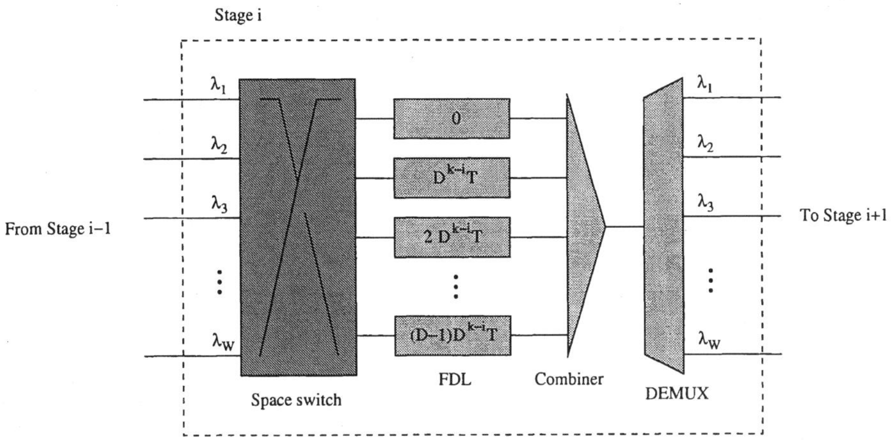


*Figure 5.5.* Stage $i$ of multi-stage switch with $W$ wavelengths and $D$ FDLs per stage.

packet must be sent in order to realize some arbitrary delay through the switch. Finally, the output part of the switch (not shown in the figure) consists of a switch that connects any FDL and any wavelength to any of the output ports. A similar multi-stage switch with feed-forward buffers was proposed in [20].

**Data vortex: multi-stage switch without buffers.** The data vortex architecture was recently proposed in [32] to implement a large-scale switch with low latency. The objective was to minimize the number of switching operations and to eliminate the optical buffers, resulting in a cost-effective and practical implementation. The architecture of the data vortex switch is novel, consisting of a group of routing nodes which lie on a collection of concentric cylinders. The switch operates in synchronous mode, and it employs a number of sophisticated techniques such as a hierarchical routing structure and distributed traffic control. For details on the architecture and operation of the data vortex switch, the reader is referred to [32].

## 5.4 Testbeds and Experimental Efforts

A number of experimental projects have been carried out in the area of optical packet switching over the last decade. We now briefly describe a representative set of these efforts; for details and a comprehensive description of the results of these projects, the reader is referred to the relevant publications.

The KEOPS (keys to optical packet switching) project [18, 29], was funded by the Advanced Communications Technology and Service (ACTS) program, a research program of the European Union, from 1994 to 1998. Partners of KEOPS included companies and universities from Europe. Building upon the results of the previous ATM optical switching (ATMOS) project, the objective of KEOPS was to analyze, develop, and demonstrate bit-rate-transparent all-optical packet switching for future all-optical networks. Two of the major results of the project were the demonstration of a 4×4 wavelength-routing switch operating at 2.5Gbps and a 16×16 broadcast-and-select switch operating at 10Gbps.

The WASPNET (wavelength switched packet network) project [21, 25], funded by the U.K. Engineering and Physical Sciences Research Council (EPSRC) starting in July 1997, is a collaboration between three British Universities, the Universities of Essex, Bristol and Strathclyde, and three industrial partners, BT Laboratories, Marconi Communications, and Fujitsu Telecommunications. The objective of the project is to understand the advantages and potential of optical packet switching compared to the conventional electronic packet switching. A 8×8 wavelength-routing switch [21], and a cascade of 14 switches operating at 2.5Gbps [21] were demonstrated.

The DAVID (data and voice integration over WDM) project, was funded by the Information Society Technology (IST) Program, another research program of the European Union, from 1998 to 2002. DAVID was pursued by a fellowship of major operators, manufacturers, as well as leading universities, and research organizations from all over Europe. The main objective was to propose a packet-over-WDM network solution, including traffic engineering capabilities and network management, covering the entire area from MAN to WAN. For additional information and a list of relevant publications, see [13].

All-optical label swapping (AOLS) [4], is a new approach to routing packets in optics that combines optical network traffic engineering techniques with photonic packet switching technologies. AOLS is being developed by the University of California at Santa Barbara in collaboration with several companies, and is funded by a DARPA NGI grant and by the DARPA sponsored Center for Multidisciplinary Optical Switching Technology (MOST). Using technology based on XPM wavelength converters, the project has demonstrated AOLS with variable-length packets at 80Gbps and optical labels at 10 Gbps [5].

## 5.5 Concluding Remarks

The concept of optical packet switching, which seeks to replace the electronic switching functions by optical ones, represents a paradigm shift with the potential to revolutionize networking. Before a practical OPS network becomes a reality however, a number of technological issues must be addressed, as we discussed in this chapter. Nevertheless, given the ongoing research activities in this area, it is reasonable to expect that the key technological challenges will be eventually overcome, and some form of OPS will become possible within the next decade.

As optical technology advances and the OPS vision comes closer to reality, a number of other challenges will emerge. The constraints and new realities imposed by the optical layer and WDM transmission and switching technology will certainly affect our long-held assumptions regarding fundamental networking issues such as routing, control, packet transport, etc., which have been developed for mostly opaque electronic networks. As we carefully rethink these issues in the context of transparent OPS networks, protocol and algorithm design will certainly evolve to better accommodate the OPS technology, creating the need for proof-of-concept systems and testbeds of realistic size in which to experimentally study these new solutions. Migration strategies will also need to be developed for the network infrastructure to make the transition from electronic to optical packet switching, as well as from other forms of optical switching (e.g., wavelength routing or OBS) to packet switching. In the years ahead, as research in OPS shifts from components to integrated systems (i.e., OPS nodes and networks), we can expect a wide range of exciting research opportunities requiring interdisciplinary approaches that combine expertise in networking and optical engineering.

# References

[1] Baldine, I., Cassada, M., Bragg, A., Karmous-Edwards, G., and Stevenson, D. (2003). Just-in-time optical burst switching implementation in the ATDnet all-optical networking testbed. In *Proceedings of Globecom 2003*.

[2] Baldine, I., Rouskas, G. N., and Stevenson, D. (2002). `JumpStart`: A just-in-time signaling architecture for WDM burst-switched networks. *IEEE Communications*, 40(2):82–89.

[3] Blumenthal, D., Ikegami, T., Prucnal, P. R., and (editors), L. T. (1999). Special issue on photonic packet switching technologies, techniques, and systems. *Journal of Lightwave Technology*, 17(12).

[4] Blumenthal, D. J. (2001). Photonic packet switching and optical label swapping. *Optical Networks Magazine*, 2(6):54–65.

[5] Blumenthal, D. J., Bowers, J. E., *et al.* (2003). Optical signal processing for optical packet switching networks. *IEEE Communications Magazine*, 41(2):23–29.

[6] Callegati, F. (2000). Optical buffers for variable length packets. *IEEE Communications Letters*, 4(9):292–294.

[7] Callegati, F., Corazza, G., and Raffaelli, C. (2002). Exploitation of DWDM for optical packet switching with quality of service guarantees. *IEEE Journal on Selected Areas in Communications*, 20(1):190–200.

[8] Chia, M. C., Hunter, D. K., and *et. al.* (2001). Packet loss and delay performance of feedback and feed-forward arrayed-waveguide gratings-based optical packet switches with WDM inputs-outputs. *Journal of Lightwave Technology*, 19(9):1241–1253.

[9] Chlamtac, I. *et al.*, (1996). CORD: Contention resolution by delay lines. *IEEE Journal on Selected Areas in Communications*, 14(5):1014–1029.

[10] Chu, P. B., Lee, S.-S., and Park, S. (2002). MEMS: The path to large optical crossconnects. *IEEE Communications*, 40(3):80–87.

[11] Danielsen, S. L. *et al.* (1998a). Analysis of a WDM packet switch with improved performance under bursty traffic conditions due to tunable wavelength converters. *IEEE/OSA Journal of Lightwave Technology*, 16(5):729–735.

[12] Danielsen, S. L. *et al.* (1998b). Optical packet switched network layer without optical buffers. *IEEE Photonics Technology Letters*, 10(6):896–898.

[13] DAVID Project (2000). http://david.com.dtu.dk.

[14] Dorren, H. J. S., Hill, M. T., and *et. al.* (2003). Optical packet switching and buffering by using all-optical signal processing methods. *Journal of Lightwave Technology*, 21(1):2–12.

[15] El-Bawab, T. and Shin, J.-D. (2002). Optical packet switching in core networks: Between vision and reality. *IEEE Communications Magazine*, 40(9):60–65.

[16] Elmirghani, J. M. H. and Mouftah, H. T. (2000). All-optical wavelength conversion: Technologies and applications on DWDM networks. *IEEE Communications Magazine*, 38(3):86–92.

[17] Eramo, V. and Listanti, M. (2000). Packet loss in a bufferless optical WDM switch employing shared tunable wavelength converters. *Journal of Lightwave Technology*, 18(12):1818–1833.

[18] Guillemot, C. *et al.* (1998). Transparent optical packet switching: The European ACTS KEOPS project approach. *Journal of Lightwave Technology*, 16(12):2117–2134.

[19] Hunter, D. K., Chia, M. C., and Andonovic, I. (1998a). Buffering in optical packet switches. *Journal of Lightwave Technology*, 16(12):2081–2094.

[20] Hunter, D. K., Cornwell, W. D., Gilfedder, T. H., Frenzen, A., and Andonovic, I. (1998b). SLOB: A switch with large optical buffers for packet switching. *Journal of Lightwave Technology*, 16(10):1725–1736.

[21] Hunter, D. K. *et al.* (1999). WASPNET: A wavelength switched packet network. *IEEE Communications Magazine*, 37(3):120–129.

[22] Li, B., Qin, Y., Cao, X., and Sivalingam, K. M. (2001). Photonic packet switching: Architecture and performance. *Optical Networks Magazine*, 2(1):27–39.

[23] McGreer, K. A. (1998). Arrayed waveguide gratings for wavelength routing. *IEEE Communications Magazine*, 36(12):62–68.

[24] O'Mahony, M. J., Simeonidou, D., Hunter, D. K., and Tzanakaki, A. (2001). The application of optical packet switching in future communication networks. *IEEE Communications Magazine*, 39(3):128–135.

[25] O'Mahony, M. J. *et al.* (2001). An optical packet switched network (WASPNET)- concept and realisation. *Optical Networks Magazine*, 2(6):46–53.

[26] Papadimitriou, G. I., Papazoglou, C., and Pomportsis, A. S. (2003). Optical switching: Switch fabrics, techniques, and architectures. *Journal of Lightwave Technology*, 21(2):384–405.

[27] Qiao, C. and Yoo, M. (1999). Optical burst switching (OBS)-A new paradigm for an optical Internet. *Journal of High Speed Networks*, 8(1):69–84.

[28] Ramamurthy, B. and Mukherjee, B. (1998). Wavelength conversion in WDM networking. *IEEE Journal Selected Areas in Communications*, 16(7):1061–1073.

[29] Renaud, M., Masetti, F., Guillemot, C., and Bostica, B. (1997). Network and system concepts for optical packet switching. *IEEE Communications Magazine*, 35(4):96–102.

[30] Tancevski, L., Ge, A., Castanon, G., and Tamil, L. S. (1999). A new scheduling algorithm for asynchronous, variable length IP traffic incorporating void filling. In *Proceedings of OFC*.

[31] Tancevski, L., Yegananarayanan, S., Castagnon, G., Tamil, L., Masetti, F., and McDermott, T. (2000). Optical routing of asynchronous variable length packets. *IEEE Journal on Selected Areas in Communications*, 18(10):2084–2093.

[32] Yang, Q., Bergman, K., Hughes, G. D., and Johnson, F. G. (2001). WDM packet routing for high-capacity data networks. *Journal of Lightwave Technology*, 19(10):1420–1426.

[33] Yao, S., Dixit, S., and Mukherjee, B. (2000). Advances in photonic packet switching: An overview. *IEEE Communications Magazine*, 38(2):84–94.

[34] Yao, S., Mukherjee, B., Yoo, S. J. B., and Dixit, S. (2003). A unified study of contention-resolution schemes in optical packet-switched networks. *Journal of Lightwave Technology*, 21(3):672–683.

[35] Yao, S., Yoo, S. J. B., Mukherjee, B., and Dixit, S. (2001). All-optical packet switching for metropolitan area networks: Opportunities and challenges. *IEEE Communications Magazine*, 39(3):142–148.

Chapter 6

# WAVEBAND SWITCHING: A NEW FRONTIER IN OPTICAL WDM NETWORKS

Xiaojun Cao[1], Vishal Anand[2], Yizhi Xiong[1], and Chunming Qiao[1]

[1] *Department of Computer Science and Engineering, State University of New York at Buffalo, Amherst, NY 14226*

[2] *Department of Computer Science, State University of New York, College at Brockport, Brockport, NY*

Email: xcao2@cse.buffalo.edu, vanand@brockport.edu, yxiong@cse.buffalo.edu, qiao@computer.org

**Abstract** The rapid advances in dense wavelength division multiplexing (DWDM) technology with hundreds of wavelengths per fiber and world-wide fiber deployment have brought about a tremendous increase in the size (i.e., the number of ports) of photonic cross-connects, as well as in the cost and difficulty associated with controlling such large cross-connects. Waveband switching (WBS) has attracted attention for its practical importance in reducing the port count, associated control complexity, and cost of photonic cross-connects. In this chapter, we show that WBS is different from traditional wavelength routing, and thus techniques developed for wavelength-routed networks (including, for example, those for traffic grooming) cannot be directly applied to effectively address WBS-related problems. We describe a Three-layer multi-granular optical cross-connect (MG-OXC) architecture for WBS. By using this MG-OXC in conjunction with intelligent WBS algorithms, we show that one can achieve considerable savings in the port count. We present various WBS schemes and lightpath grouping strategies. Finally we discuss issues related to waveband conversion and failure recovery in WBS networks.

**Keywords:** Wavelength division multiplexing, Waveband switching, Multi-granular optical networks

## 6.1 Introduction

Optical networks using wavelength division multiplexing (WDM) technology, which divides the enormous fiber bandwidth into a large number of wavelengths, is a key solution to keep up with the tremendous growth in data traffic demand. However, as the WDM transmission technology matures and fiber deployment becomes ubiquitous, the ability to manage traffic in a WDM network is becoming increasingly critical and complicated. In particular, the rapid advance and use of dense WDM technology has brought about a tremendous increase in the size of photonic (both optical and electronic) cross-connects. The number of ports in a cross-connect is the most significant contributor to the cross-connect size and hence, the cost. Furthermore, owing to their large size the control and management of these cross-connects is also becoming formidable task. Hence, despite the remarkable technological advances in building photonic cross-connect systems and associated switch fabrics, the high cost (both capital and operating expenditures) and unproven reliability of huge switches have hindered their deployment.

The concept of *wavelength band switching* (WBS) (or simply waveband switching), has been proposed to reduce this complexity to a reasonable level. The main idea of WBS is to group several wavelengths together as a band and switch the band (optically) using a single port. In this way, not only the size of digital cross-connects or DXCs (e.g., the OEO grooming switches) can be reduced because bypass (or express) traffic can now be switched optically, but also the size of optical cross-connects (OXCs) that traditionally switch at the wavelength level can be reduced because of the bundling of lightpaths into bands in WBS networks. In this chapter, we focus on the use of WBS to reduce the size of the multi-granular optical cross-connect (MG-OXC) [1, 2, 3, 4, 5, 6], which is a part of the multi-granular photonic cross-connect (see Figure 6.1 for an example).

WBS differs from conventional wavelength routing in several ways, one for example is that each has different objectives. Accordingly, techniques developed for wavelength-routed networks (including for example, those for traffic grooming) cannot be directly applied to effectively address WBS-related problems. More specifically, in networks employing ordinary-OXC, the routing and wavelength assignment (RWA) problem is to find a route for a lightpath and assign a wavelength to it. One of the key objectives of the traditional RWA algorithms is to minimize the total number of wavelength-hops (WHs) or the maximum number of wavelengths required to satisfy a given set of lightpath requests, which is known to be NP-complete [7, 8, 9]. In this chapter, we study the optimal WBS problem, with its main objective being to route lightpaths and assign appropriate wavelengths to them so as to minimize the total number of ports required by the MG-OXCs. As to be shown, even though traditional RWA is still an important component of WBS, new waveband assignment algorithms need to be developed in order to effectively achieve the objective.

The rest of the chapter is organized as follows. We first describe a Three-layer MG-OXC architecture for WBS, and presents various WBS schemes and lightpath grouping strategies. We then explain how WBS differs from wavelength routing. Next, we focus on the design of WBS algorithms and present the results of our simulation of the same. Finally, we conclude this chapter.

## 6.2 Multi-granular Optical Cross-connect Architecture

In wavelength-routed networks (WRNs) with ordinary-OXCs (i.e., the single-granular OXCs) that switch traffic only at the wavelength level, wavelengths either terminate at or transparently pass-through a node, each requiring a port. However, in WBS networks, several wavelengths are grouped together as a band, and switched as a single entity (i.e., using a single port) whenever possible. A band is de-multiplexed into individual wavelengths if and only if necessary, for example, when the band carries at least one lightpath which needs to be dropped or added. WBS networks employ MG-OXCs to not only *switch* traffic at multiple levels or granularities such as fiber, band, and wavelength (and DXCs to switch traffic at sub-wavelength level), but to also *add* and *drop* traffic at multiple granularities. Traffic can be transported from one level to another via multiplexers and demultiplexers within the MG-OXC.

### 6.2.1 A Three-layer Multi-granular Optical Cross-connect

The MG-OXC is a key element for routing high speed WDM data traffic in a multi-granular optical network. While reducing its size has been a major concern, it is also important to devise node architectures that are flexible (reconfigurable) yet cost-effective. Figure 6.1 shows a typical MG-OXC considered in [3, 10], which includes the fiber cross-connect (FXC), band cross-connect (BXC) and wavelength cross-connect (WXC) layers.

As shown in Figure 6.1, the *WXC, BXC* layers consist of cross-connect(s) and multiplexer(s)/demultiplexer(s). The *WXC* layer includes a wavelength cross-connect that is used to switch lightpaths. To add/drop wavelengths from the *WXC* layer, we need $W_{add}/W_{drop}$ ports. In addition, band-to-wavelength ($BTW$) demultiplexers are used to demultiplex bands to wavelengths and wavelength-to-band ($WTB$) multiplexers are used to multiplex wavelengths to bands. At the *BXC* layer, the band cross-connect, $B_{add}$ and $B_{drop}$ ports are used for bypass bands, added bands and dropped bands respectively. The fiber-to-band ($FTB$) demultiplexers and band-to-fiber ($BTF$) multiplexers are used to demultiplex fibers to bands, and multiplex bands to form fibers respectively. Similarly, fiber cross-connect $/F_{add}/F_{drop}$ ports are used to switch/add/drop fibers at the *FXC* layer. In order to reduce the number of ports, the MG-OXC switches a fiber using one port (space switching) at the FXC cross-connect if none of its wavelengths is used to add or drop a lightpath. Otherwise, it will demultiplex the fiber into bands, and switch an entire band using one port at the BXC cross-connect if none of its wavelengths needs to be added

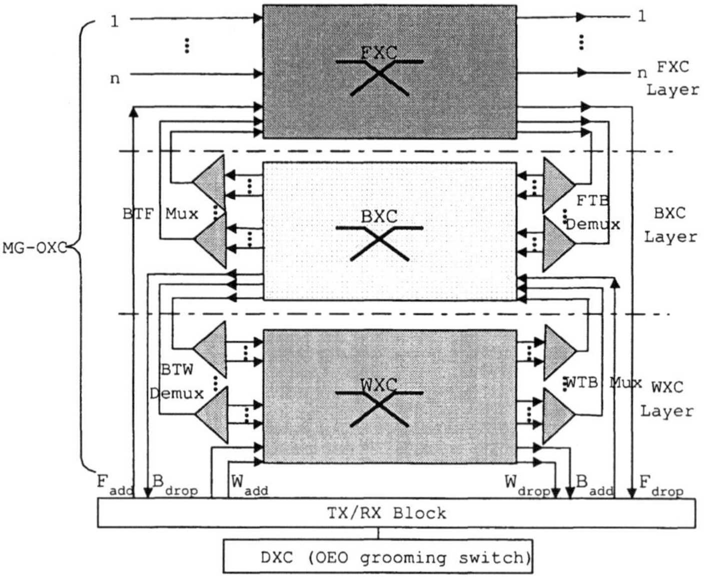


*Figure 6.1.* Three-layer multi-granular photonic cross-connect.

or dropped. In other words, only the band(s) whose wavelengths need to be added or dropped will be demultiplexed, and only the wavelengths in those bands that carry bypass traffic need to be switched using the WXC. This is in contrast to the ordinary-OXCs, which need to switch every wavelength individually using one port.

With this architecture, it is possible to dynamically select fibers for multiplexing/demultiplexing from FXC layer to the BXC layer, and bands for multiplexing/demultiplexing from BXC to the WXC layer. For example, at the $FXC$ layer, as long as there is a free $FTB$ demultiplexer, *any* fiber can be demultiplexed into bands. Similarly, at the $BXC$ layer any band can be demultiplexed to wavelengths using a free $BTW$ demultiplexer by appropriately configuring the FXC, BXC cross-connects and associated demultiplexers. On the other hand, in the architecture in [1], these configurations are fixed, in that only certain fixed fibers (bands) can be demultiplexed. This dynamic grouping architecture is more flexible (but more complicated), as it allows a complete dynamic reconfiguration of the fibers, bands, and wavelengths for drop, add or bypass.

## 6.2.2 Port Count Reduction: An Example

We use an example to illustrate the working of the Three-Layer MG-OXC. When counting the number of ports, we will only focus on the input-side of the MG-OXC (due to the symmetry of the MG-OXC architecture), which consists of locally added traffic and traffic coming into the MG-OXC node from all other nodes (i.e., the bypass and locally dropped traffic). Assume there are 10 fibers, each having 100 wavelengths, and one wavelength needs to be dropped and one to be added at a node. The total number of ports required at the node when using an ordinary-OXC is 1000 for incoming wavelengths (including 999 for bypass and 1 for drop wavelength), plus 1 for add wavelength for a total of 1001. However, if the 100 wavelengths in each fiber are grouped into 20 bands, each having 5 wavelengths, then using a MG-OXC as in Figure 6.1, only one fiber needs to be demultiplexed into 20 bands (using a 11-port FXC). Hence, only one of these 20 bands needs to be demultiplexed into 5 wavelengths (using a 21-port BXC). Finally, one wavelength is dropped and added (using a 6-port WXC). Accordingly, the MG-OXC has only $11 + 21 + 6 = 38$ ports, which is about 30 times reduction in port count.

## 6.3 Waveband Switching

In this section, we introduce various WBS schemes and lightpath grouping strategies, and summarize the major benefits of using MG-OXCs for WBS.

### 6.3.1 Waveband Switching Schemes

We first classify *WBS schemes* into two variations depending on whether the number of bands in a fiber ($\mathcal{B}$) is fixed or variable as in Figure 6.2. Each variation is further classified according to whether the number of wavelengths in a band (denoted by $\mathcal{W}$) is fixed or variable. For a given fixed value of $\mathcal{W}$, the set of wavelengths in band can be further classified depending on whether they are pre-determined (e.g., consists of consecutively numbered subset of wavelengths) or can be adaptive (dynamically configured). For example, one variation, could be to allow a variable number of wavelengths in a band at different nodes, with these wavelengths being chosen randomly (not necessarily consecutively). Such a variation may result in more flexibility (efficiency) in using MG-OXC than the variation shown shaded, on the other hand, the MG-OXC (especially its BXC) required to implement this variation may be too complex to be feasible with the current and near-future technology.

Hereafter, we concentrate on the variation shown shaded in Figure 6.2, where each fiber has a fixed number of bands and each band has a fixed number as well as a fixed set of wavelengths, though the principles to be discussed can be extended to other WBS variations as well.

### 6.3.2 Lightpath Grouping Strategy

The following grouping strategies can be used to group lightpaths into wavebands.

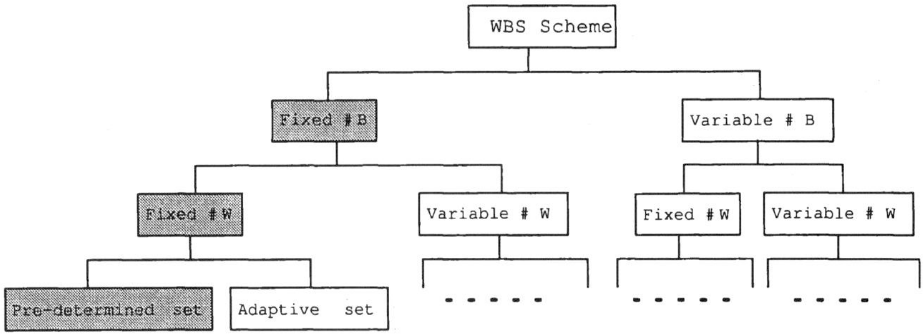


*Figure 6.2.* Classification of the waveband switching schemes.

(1) *end-to-end grouping*: grouping the traffic (lightpaths) with same source-destination (s-d) only;

(2) *one-end-grouping*: grouping the traffic between the same source (or destination) nodes and different destination (or source) nodes;

(3) *subpath grouping*: grouping traffic with common subpath (from any source to any destination).

We can see that Strategy 3 is the most powerful (in terms of being able to maximize the benefits of WBS) although it is also the most complex to use in WBS algorithms.

### 6.3.3 Major Benefits of WBS Networks

From the previous discussion (and performance results to be described later), we see that WBS in conjunction with MG-OXCs can bring about tremendous benefits in terms of reducing the size (i.e., the number of ports) of optical cross-connects, which in turn reduces the size of the OEO grooming switch, as well as the cost and difficulty associated with controlling them. In addition to reducing the port count (which is a major factor contributing to the overall cost of switching fabrics), the use of bands reduces the number of entities that have to be managed in the system, and enables hierarchical and independent management of the information relevant to bands and wavelengths. This translates into reduced size (footprint), power consumption and simplified network management. Moreover, relatively small-scale modular switching matrices are now sufficient to construct large-capacity optical cross-connects, making the system more scalable. With WBS, some or most of the wavelength paths (or lightpaths) do not have to pass through individual wavelength filters, thus simplifying the multiplexer and demultiplexer design as well. In fact, cascading of FTB and BTW demultiplexers has been shown to be effective in reducing cross-talk [2], which is critical in building large capacity backbone networks. Finally, all of these also result in reduced complexity of controlling the switch matrix, provisioning and providing protection/restoration.

## 6.4 Waveband Switching Versus Wavelength Routed Networks

Although there has been a tremendous amount of work on WRNs, and wavelength routing is still fundamental to a WBS network, the *objective* and *techniques* used for WBS are quite different. For example, a common objective in designing (dimensioning) a WRN is to reduce the number of required wavelengths or the number of used WHs, see [8] for example. However, in WBS networks, the objective is to minimize the number of ports required by the MG-OXCs. As to be shown, minimizing the number of wavelengths or WHs does *not* lead to the minimization of the port count of the MG-OXCs in WBS networks, and even a simple WBS algorithm is not a trivial extension of traditional RWA algorithms. In fact, our studies have indicated that when using the traditional optimal RWA algorithm (based on an Integer Linear Programming) with a best-effort lightpath grouping heuristic can *backfire* (i.e., result in an increase instead of decrease in the number of ports), and that an ideal WBS algorithm may need to trade a slight increase in number of wavelengths (or WHs) for a much reduced port count [10]. While many optimization problems (e.g., optimal RWA) in WRNs are already NP-complete, some of the optimization problems in WBS networks have more constraints and accordingly are even harder to solve in practice.

Due to the differences in the objectives, techniques developed for WRNs (including for example, those for traffic grooming) cannot be directly applied to effectively address WBS-related problems. For example, techniques developed for traffic grooming in WRNs, which are useful mainly for reducing the electronics (e.g., SONET Add-Drop Multiplexers) and/or number of wavelengths required, cannot be directly applied to effectively group wavelengths into bands. This is because in WRNs, one can multiplex just about any set of lower bit rate (i.e., sub-wavelength) traffic such as STS-1s into a wavelength, subject only to the constraint that the total bit rate does not exceed that of the wavelength. However, in WBS networks, there is at least *one more constraint* that is only the traffic carried by a fixed set of wavelengths (typically consecutive) can be grouped into a band.

### 6.4.1 Wavelength Conversion

There has been a significant amount of research on the benefit of wavelength conversion in WRNs, but the benefits of wavelength conversion in WBS networks with MG-OXCs needs further investigation. The following example shows that while in WRNs with full wavelength conversion, wavelength assignment is trivial, *in WBS networks, one must assign wavelengths judiciously* in order to reduce the port count of MG-OXCs. In the example shown in Figure 6.3, assume that there is one fiber on each link with two bands, each having two wavelengths (i.e., $\{\lambda_0, \lambda_1\} \in b_0, \{\lambda_2, \lambda_3\} \in b_1$). However, wavelength $\lambda_2$ is not available on any of the links shown. In addition, there are three existing lightpaths, one from node 1 to node 5 using $\lambda_0$, the second from node 2 to node 4 using $\lambda_3$, and the third from node 6 to node 4 using $\lambda_3$. Hence, the only wavelengths available on the link from node 4 to node 5 are $\lambda_1$ and $\lambda_3$.

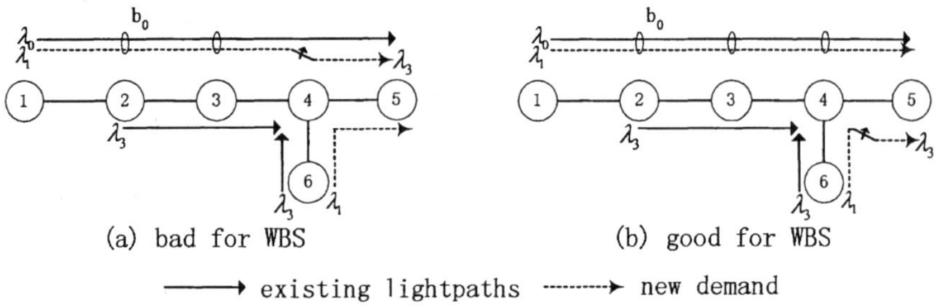


*Figure 6.3.* Wavelength assignment & schemes with full wavelength conversion.

Now assume that a new lightpath from node 6 to 5 is assigned wavelength $\lambda_1$ on both the links from node 4 to node 5 and node 6 to node 4 as shown in Figure 6.3(a). As a result, another new lightpath from node 1 to node 5 must then use $\lambda_1$ on links from node 1 to node 4, and then be converted to $\lambda_3$ on the link from node 4 to node 5.

Alternately, as shown in Figure 6.3(b), one can assign $\lambda_3$ to the first new lightpath on the link from node 4 to node 5, and assign $\lambda_1$ to the second new lightpath all the way from node 1 to node 5. In a WRN, this alternative does not result in much difference at all as it also requires a wavelength conversion at node 4. However, in a WBS network using MG-OXCs, this alternative will require fewer ports. The reason is that in Figure 6.3(b), band $b_0$ no longer needs to be demultiplexed at node 4. Note that, performing a wavelength conversion to the first new lightpath does not increase the port count because even in Figure 6.3(a), band $b_0$ on the fiber from node 6 to node 4 carrying the first new lightpath needs to be demultiplexed into wavelengths so that its $\lambda_1$ can be multiplexed with $\lambda_0$ on the link from node 4 to node 5. Therefore, in WBS networks, although wavelength conversion does facilitate wavelength grouping (or banding), but performing wavelength conversion requires each fiber or band to be demultiplexed first into wavelengths, thus potentially increasing the number of ports needed. In other words, even with full wavelength conversion, efficient WBS algorithms are still necessary to ensure the reduction in port count.

### 6.4.2 Waveband Conversion

We can see that even if wavelength conversion itself would cost nothing, in order to minimize the port count of MG-OXCs, one can no longer use wavelength conversion *freely* to make up for careless wavelength assignment as is possible in WRNs with full wavelength conversion capability. For this reason, we should also explore the use of *waveband conversion*, which can be accomplished using novel technologies [11] without having to demultiplex the band into individual wavelengths when doing conversion in WBS networks. Having

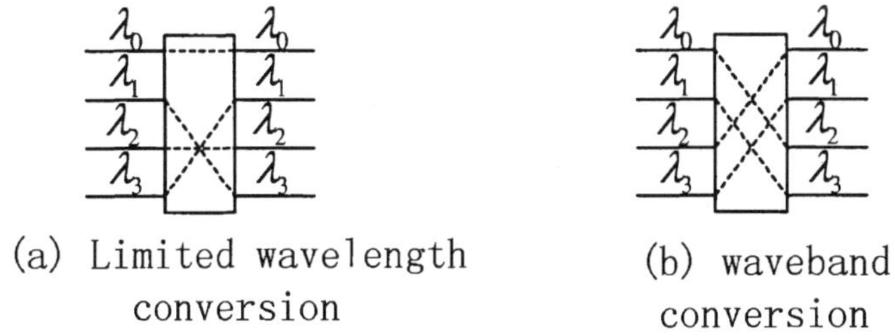


*Figure 6.4.* Limited wavelength conversion and waveband conversion.

waveband conversion is *similar* to but *not identical* to having limited wavelength conversion in WRNs. More specifically, not only a given wavelength in a band $b_i$ can be converted to a corresponding wavelength only in another band $b_j$, but also all other wavelengths in band $b_i$ have to be converted to their corresponding wavelengths in $b_j$ *at the same time*. For example, if we assume there are 2 wavelengths in each band (i.e., $\{\lambda_0, \lambda_1\} \in b_0, \{\lambda_2, \lambda_3\} \in b_1, \{\lambda_4, \lambda_5\} \in b_2, \ldots$). Then with waveband conversion, converting band $b_0$ to bands $b_1$ or $b_2$ is similar to having limited conversion, i.e., $\lambda_0$ can only be converted to $\lambda_2$ or $\lambda_4$, while $\lambda_1$ can only be converted to $\lambda_3$ and $\lambda_5$. On the other hand, the difference is that, with waveband conversion, we are now forced to convert $\lambda_0$ to $\lambda_2$ and *also* $\lambda_1$ to $\lambda_3$ at the same time as shown in Figures 6.4(a) and 6.4(b).

### 6.4.3 Waveband Failure Recovery in MG-OXC networks

Protection and restoration schemes for failure recovery from a broken fiber link or an OXC node (or in general a failed Shared Risk Link Group or SRLG) have been studied extensively in WRN. However, previous research has only examined recovery from such a failure at either the fiber or wavelength level in WRNs, and studied the tradeoffs involved in recovery at these two different levels.

With the introduction of multi-granular WBS networks, a waveband may fail because of a malfunctioning port at the BXC layer, a broken waveband multiplexing/demultiplexers or waveband converter. If the other bands in the same fiber are not affected by the failure, simply recovering the traffic carried by the affected band can be more bandwidth efficient (or more likely to succeed in restoring the traffic) than recovering the traffic carried by the entire fiber (as if the fiber is cut) although the latter is more simple and has a faster response/restoration time. Even when a fiber is cut, treating the traffic carried by one band as a basic unit for recovery can achieve a useful balance between treating the entire fiber or each individual wavelength as a basic recovery unit. Next, we propose new ways to recover from fiber link failures in WBS networks at the band level as opposed to the fiber or wavelength level.

While recovering at the fiber level is done via link protection/restoration, recovering at the wavelength level is often done via path protection (where an entire lightpath is routed from the source). To recover at the band level, it may be useful to first define *band-segment* or BS of a given band $b_i$ to be the portion of fiber route between two MG-OXCs such that $b_i$ is formed (e.g., multiplexed from wavelengths using a WTB) at the first MG-OXC and then demultiplexed into wavelengths at the second MG-OXC (e.g., using a BTW). That is, within an BS, the lightpaths carried in the band are not switched individually. Two examples of *active* (also called primary or working) band-segments are shown in Figure 6.5. The first, denoted by ABS0, spans from node 1 to node 3 via node 2, carrying two active lightpaths AP0 and AP1 (the former is dropped at node 3). The second, denoted by ABS1, spans from node 3 to node 4 carrying two active lightpaths AP1 and AP2 (the latter is added at node 3). Based

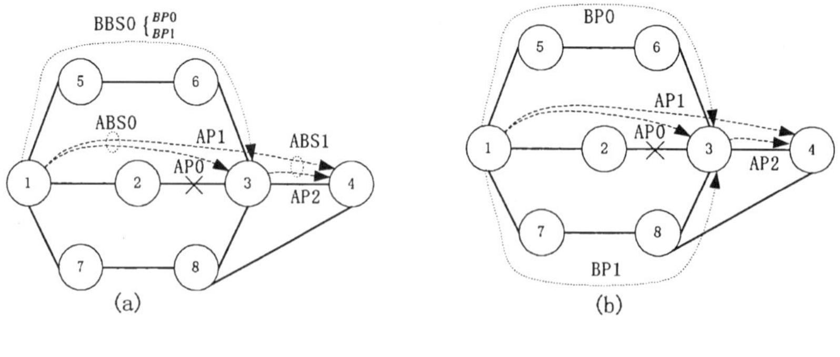


*Figure 6.5.* Recovery schemes using WBS.

on the concept of band-segment (BS), failure recovery can be accomplished in two ways as shown in Figure 6.5(a) and (b), respectively. The first approach is to recover the affected ABS0 as a basic unit using one backup (or alternate) BS, denoted by BBS0 (which includes backup lightpath BP0 and lightpath segment BP1) as shown in Figure 6.5(a). The second approach is to recover each individual lightpath/lightpath-segment in the affected band-segment as a unit. More specifically, if only the lightpaths with same source and destination are grouped into a band, it is convenient to protect all the lightpaths in a waveband segment. Otherwise, a lightpath may transit one or more band-segment along its route as AP1 does in Figure 6.5, which reduces the port count but complicates issues such as fault localization.

Similarly, the use of waveband conversion and/or wavelength conversion to recover from waveband failures is *unique* to WBS networks and becomes interesting. For example, in WRNs, one cannot merge the traffic carried by two or more wavelengths without going through OEO conversions (one may consider traffic grooming as a way to merge wavelengths through OEO conversion). However, in WBS networks we may use a new recovery technique that *merges* the critical traffic carried in a band affected by a waveband (level) failure with the traffic carried by an unaffected band, without having to go through any OEO conversions. For example, assume that a fiber has two bands $b_0$ and $b_1$, each with 3 wavelengths such that $\lambda_0$ and $\lambda_1$ are used in $b_0$, so is $\lambda_5$ in $b_1$ as in Figure 6.6(a). Further assume that band $b_0$ is affected by a *band failure*. Instead of having to re-route the traffic carried by band $b_0$ using a backup band along a link-disjoint path, one may use a technique which we call *band-merging*, whereby the traffic carried by wavelengths $\lambda_0$ and $\lambda_1$ can be restored on their corresponding wavelengths in $b_1$ (i.e., $\lambda_3$ and $\lambda_4$, respectively). Note that, the traffic carried on $\lambda_5$ should remain intact as a result of band-merging as its corresponding wavelength $\lambda_2$ in $b_0$ is inactive. The band-merging technique can be implemented by simply converting $\lambda_0$ and $\lambda_1$ to $\lambda_3$ and $\lambda_4$, respectively, it may also be implemented by using a novel device operating under a principle similar to that of waveband conversion, which can avoid demultiplexing bands $b_0$ and $b_1$, as required by wavelength conversion.

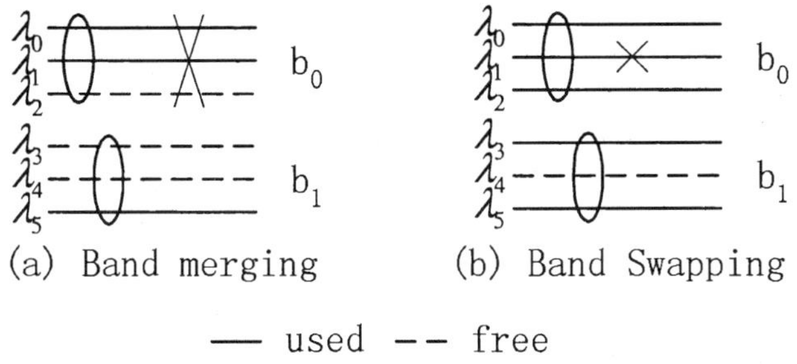


*Figure 6.6.* Band merging and band swapping in failure recovery.

As an another example (see Figure 6.6(b)), assume that all wavelengths except $\lambda_4$ are used, and that $\lambda_1$ (in $b_0$) alone is affected by a *wavelength failure.* To recover from such a failure using the spare bandwidth on $\lambda_4$, one may convert $\lambda_1$ to $\lambda_4$ at a node prior to the fault, but this requires both bands to be demultiplexed at this node. To avoid demultiplexing of the bands and preserve the wavelength grouping, a new technique called *band-swapping* which converts band $b_0$ to $b_1$ and $b_1$ to $b_0$, can be used to recover from the failure.

## 6.5 Algorithms and Performance of WBS Networks

In this section, we first introduce an ILP model, which can be solved to minimize either the total port count or the maximum port count over all Three-layer MG-OXC nodes in the network to satisfy a given set of traffic demands on a given network topology, and heuristic-based solutions for efficient WBS. We then compare the numerical results of the heuristics and ILP model.

### 6.5.1 ILP Model for WBS

This section formulates the WBS scheme using integer linear programming (ILP). The ILP formulation is for multi-fiber networks, and is more general than existing ILP models, including that for the single-fiber case in [1].

**Notations.** The following notations are used in the ILP model.

- $I_n$: Set of input fibers at node $n$ (excluding those for local add);
- $\mathcal{I}^f_{n,m}$: Input fiber $f$ at node $n$, connected to node $m$. So $I_n = \bigcup_{m,f} \mathcal{I}^f_{n,m}$;
- $O_n$: Set of output fibers at node $n$ (excluding those for local drop);
- $\mathcal{O}^f_{n,m}$: Output fiber $f$ at node $n$, connected to node $m$. So $O_n = \bigcup_{m,f} \mathcal{O}^f_{n,m}$;
- $A_n$: Set of local add fibers at node $n$, including those used at the ports of WXC, BXC and FXC layer;
- $D_n$: Set of local drop fibers at node $n$, including those used at the ports of WXC, BXC and FXC layer;
- $IA_n$: $I_n \bigcup A_n$. This set includes the set of all incoming fibers (local and non-local) at node $n$;
- $OD_n$: $O_n \bigcup D_n$. This set includes the set of all outgoing fibers (local and non-local) at node $n$;
- $\mathcal{L}_b$: Set of wavelengths in band $b$;

- $\mathcal{F}$: Number of fibers per link that can be used for each direction;
- $\mathcal{K}$: Number of wavelengths per fiber;
- $\mathcal{B}$: Number of wavelength bands per fiber;
- $\mathcal{W}$: Number of wavelengths per wavelength band ($\mathcal{K} = \mathcal{B} \times \mathcal{W}$);
- P: Set of node pairs having non-zero traffic demand. Each node pair can be denoted by $p = (p.src, p.dest)$, where $p.src$ and $p.dest$ represent the source and destination nodes of one or more request lightpaths, respectively;
- T[p]: Traffic matrix whose element $t_p$ is an integer, representing the traffic demand (i.e., the number of lightpaths) of the node pair $p$.

**ILP Variables.** To facilitate the presentation and understanding of our ILP model, we first define the variable $V_{i,o,p}^{n,w}$, which describes the lightpaths between node pairs in the network.

- $V_{i,o,p}^{n,w}$: 1 if at node $n$ there is a lightpath between node pair $p$ using wavelength $w$ on an incoming fiber $i$ to outgoing fiber $o$, and 0 otherwise;

To describe the drop/bypass/add traffic (lightpath) at different layers of an MG-OXC node, the following three variables: $W_{i,o}^{n,w}$, $B_{i,o}^{n,b}$ and $F_{i,o}^{n}$ are used. In the above $I_n \bigcup A_n$ is called an incoming fiber and $O_n \bigcup D_n$ is called an outgoing fiber. More specifically, when $i \in I_n, o \in O_n$, these variables represent bypass traffic; add traffic when $i \in A_n, o \in O_n$, and drop traffic when $i \in I_n, o \in D_n$ (note that the case when $i \in A_n, o \in O_n$ does not make sense).

- $W_{i,o}^{n,w}$: 1 if node $n$ has a lightpath using wavelength $w$ on an incoming fiber $i$ through the *WXC* layer onto an outgoing fiber $o$, and 0 otherwise;
- $B_{i,o}^{n,b}$: 1 if node $n$ has a set of lightpaths using waveband $b$ ($b \in [1, 2, \ldots, \mathcal{B}]$) on an incoming fiber $i$ through the *BXC* layer onto an outgoing fiber $o$, and 0 otherwise;
- $F_{i,o}^{n}$: 1 if node $n$ has a set of lightpaths using an incoming fiber $i$ through the *FXC* layer onto an outgoing fiber $o$, and 0 otherwise;

The following four additional variables are also defined for describing the multiplexing/demultiplexing at the *FXC, BXC* and *WXC* layers.

- $FTB_i^n$: 1 if input fiber $i$ ($i \in I_n$) needs to be demultiplexed into bands at node $n$, and 0 otherwise;
- $BTW_i^{n,b}$: 1 if band $b$ on input fiber $i$ ($i \in I_n$) needs to be demultiplexed into wavelengths at node $n$, and 0 otherwise;

- $BTF_o^n$: 1 if a band needs to be multiplexed onto an output fiber $o$ ($o \in O_n$) at node $n$, and 0 otherwise;
- $WTB_o^{n,b}$: 1 if a wavelength needs to be multiplexed on to band $b$ of an output fiber $o$ ($o \in O_n$) at node $n$, and 0 otherwise;

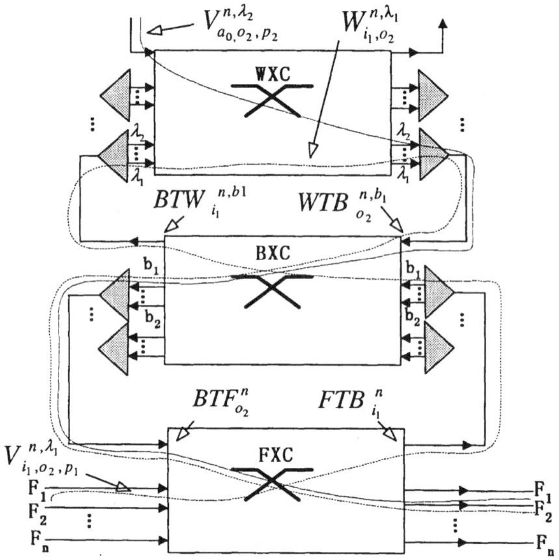


*Figure 6.7.* An example of wavebanding at node $n$.

As a consequence of multiplexing/demultiplexing, we need to use multiplexer/demultiplexer port(s) at the respective layers. Figure 6.7 shows one such example involving two lightpaths, one for node pair $p_1$ using $\lambda_1$ on input fiber 1 and the other for node pair $p_2$ using $\lambda_2$ to be added locally. Using the MG-OXC, the two lightpaths are grouped together in the same band of the same output fiber (e.g., fiber 2). By definition, we have $V^{n,\lambda_1}_{i_1,o_2,p_1} = V^{n,\lambda_2}_{a_0,o_2,p_2} = 1$. For this, input fiber 1 (containing the lightpath for $p_1$) has to be demultiplexed into band $b_1$ (and other bands) using a FTB demultiplexer (hence, $FTB^n_{i_1} = 1$). Band $b_1$ then has to be further demultiplexed into $\lambda_1$ and other wavelengths (hence, $BTW^{n,b_1}_{i_1} = 1$) to switch the lightpath for $p_1$ (hence, $W^{n,\lambda_1}_{i_1,o_2} = 1$). The second lightpath for $p_2$ is added into band $b_1$ using a WTB multiplexer (hence, $WTB^{n,b_1}_{o_2} = 1$). Now that the two lightpaths are in the same band, the band is multiplexed onto a fiber using a BTF multiplexer (hence, $BTF^n_{o_2} = 1$), and then transmitted onto output fiber 2.

**Objective Function.** Let $WXC_n$, $BXC_n$ and $FXC_n$ be the number of ports at *WXC,BXC* and *FXC* layers at node $n$, respectively. There are two reasonable objectives. The first is to minimize the total cost associated with the MG-OXC ports in the network, that is,

$$minimize\ [\alpha \times \sum_n WXC_n + \beta \times \sum_n BXC_n + \gamma \times \sum_n FXC_n] \quad (6.1)$$

where $\alpha, \beta$ and $\gamma$ are the coefficients or weights corresponding to the cost of each port at the *WXC,BXC* and *FXC* layer, respectively. When $\alpha = \beta = \gamma = 1$, the objective becomes to minimize the total number of MG-OXC ports in the network, which is the sum of the port count at FXC, BXC and WXC layers respectively.

The second objective is to minimize the maximum cost at each node over all nodes. This can be formulated as:

$$minimize\ \max_n(\alpha \times WXC_n + \beta \times BXC_n + \gamma \times FXC_n) \quad (6.2)$$

When $\alpha = \beta = \gamma = 1$, this becomes equal to minimizing the maximum port count (node size) over all the nodes in the network.

**Constraints.** The following are the various constraints that need to be satisfied by the ILP for WBS.
For *routing and wavelength assignment*, the following constraints on traffic flows, wavelength-capacity and wavelength-continuity are similar to those in the traditional RWA ILP formulations.

$$\begin{cases} \sum\limits_{i \in A_n, o \in O_n} V_{i,o,p}^{n,w} = \sum\limits_{i \in I_n, o \in D_n} V_{i,o,p}^{n,w} = 0 & n \neq p.src, p.dest, \forall w \quad \text{(i)} \\ \sum\limits_{w, i \in A_n, o \in O_n} V_{i,o,p}^{n,w} = t_p & n = p.src, \quad \text{(ii)} \\ \sum\limits_{w, i \in I_n, o \in D_n} V_{i,o,p}^{n,w} = t_p & n = p.dest, \quad \text{(iii)} \end{cases} \quad (6.3)$$

$$\sum_{p, o \in OD_n} V_{i,o,p}^{n,w} \leq 1 \ \forall w, i \in I_n; \quad (6.4)$$

$$\sum_{p, i \in IA_n} V_{i,o,p}^{n,w} \leq 1 \ \forall w, o \in O_n; \quad (6.5)$$

$$\sum_{i \in IA_m, o \in \mathcal{O}_{m,n}^f} V_{i,o,p}^{m,w} - \sum_{i \in \mathcal{I}_{n,m}^f, o \in OD_n} V_{i,o,p}^{n,w} = 0 \ \forall m, n, p, w, f; \quad (6.6)$$

Equation (6.3) is the traffic flow constraint; Equations (6.4) and (6.5) are the wavelength capacity constraint; Equation (6.6) is the wavelength continuity constraint.

For *waveband switching*, we need the following additional constraints.

$$1 \geq F^n_{i,o} + B^{n,b}_{i,o} + W^{n,w}_{i,o} \geq \sum_p V^{n,w}_{i,o,p} \; \forall w \in \pounds_b, i \in IA_n, o \in OD_n; \quad (6.7)$$

$$1 \geq F^n_{i,o} + \sum_{p,o_1 \neq o} V^{n,w}_{i,o_1,p}, \quad 1 \geq F^n_{i,o} + \sum_{p,i_1 \neq i} V^{n,w}_{i_1,o,p} \; \forall w, i, o; \quad (6.8)$$

$$1 \geq B^{n,b}_{i,o} + \sum_{p,o_1 \neq o} V^{n,w}_{i,o_1,p}, \quad 1 \geq B^{n,b}_{i,o} + \sum_{p,i_1 \neq i} V^{n,w}_{i_1,o,p} \; \forall i, o, w \in \pounds_b; \quad (6.9)$$

Constraints (6.7), (6.8) and (6.9) ensure that if a lightpath uses wavelength $w$ belonging to band $b$ of incoming fiber $i$ and outgoing fiber $o$ (i.e., $\sum_p V^{n,w}_{i,o,p} = 1$), then at node $n$,

- exactly one of FXC, BXC and WXC cross-connect port will be used for switching this lightpath when it is a bypass (i.e., $i \in I_n, o \in O_n$) or
- exactly one of $F_{add}$, $B_{add}$ and $W_{add}$ port will be used for adding this lightpath when it is added (i.e., $i \in A_n, o \in O_n$) or
- exactly one of $F_{drop}$, $B_{drop}$ and $W_{drop}$ port will be used for dropping this lightpath when it is dropped (i.e., $i \in I_n, o \in D_n$).

$$BTF^n_o \geq WTB^{n,b}_o \geq W^{n,w}_{i,o} \; \forall w \in \pounds_b, o \in O_n, i \in IA_n; \quad (6.10)$$

The above constraint ensures that a wavelength $w$ at node $n$ switched or added at the *WXC* layer has to pass a WTB multiplexer to the *BXC* layer. At the same time, every band from a WTB multiplexer has to pass a BTF multiplexer before it can leave node $n$. Similarly, Equation (6.11) below specifies that a wavelength $w$ switched or dropped at the *WXC* layer has to come from *BXC* layer using a BTW demultiplexer, and in addition every band demultiplexed by BTW can only come from a FTB demultiplexer.

$$FTB^n_i \geq BTW^{n,b}_i \geq W^{n,w}_{i,o} \; \forall w \in \pounds_b, o \in OD_n, i \in I_n; \quad (6.11)$$

Finally, any bypass or add bands should pass a BTF multiplexer as specified in equation (6.12) and similarly, any drop or bypass band can only come from a FTB demultiplexer as specified in Equation (6.13) .

$$BTF_o^n \geq B_{i,o}^{n,b}\ \forall o \in O_n, i \in IA_n; \tag{6.12}$$

$$FTB_i^n \geq B_{i,o}^{n,b}\ \forall o \in OD_n, i \in I_n; \tag{6.13}$$

For *port numbers*, the following constraints specify the minimum number of ports required at each layer of the MG-OXC.

$$WXC_n = \sum_{i \in IA_n, o \in OD_n, w} W_{i,o}^{n,w}\ \forall n; \tag{6.14}$$

$$BXC_n = \sum_{i \in IA_n, o \in OD_n, b} B_{i,o}^{n,b} + \sum_{o \in O_n, b} WTB_o^{n,b} + \sum_{i \in I_n, b} BTW_i^{n,b}\ \forall n; \tag{6.15}$$

$$FXC_n = \sum_{i \in IA_n, o \in OD_n} F_{i,o}^n + \sum_{o \in O_n} BTF_o^n + \sum_{i \in I_n} FTB_i^n\ \forall n; \tag{6.16}$$

For the *WXC* layer, the number of input-side ports include the bypass, add/drop lightpaths as specified in (6.14). The number of input-side ports needed at the *BXC* layer is the sum of the number of wavebands $B_{i,o}^{n,b}$ (BXC cross-connect and add/drop/bypass bands) and the number of wavebands from the WTB/BTW multiplexers/demultiplexers as in (6.15). Similarly, Equation (6.16) can be used to determine the number of ports at the *FXC* layer.

In short, our ILP model (and heuristics to be described next) considers the design of MG-OXC nodes (i.e., the number of ports allocated at each of the layers) with the objective to minimize either the total port count or the maximum port count over all MG-OXC nodes in the network given a set of traffic demands to be satisfied on a given network topology, wherein each link in the network may have single or multiple fibers.

Note that if we eliminate the *FXC* and *BXC* layers (i.e., by setting the corresponding variables to 0) from the MG-OXC, the above ILP formulation with objective in Equation (6.1) will minimize the total number of ports, which is equivalent to minimizing WHs using ILP for optimal RWA in WRNs. In fact, above optimal WBS algorithm based on an ILP model includes the well-known optimal RWA problem as a special instance [9]. For example, the problem of minimizing the number of wavelength-hops in WRNs (with K wavelengths on each link) is identical to the problem of minimizing the port count in WBS networks (where either $\mathcal{B} = 1$ & $\mathcal{W} = \mathcal{K}$ or $\mathcal{B} = \mathcal{K}$ & $\mathcal{W} = 1$). Since the optimal WBS problem (using ILP) is not solvable for larger networks, we need to develop efficient heuristic WBS algorithms.

### 6.5.2 Waveband Oblivious RWA (WBO-RWA)

To study the relationship between WBS and traditional RWA, we use ILP formulations for RWA [8] that minimizes the total number of used WHs. We then group the assigned wavelengths into bands and calculate the number of required ports. Since the RWA is done completely oblivious to the existence of wavebands, this algorithm will be referred to as waveband oblivious routing and wavelength assignment (WBO-RWA).

### 6.5.3 Balanced Path routing with Heavy-Traffic first waveband assignment (BPHT)

Intuitively, to maintain *wavelength-continuity* in wavelength routed optical networks without wavelength conversion, it is better to assign wavelengths to longer paths (in terms of hops) first. Further, to reduce the number of ports in MG-OXC, it is better to assign paths that have maximum number of links in common, wavelengths in the same fiber (and band), thus increasing the probability of switching the whole fiber (and band) by just using a single FXC (and BXC) port. The following is our three-stage heuristic algorithm called Balanced Path routing with Heavy-Traffic (BPHT) first waveband assignment, which tries to maximize the reduction in the MG-OXC size using the above ideas.

**Stage 1: Balanced Path Routing.** In this stage, we use the following steps to achieve load balanced routing.

- Find K-shortest routes for every node pair $(s, d)$ with non-zero traffic demand, and order them from the shortest to the longest (in terms of hop number) as $P^1_{s,d}, P^2_{s,d}, \cdots, P^k_{s,d}$. Let the number of hops of the shortest route be $H_{s,d}$.

- Define the load on every link $l$ to be the number of routes already using link $l$ (initially, this is 0). Let $C$ be the maximum link load over all the links.

- Use $C$ to achieve load balanced routing, starting with the node pair $(s, d)$ with the largest $H_{s,d}$ value over all node pairs, to determine the route for each node pair. More specifically, for the K-shortest routes $P^i_{s,d}$ of the selected node pair $(s, d)$, where $i = 1, 2, \cdots, k$, we compute $C$ and pick one of the routes that minimizes $C$. If more than one routes, say $P^i_{s,d}$ and $P^j_{s,d}$, have the same minimum $C$, the shortest one (i.e., $P^i_{s,d}$, if $i < j$) will be used as the route for $(s, d)$. That is, all the lightpaths from $s$ to $d$ will take this route. After the route for $(s, d)$ is chosen, the process continues to choose one route for each of the remaining node pairs, starting with the one having the largest number of hops along the shortest path, until every node pair with non-zero traffic demand is assigned a route.

**Stage 2: Wavelength Assignment.** Based on the observation that bypass traffic, which goes through two or more hops accounts for $60\% - 80\%$ of the total traffic in the backbone, we assign the wavelengths to those bypass lightpaths first. At the same time, we also want to give preference to the lightpaths that overlap with many other (shorter) lightpaths in order to maximize the advantage of wavebanding.

The following steps are used to assign wavelengths to all the lightpath demands once the routing is done in Stage 1. To maximize the benefit of WBS in multi-fiber networks, we introduce a new waveband assignment algorithm, called waveband assignment for multi-fiber WBS (*WA-MF-WBS*, see Step (D) below).

(A) For every node pair $(s, d)$, whose route is determined as $s = s_0 \rightarrow s_1 \rightarrow s_2 \ldots s_{n-1} \rightarrow s_n = d$ in Stage 1, define a set $Q_d^s$, which includes all node pairs $(s_i, s_j)$, whose route is $s_i, s_{i+1}, \ldots, s_j$, as determined in Stage 1, where $0 \le i \le n-2$, and $i+2 \le j \le n$. Note that it is possible that the route chosen for $(s_i, s_j)$ in Stage 1 is not a sub-path of the route chosen for $(s, d)$, in which case, $(s_i, s_j)$ will not belong to $Q_d^s$.

(B) Calculate the weight (similar to the concept of wavelength-hops) for each set $Q_d^s$ as $W_{sd} = \sum_{p \in Q_d^s} h_p \times t_p$, where $p = (s_i, s_j) \in Q_d^s$, $h_p$ is the number of hops and $t_p$ is the required number of lightpaths from $s_i$ to $s_j$;

(C) Find the set $Q_d^s$ with the largest $W_{sd}$.

(D) Call set $Q_d^s$ as $\mathcal{L}$, and assign wavelengths to $\mathcal{L}$ as follows.
**i.** Suppose that the longest path in $\mathcal{L}$ is as follows: $s_0 \rightarrow s_1 \rightarrow s_2 \ldots s_{n-1} \rightarrow s_n$. Let $s = s_0$ and $d = s_n$ (which is the case initially based on the definition of $Q_d^s$). Assign wavelengths to the requested lightpaths for the node pair $(s, d)$ by trying to group them into the same fiber, and within each fiber, into the same band(s).

More specifically, for each fiber, let $0 \le w \le \mathcal{K} - 1$ and $0 \le b \le \mathcal{B} - 1$ be the index of wavelength and band respectively, starting from which, an available wavelength and band will be searched in order to fulfill new lightpath requests; In addition, let $0 \le f \le \mathcal{F} - 1$ be the index of the fiber currently under consideration (i.e., whose wavelengths may be used for new lightpaths). Initially, $f = 0$ and $w = b = 0$ for all fibers. The following algorithm *WA-MF-WBS* assigns wavelengths to the lightpaths for a specified node pair $p$.

**Algorithm: WA-MF-WBS**

**while** $t_p > \mathcal{W}$ **do**

Find a fiber starting from index $f$ that has as many free bands as possible (say $a \le \lfloor \frac{t_p}{\mathcal{W}} \rfloor$ ) {
Call the found fiber $g$, where $g$ may or may not be the same as $f$;
Assign the bands in fiber $g$ to the $a \cdot \mathcal{W}$ lightpaths for $p$;
$t_p = t_p - a \cdot \mathcal{W}$;
Set $f = g$, and update $w$ and $b$ for fiber $g$ accordingly;
}

**end while**

**while** $t_p > 0$ **do**

Find a fiber ($g$), starting from index $f$, that has at least one free wavelength;
Assign a free wavelength ($x$), starting from index $w$, to a lightpath for $p$, where $x$ is most likely to be $w$;
$t_p = t_p - 1$;
Set $f = g$, and $w = x + 1$. Also, update $b$ for fiber $g$ accordingly;

**end while**

**ii.** Use *WA-MF-WBS* to assign wavelengths to the requested lightpaths for ($s$,$s_j$) starting with the largest $j$ (i.e., $j = n-1, n-2, \ldots, 2$).

**iii.** Use *WA-MF-WBS* to assign wavelengths to the requested lightpaths for ($s_i$, $d$) starting with the smallest $i$ (i.e., $i = 1, 2, \ldots, n-2$).

**iv.** If there are still node pairs $(s_i, s_j) \in Q_d^s$ that have not been considered, repeat from Step (D) by treating $s_i$ with the smallest $i$ as $s$, and $s_j$ with the largest $j$ as $d$. Otherwise go to Step (E).

(E) Recompute the weight for those node pairs whose routes use any part of the route used by node pair $(s, d)$. For each fiber, re-adjust $b$ and $w$ to be the "next" waveband and the first wavelength in the next waveband, respectively, so as to prevent the lightpaths of the next node pair set (e.g., $Q_{d'}^{s'}$) from using the same bands as the lightpaths of $Q_d^s$ (thus reducing the need to demultiplex and multiplex the lightpaths belonging to these two sets when they merge and diverge). More specifically, set $b = (b+1)$ *mod* $\mathcal{B}$, and $w = b \times \mathcal{W}$, and then go to step $(C)$. Repeat until all the bypass (multi-hop) lightpath demands are satisfied.

For example, suppose that the lightpaths numbered from 1 to 6 are routed as in Figure 6.8(a) as dictated by the load balancing routing algorithm. Accordingly, the set $Q_{D_1}^{S_0}$ consists of the 4 lightpaths: 1, 2, 3 and 4, and the weight of the set is 14 as shown (note that $t_p = 1$ for all p). Similarly, the set $Q_{D_2}^{S_6}$ consists of the 3 lightpaths: 4, 5 and 6, and its

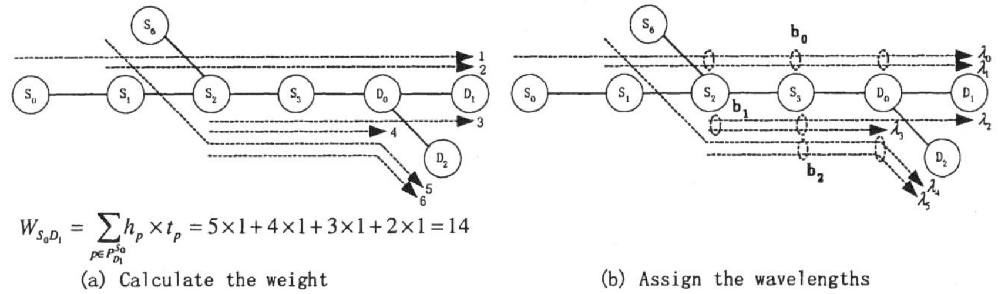


(a) Calculate the weight (b) Assign the wavelengths

*Figure 6.8.* An example illustrating step (C) and (D) in stage 2 of BPHT.

corresponding weight is 9. Hence, the wavelengths $\lambda_0$ to $\lambda_3$ will be assigned to the lightpaths in set $Q_{D_1}^{S_0}$ first. Note that after lightpath 4 is satisfied, set $Q_{D_2}^{S_6}$ includes only lightpaths 5, and 6 and its weight needs to be updated to 7. Since now $Q_{D_2}^{S_6}$ is the set with maximum weight, wavelengths $\lambda_4$ and $\lambda_5$ will be assigned to the lightpaths in the set $Q_{D_2}^{S_6}$ as in Figure 6.8(b).

(F) Finally, assign wavelengths to lightpaths between two nodes separated by only one hop, starting with the node pair having the largest lightpath demand.

**Stage 3: Waveband Switching.** Once the wavelength assignment is done, WBS can be performed in a fairly straight-forward way. Basically, we switch as many fibers using FXCs as possible; and then as many wavebands using BXCs as possible. The remaining lightpaths are then individually switched at the WXC layer. The total number of ports used at a given node can then be determined as discussed at previous section.

Ideally, BPHT will group traffic from the same source to the same destination, and most of the traffic that has common intermediate links. One of the variations of BPHT (in Stage 1) is to balance the amount of traffic (in terms of the actual number of lightpaths instead of just one route for each node pair) on every link. Another variation is to assign wavelengths to lightpaths with the largest hop count or those for node pairs with the largest weighted traffic demand (i.e., $h_p \times t_p$) first (assuming e.g., shortest-path routing) in Stage 2. In our experiments, the overall performance of BPHT is the best, and hence, only show the results of the comparison of BPHT and WBO-RWA.

### 6.5.4 Numerical Results

In this section, we compare WBS networks employing MG-OXCs with WRNs employing ordinary-OXCs, and the corresponding WBS algorithms.

**Performance Metrics.** We define the following three *performance metrics*. Each metric is a function of a WBS algorithm, denoted by "a".

- **Total port number ratio T(a):** $\frac{Total(FXC_n+BXC_n+WXC_n)\, used\, by\, WBS\, algorithm\, 'a'}{Total(OXC_n)\, of\, ordinary-OXC}$
- **Max port number ratio M(a):** $\frac{max(FXC_n+BXC_n+WXC_n)\, used\, by\, WBS\, algorithm\, 'a'}{max(OXC_n)\, of\, ordinary-OXC}$
- **Used wavelength-hop ratio W(a):** $\frac{wavelength-hops\, used\, by\, WBS\, algorithm\, 'a'}{wavelength-hops\, used\, by\, optimal\, RWA\, without\, WBS}$

Note that by definition W(WBO-RWA)=1.

**Results for a Six-node Network.** We first compare the results of optimal WBS obtained from our ILP formulations with that of BPHT and WBO-RWA. The results presented below are for a randomly generated six-node network topology with $F$ bidirectional fibers per link.

Table 6.1 shows the performance ratios for optimal WBS (based on ILP), WBO-RWA and BPHT in a random six-node network for three different representative random traffic patterns where the total lightpaths (i.e., $\sum_p t_p$) is 25, 31 and 53, respectively. As the basis for the comparison, the last row (OXC) indicates the minimum total number of ports required when ordinary-OXCs without WBS are used. The rows T(a), M(a) and W(a) represent the respective performance ratios.

*Table 6.1.* Results for a six-node network with random topology

| | Optimal WBS | | | WBO-RWA | | | BPHT | | |
|---|---|---|---|---|---|---|---|---|---|
| $\sum t_p$ | 25 | 31 | 53 | 25 | 31 | 53 | 25 | 31 | 53 |
| T(a) | 0.48 | 0.42 | 0.51 | 1.23 | 0.84 | 1.26 | 0.54 | 0.43 | 0.56 |
| M(a) | 0.69 | 0.50 | 0.73 | 1.44 | 1.19 | 1.50 | 0.63 | 0.50 | 0.69 |
| W(a) | 1.02 | 1.02 | 1.01 | 1.00 | 1.00 | 1.00 | 1.00 | 1.02 | 1.02 |
| OXC | 71 ($\sum t_p = 25$); 83 ($\sum t_p = 31$); 142 ($\sum t_p = 53$) | | | | | | | | |

From the table, we see that the performance of BPHT is close to that of the ILP model (Optimal WBS) and much better than that of WBO-RWA. We note that the average saving when using WBS is 53% for optimal WBS and 49% for BPHT (compared to the total ports required when using ordinary-OXCs). In addition, in the process of trying to reduce the total number of ports, both our ILP solution and heuristic, BPHT have W(a)>1, that is, use more wavelength-hop (WH) than the ILP solution for traditional RWA (i.e., WBO-RWA). This can be explained as follows: sometimes, to reduce port count, a longer path that utilizes a wavelength in a band may be chosen even though a shorter path (that cannot be packed into a band) exists. In other words, minimizing the number of ports at MG-OXC does not necessarily imply minimizing the number of WHs (even though minimizing WHs in networks without MG-OXC is equivalent to minimizing the number of ports). In fact, there is a trade-off between the required number of WHs and ports.

**Results for the NSF Network.** In this section we show the results of the comparison of algorithms BPHT and WBO-RWA in a multi-fiber NSF network.

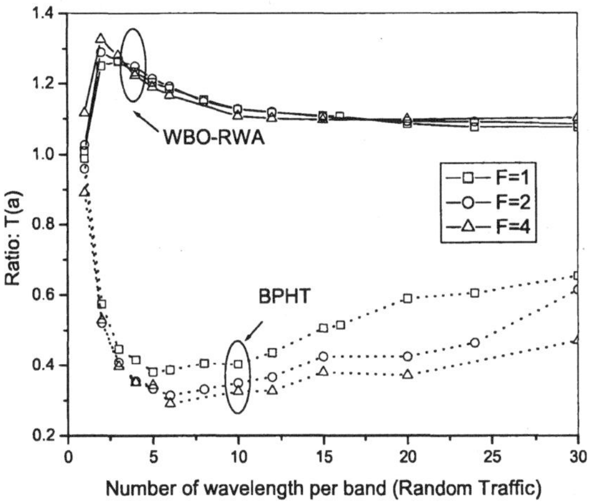


*Figure 6.9.* Total port number ratio.

Figures 6.9 and 6.10 illustrate how the ratios $T(a)$ and $M(a)$ vary with changing waveband granularity (i.e., number of wavelengths in a band) and number of fibers but a fixed number of total wavelengths per link (i.e., $\mathcal{F} \times \mathcal{B} \times \mathcal{W} = 240$) and a fixed traffic load (i.e., the total traffic does not change with $\mathcal{F}$ or $\mathcal{B}$ or $\mathcal{W}$) but a random pattern. From the figures, we notice that the total number of ports in the network and the maximum number of ports at a node among all nodes by using BPHT is much less than those from WBO-RWA, and heuristic WBO-RWA requires more ports at MG-OXC than using ordinary OXCs (as T(WBO-RWA) $>$ 1). Interestingly, the curves for BPHT in Figures 6.9 and 6.10 also indicate that with an appropriate waveband granularity ($\mathcal{W} \simeq 6$), BPHT performs the best in terms of both $T(a)$ and $M(a)$, achieving a savings of nearly 70% in number of ports when using MG-OXCs instead of ordinary OXCs.

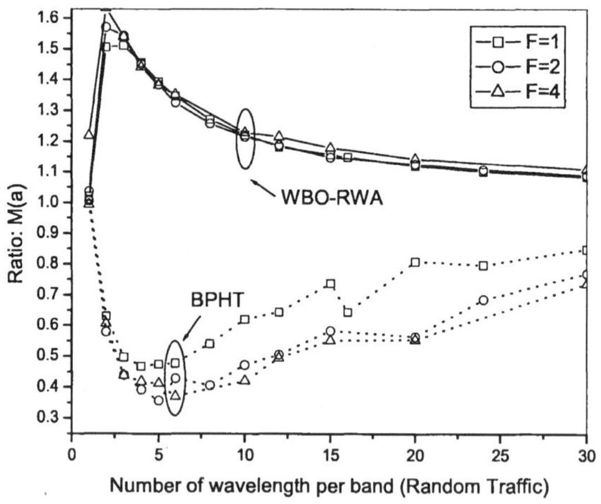


*Figure 6.10.* Max port number ratio.

More specifically, we notice that multi-fiber MG-OXC networks perform better than single-fiber MG-OXC networks, as they can achieve a larger reduction in port count when using BPHT. This is because with multiple fibers (e.g., $\mathcal{F} = 4$) there is a higher probability to switch lightpaths as a group (whole fiber or band). In single-fiber networks the advantage of having a FXC layer and fiber switching is not evident [1]. On the other hand, the situation is slightly reversed for WBO-RWA, since WBO-RWA does not appropriately consider band or fiber switching, the wavelength assignment is done in manner unsuitable for reducing port count. Hence the benefit of multi-fiber in reducing port count does not show up in WBO-RWA algorithm.

## 6.6 Concluding remarks

It is well known that optical cross-connects can reduce the size, the cost, and control complexity of electronic (e.g., OEO grooming switches) cross-connects. Waveband switching (WBS) is a key technique to reduce the cost and complexity associated with current optical networks with large photonic cross-connects (both electrical and optical cross-connects). Since techniques developed for wavelength-routed networks cannot be efficiently applied to WBS networks, new techniques are necessary to efficiently address WBS-related issues such as lightpath routing, wavelength assignment, lightpath grouping, waveband conversion and failure recovery. In this chapter, we have provided a comprehensive overview of the issues associated with WBS. In particular, we have classified the WBS schemes into several variations and described a Three-layer multi-granular optical cross-connect (MG-OXC) architecture. We show that WBS networks using MG-OXCs can have a much lower port count when compared to traditional wavelength routed networks using ordinary-OXCs. For example, the WBS algorithm Balanced Path routing with Heavy-Traffic (BPHT) uses about 50% and 70% fewer total ports in single and multi-fiber networks, than using ordinary-OXCs in traditional wavelength routed networks. Issues pertaining to the design and comparison of different MG-OXC architectures (e.g., the Single-Layer and Multi-Layer MG-OXC architectures), the impact of waveband conversion and survivability in WBS networks need further investigation.

## References

[1] X. Cao, Y. Xiong, V. Anand, and C. Qiao, "Wavelength band switching in multi-granular all-optical networks," in *SPIE's Proc. vol. 4874, OptiComm'02, Boston Massachusetts*, 2002, pp. 198–210.

[2] K. Harada, K. Shimizu, T. Kudou, and T. Ozeki, "Hierarchical optical path cross-connect systems for large scale WDM networks," in *Proceedings OFC*, 1999, p. WM55.

[3] L. Noirie, M. Vigoureux, and E. Dotaro, "Impact of intermediate grouping on the dimensioning of multi-granularity optical networks," in *Proceedings OFC*, 2001, p. TuG3.

[4] O. Gerstel, R. Ramaswami, and W. Wang, "Making use of a two stage multiplexing scheme in a WDM network," in *Proceedings OFC*, 2000, p. ThD1.

[5] R. Lingampalli and P. Vengalam, "Effect of wavelength and waveband grooming on all-optical networks with single layer photonic switching," in *Proceedings OFC*, 2002, p. ThP4.

[6] M. Lee, J. Yu, Y. Kim, C. Kang, and J. Park, "Design of hierarchical crossconnect WDM networks employing a two-stage multiplexing scheme of waveband and wavelength," *IEEE JSAC*, vol. 20, no. 1, pp. 166–171, Jan. 2002.

[7] D. Banerjee and B. Mukherjee, "A practical approach for routing and wavelength assignment in large wavelength-routed optical networks," *IEEE JSAC*, vol. 14, pp. 903–908, Jun. 1996.

[8] B. Mukherjee, D. Banerjee, S. Ramamurthy, and A. Mukherjee, "Some principles for designing a wide-area optical network," *IEEE/ACM Tran. on Networking*, vol. 4, pp. 684–696, Oct. 1996.

[9] I. Chlamtac, A. Ganz and G. Karmia, "Lightpath communications: An approach to High Bandwidth Optical WAN's," *IEEE Tran. on Comm.*, vol. 40, no. 2, pp. 1171–1182, Jul. 1992.

[10] X. Cao, V. Anand, Y. Xiong, and C. Qiao, "Performance evaluation of wavelength band switching in multi-fiber all-optical networks," in *Proceedings of INFOCOM '03*, 2003, vol. III.

[11] H. E. Escobar and L. R.Marshall, "All-optical wavelength band conversion enables new scalable and efficient optical network architectures," in *Proceedings OFC*, 2002, p. WH2.

Chapter 7

# OPTICAL BURST SWITCHING: QUALITY OF SERVICE, MULTICAST, AND OPERATION AND MAINTENANCE

Hakki Candan Cankaya[1] and Myoungki Jeong[2]

[1] *Network Strategy Group, Alcatel USA*

[2] *Telecom R&D Center, Samsung Electronics*

Email: hakki.cankaya@alcatel.com, myoungki.jeong@samsung.com

**Abstract** A new switching technique, called Optical Burst Switching (OBS), has been proposed as one of the candidate technologies for future optical networks with higher bandwidth efficiency. Many issues of OBS have been researched and some solutions of them are published by both academia and industry. In this chapter, after a brief review of OBS concepts, we present some of our experiences that we have obtained from research on OBS related projects in Alcatel research and network strategy group. These will include multi-Terabit IP Optical Router (TIPOR), the first publicly demonstrated prototype, that is based on burst switching technology, preemptive scheduling technique with QoS support, tree- shared multicasting, and operation and maintenance framework for OBS networks.

**Keywords:** Optical burst switching networks, Tree-shared multicast, Data channel scheduling, Operation and maintenance, Quality of service.

## 7.1 Introduction

Optical Burst Switching (OBS) [1][2] has been proposed as a switching technology to transport traffic directly over WDM with a granularity between Optical Wavelength Switching (OWS) and Optical Packet Switching (OPS), taking into consideration the state of the art of optics and electronics. It has been considered as one of the candidate technologies for future optical networks with higher bandwidth-efficiency [3]. In OBS, packets which are destined to the same OBS egress edge node are assembled (i.e., burstification) into a bigger data entity, called data burst (DB), and are switched through an

OBS network in an optically transparent manner. To make it happen, a control packet called Burst Header Packet (BHP) containing necessary information for such as switching and scheduling is formed for each DB. Then the BHP travels ahead of its associated DB in time, and is processed electronically at every node along the path to the destination (egress edge node) in the OBS network so that its associated DB can be switched transparently at each node along the path in the optical domain. One of the important features of OBS is the spatial and temporal separation of DB and BHP. In the context of spatial separation, DB and its BHP travel on different channels (i.e., wavelengths); therefore, a link in OBS networks consists of data channels and control channel(s). For the temporal separation, BHP is transmitted earlier than its associated DB in time by some amount, and is processed electronically at each node along the path to the (destination) egress-edge router. These unique features of OBS require a strict coordination between DB and BHP and make it a challenging task [4]. Erroneous event detection with this existing optical transparency in data plane is another such requirement. Given that the number of control channels are fewer than the number of data channels and guard bands (GBs) for DBs on data channels are necessary against jitter and other imperfections in such as transmission and switching, it is crucial to reduce these overheads caused by the volume of control traffic (i.e., the number of BHPs on control channels, etc.) and GBs on data channels. As a result, scheduling, operation and maintenance, multicasting and many other issues are needed to be studied for OBS under these requirements.

OBS has been in the radar screen of the industry as being one of the future possibilities of revenue, and it has been extensively studied in detail by academia as well. For example, Alcatel implemented and publicly demonstrated, as a proof of concept, a scaleable multi-Terabit IP Optical Router (TIPOR) prototype that is based on burst switching technology with an optical packet switching fabric [5]. In the rack-mounted prototype, four functions were implemented on three different modules. Burstification of packets with contention resolution and scheduling functions were built on a Burst Card (BC) with 10 Gbit/s throughput which is also scaleable up to 160 Gbit/s. On the ingress side of this module, 9 Kbit/s bursts are generated from ATM cells. The module also includes a two-level scheduler that finds the maximal matching between input and output port servers for each switching slot. On the egress side, bursts are simply disassembled back to cells with error-check. The prototype was designed to contain up to 16 Burst Cards. Framer/transceiver board/module implements the switching matrix interfacing functionality. It speeds-up the 2.5 Gbits/s burst flow to 10 Gbits/s and generates the optical packet frame with guard band. It contains E/O and O/E conversion sub-modules for both ingress and egress sides. Fast optical packet switching functionality was also implemented on an optical switching matrix module [6]. That employed two-stage (fiber and wavelength) switch design and a broadcast-and-select switch structure. The architecture is scaleable up to 2.5 Tbit/s and beyond.

In this chapter, we present some of our experiences that we have obtained from research on OBS related projects in Alcatel research and network strategy group. In the next section, we briefly describe OBS concepts, an OBS network, and its components. In the third section, we introduce a new scheduling algorithm, called partially preemptive scheduling. The new scheduling technique also supports Quality of Service (QoS). We provide a set of results from this study which shows up to 40% improvement in data channel utilization. In Section 7.4, we describe an efficient multicasting method, called tree-shared multicasting, and provide a set of results showing significant improvement in overhead reductions of OBS. In Section 7.5, we discuss some OAM extensions for OBS networks and present a high-level design of an OAM unit for OBS core nodes. Section 7.6 concludes the chapter.

## 7.2 Optical Burst Switching Networks

Figure 7.1 depicts a generic OBS network which is made up of core nodes, edge nodes, and WDM links [1]. In an ingress edge node, arriving packets from packet sources are assembled into DBs (a jumbo packet) based on their destination edge node addresses and other attributes like QoS parameters. The burstification can be performed by using buffers on a temporal and/or spatial basis. In the temporal burstification, there is a time-out parameter. After each time-out occurrence, a DB is formed. In the spatial burstification, on the other hand, a DB is formed after it reaches a certain size, which represents another parameter. Any combination of these two parameters can also be studied. These burstification parameters can be fixed or adaptive to traffic dynamics for better performance. On WDM links, DBs and their header information, BHPs, are decoupled both in terms of time and space. DBs travel on data channels and their associated BHPs travel on control channel(s). Each DB and its BHP are related within a time frame called *offset time*. In essence, a BHP travels an offset time ahead and is processed electronically by a controller in a core node to configure the optical switch for the transparent switching of its DB. Therefore, a DB follows a pre-established optical path from ingress edge node to egress edge node. Once the DB arrives at egress edge node, it is disassembled into its original packets and they are forwarded to their final destinations.

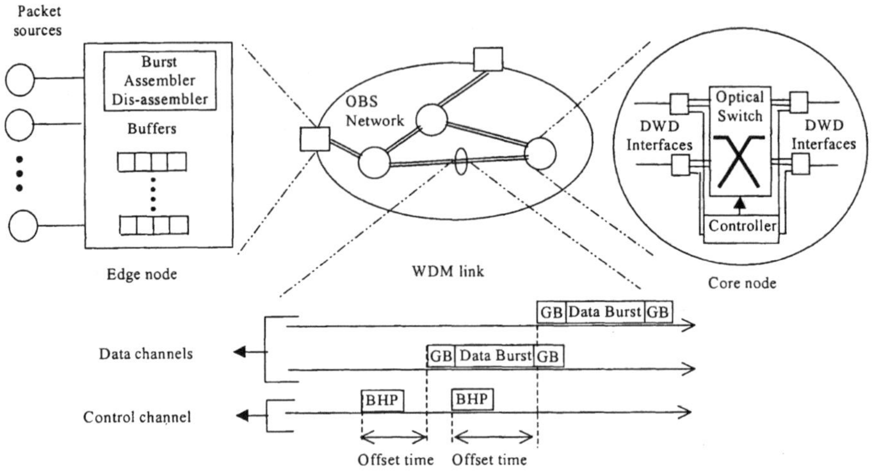


*Figure 7.1.* OBS network.

## 7.3 Preemptive Scheduling Technique with QoS Support

A number of DB scheduling techniques has been proposed for OBS so far. *Horizon scheduling* is one of the earliest [2]. In this bufferless approach, a DB is scheduled before its arrival. Indeed, it is scheduled as soon as its BHP arrives, and is processed electronically. The scheduler maintains a scheduling calendar, which includes only the latest DB on each data channel (so called *horizon* of the link). *Horizon* scheduling is simple, but it reveals low channel utilization. Later, with the introduction of a delayed reservation concept, a new era for burst scheduling has started. In *delayed reservation*, allocation of resources at the core node for switching DB is delayed until its actual arrival [1]. In this era, many protocols based on this concept, like Just-Enough-Time (JET), Just-in-Time (JIT), etc. were studied [1][7]. Delayed reservation made possible to use gaps/voids between not yet arrived but already scheduled DBs. Following this trend, void-filling scheduling was introduced to improve utilization [8] [4]. Basically this approach utilizes the gaps between already scheduled DBs and maintains a scheduling calendar which keeps track of all the scheduled DBs that are still due to arrive. Latest Available Unused-channel First with Void-Filling (LAUF-VF) [4] is an example of this kind, which aims at improving the utilization of the gaps by a specific method. However, all these scheduling techniques either with or without QoS support try to schedule a DB in its entirety; if there is even a small overlap with another DB, the recent (new incoming) DB (subject to scheduling) is dropped. There is also no preemption of an already scheduled DB if another new DB with higher priority is to be scheduled while overlapping with the former DB. In [9][10], handling DBs in parts is proposed to address the entire DB dropping problem when two DBs overlap. In these proposals the overlapping sections of DBs are considered for dropping in order to reduce packet loss.

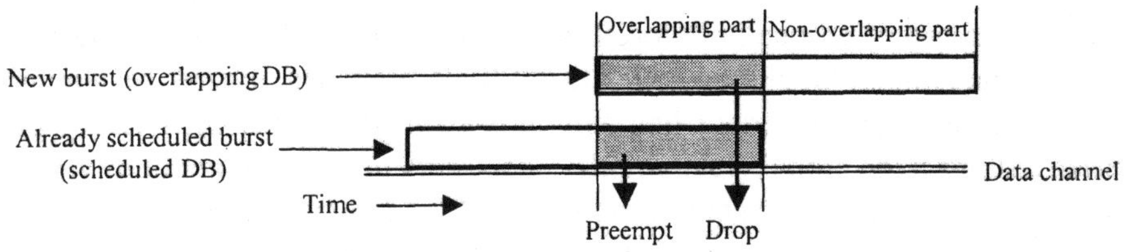


*Figure 7.2.* Overlapping bursts.

In this section, we describe a new scheduling technique, called *partially preemptive burst scheduling*, that integrates the concepts of handling DBs in parts and their preemption on a QoS supporting platform [11]. The idea of partial scheduling is to be able to handle an overlapping DB separately in two parts. These parts are overlapping part and non-overlapping part as shown in Figure 7.2. In case of an overlap between a new incoming DB and a previously scheduled DB, we call the new burst as "overlapping DB" and the already scheduled burst as "scheduled DB" as shown in Figure 7.2. Partially preemptive burst scheduling provides two options in case of a contention: (1) drop the part of the overlapping DB, (2) preempt the part of the scheduled DB in order to schedule the overlapping DB in its entirety. The options are evaluated by a QoS platform based on proportional differentiation model [12][13][14]. In this model, classes receive service in proportion to their differentiation parameters set by the network service provider. For example, let $p_i$ be a differentiation parameter and $c_i$ be a performance measure of class $i$. Then in the proportional differentiation model the following relationship holds for all pairs of service classes.

$$\frac{p_i}{p_j} = \frac{c_i}{c_j}, \text{ where } (i,j = 0..N \text{ and N is the number of classes}).$$

It is important that not only does the proportional differentiation model hold for long time scales, but also for short time scales. Long-term averages may not always be meaningful especially when the traffic is bursty [14]. Therefore, the proportional differentiation model is defined over a short monitoring time scale, with a period of $\tau$.

$\frac{\bar{c}_i(t,t+\tau)}{\bar{c}_j(t,t+\tau)} = \frac{p_i}{p_j}$, where $\bar{c}_i(t,t+\tau)$ is the average performance measured over the period $\tau$ for class $i$.

Two important features of this model are predictability and controllability. Predictability means that the differentiation among classes is consistent. In other words, higher classes are always better or at least no worse than lower classes, regardless of variations in class loads. Controllability means that network operators are able to adjust the quality spacing between classes on selected criteria (e.g. pricing, policy objectives, etc.) [15]. According to the model, selected performance measures for packet forwarding are distanced from each other in proportion to class differentiation parameters. Some of the common performance measures for packet forwarding are queuing delays, packet losses, usage, etc. In this study, we applied the proportional model to data loss and channel usage differentiation. We conducted performance evaluation and a comparison study by simulations. According to the results, the partially preemptive burst scheduling technique significantly improved dropping probability by approximately 50% and utilization by approximately 40% both at 0.8 load at a certain level of QoS.

## 7.3.1 Algorithm

In OBS core routers, scheduling of DBs is complete before they actually arrive based on the control information obtained from their corresponding BHPs, which have already arrived and converted to electronic signal. A window of short-term scheduling calendar, called "Sliding Frontier (SF)" is maintained for DB arrivals. This calendar at a core node contains information regarding DBs and BHPs arriving on all the data channels and control channels. The scheduler uses SF for making decisions.

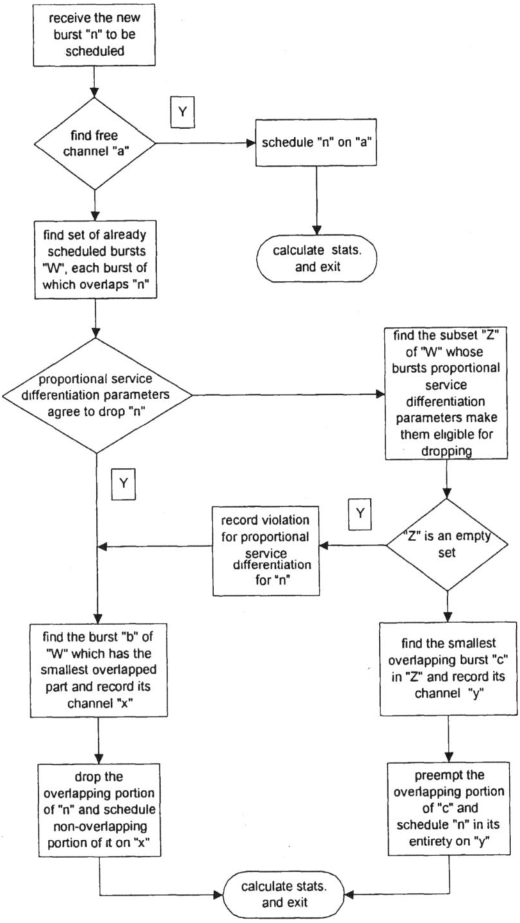


*Figure 7.3.* Flow chart of the algorithm.

In the flow chart given in Figure 7.3, the new DB "n" is first tried to be scheduled if there is any available channel. In the case that there is no channel available, then the new DB's proportional service differentiation profile is checked whether it is allowed to drop the new DB. If it is so, then the smallest overlapping is found and the new DB is partially scheduled. If not, among the already scheduled DBs, we find the ones that can be dropped considering their own profiles. Among them, the one with minimum overlapping is chosen and is preempted either in part or full in order to schedule the new DB in its entirety. In case that non of those already scheduled DBs can be dropped, a violation for the new DB's proportional service differentiation profile is recorded and the new DB is partially scheduled.

### 7.3.2 Results

In this section, we present some of the simulation results of the partially preemptive scheduling with proportional data loss only. For larger set of results including the proportional usage differentiation model, readers may want to refer to [11]. We also compare the results with proportional differentiation without partial preemption [14] along with conventional classless void-filling scheduling [4].

The simulations are performed for an OBS core-node with WDM links (with multiple channels). We assume no buffering but full wavelength conversion capability. We use four classes C0, C1, C2, C3. The C0 has the highest priority and C3 has the lowest. We let all four classes have the same arrival rate. An ON_OFF source model is used to generate DBs [11].

We set the proportion factors of four classes as follows: Class0=1.0, Class1=2.0, Class2=3.0, Class3=4.0. The channel rate is assumed to be 10Gbits/sec and mean DB length be 20Kbytes. Unless otherwise stated, we assume 4 channels in the experiments. In some experiments where we do not vary offered load, we load 80% of each channel.

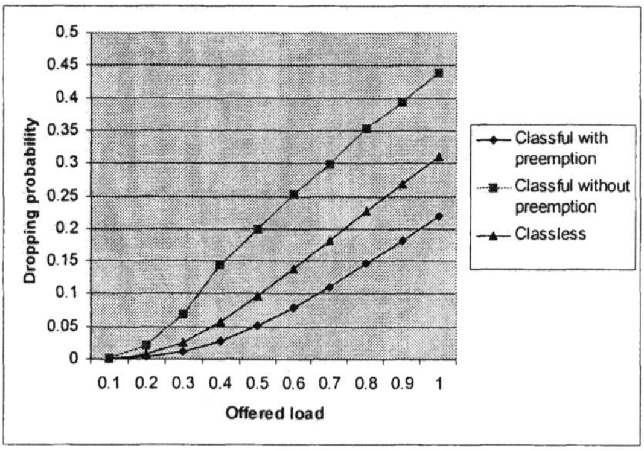


*Figure 7.4.* Dropping probability comparison for proportional loss.

In comparison in Figure 7.4, the proposed proportional loss model with partial preemption (referred to in the figure legend as "Classful with preemption") has the least dropping probability figures compared to other approaches. Proportional loss model without partial preemption (referred to in the figure legend as "Classful without preemption" ) has 55% more dropping probability at offered load of 0.8 than that of the regular void-filling without partial preemption (referred to in the figure legend as "Classless"). This increase in dropping probability is due to the tax paid for QoS support. The proposed model reduces the dropping probability more than half the value of the case without partial preemption at offered load of 0.8 while satisfying the proportional differentiation as seen in Figure 7.5.

*Figure 7.5.* Dropping ratios of partial preemption with proportional loss.

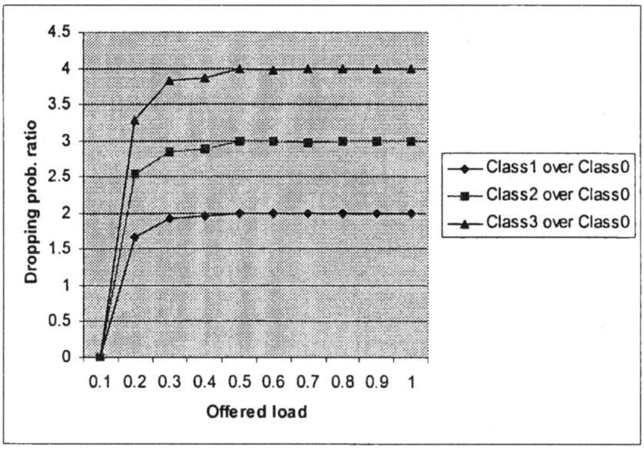


Figure 7.6 reveals that, the proposed scheduling technique can also yield the highest utilization under all load levels because of the way the algorithm run by the scheduler picks the smallest available overlap to consider for partial scheduling discussed in the previous algorithm section. The increase in utilization is almost linear with increasing offered load as the others tend to saturate. On the other hand, channel utilization in the case of proportional loss without partial preemption is lower than the case with classless void-filling because of the latter's early/intentional dropping strategy used to accomplish proportional differentiation [14]. In conclusion, at the end of this study we have learned that partial preemption can improve both the utilization and dropping probability considerably while still providing acceptable proportional differentiation.

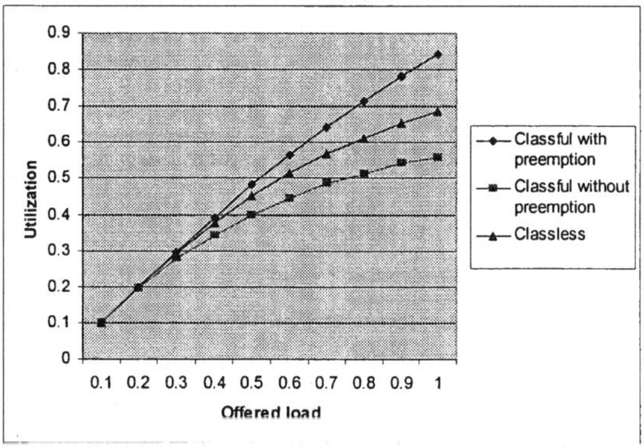


*Figure 7.6.* Utilization comparison for proportional loss.

## 7.4 Tree Shared Multicasting

As mentioned in the introduction, there are two major overheads when using OBS, namely, control packets (on out-of-band channel) and guard bands (GBs). More specifically, to send each burst, a control packet needs to be sent to set up switches and reserve bandwidth for the following burst. Given that there are a limited number of (out-of-band) control channels (e.g., a single wavelength out of all the wavelengths in a fiber) in optical WDM networks, the control channels can be a bottleneck for network performance. GBs (on data channels) are used in each burst to accommodate possible timing jitters along the path to the destination due to such as path length variations and thermal and chromatic dispersion effects, and switching transients at each node due to such as non-ideal synchronization between data burst and its control packet, and switching. Hence, it is important to use an efficient multicast scheme to reduce the overheads from the GBs and control packets when performing WDM multicast using OBS.

In this section, we describe a new multicast scheme, called Tree-Shared Multicasting (TS-MCAST) [16] in optical burst-switched WDM networks, taking into consideration overheads due to control packets and guard bands (GBs) associated with data bursts. In TS-MCAST, multicast traffic belonging to multiple multicast sessions from the same source edge node to possibly different destination edge nodes can be multiplexed together in a burst, which is delivered via a *shared* multicast tree. To support TS-MCAST, we describe three tree sharing strategies based on Equal Coverage (EC), Super Coverage (SC), and Overlapping Coverage (OC), and describe some approaches to construct shared multicast trees.

In TS-MCAST, if there is a certain degree of membership overlap or some other special relationships between multicast sessions in $H_i$ where $H_i$ be the set of multicast sessions whose source is at edge router $i$, the set $H_i$ is split into a number of disjoint subsets based on a certain strategy (to be described in the next subsection). Each subset is called Multicast Sharing Class (MSC) where as a special case, a MSC may contain only one multicast session. We assume that there is a single burst assembly queue for each MSC. IP packets belonging to all multicast sessions in a MSC are assembled together to form a burst if those arrive within a given assembly time $T_b^m$, and the burst is delivered using the ST for the MSC where either an algorithm will be used to construct a new shared tree (ST), or one of the existing multicast trees will be used as the ST. Hence, the average burst length using TS-MCAST will be longer than that without tree sharing where IP packets belonging to a multicast session are assembled to form bursts independently of other multicast sessions. This helps reduce the bandwidth waste due to GB as well as the number of control packets generated since

GB will be shared by multiple multicast sessions and there is only one control packet to generate for multiple multicast sessions in a given burst assembly time $T_b^m$. Note that when multiple multicast sessions use a single shared tree for the delivery of their multicast data, we call it *tree sharing*.

Let $N_e$ be the set of all edge routers and $N_l$ be the set of all links in the network. Let $H'_i$ be the set of MSCs at edge router $i$ where $1 \leq |H'_i| \leq |H_i|$ and $F'_{\max}$ be the maximum over all $|H'_i|$, where $i \in N_e$. Assume that the aggregated (i.e., sum) amount of the multicast traffic (determined by the number of bits assembled in a burst over a given assembly time) of the $j$th MSC, denoted by $MSC_{ij}$, is represented by $R_{|N_e| \times F'_{\max}} = [r'_{ij}]$, and define the aggregated multicast cost matrix as $S_{|N_e| \times F'_{\max}} = [s'_{ij}]$ where $s'_{ij}$ is the number of links on the shared multicast tree used for $MSC_{ij}$. Accordingly, in TS-MCAST, the average amount of bandwidth consumed by the multicast traffic per link over all multicast sessions in the network is

$$W_{TS-MCAST}^{m} = \frac{1}{|N_l|} \Big( \sum_{i \in N_e} \sum_{j \in H'_i} (r'_{ij} + \frac{G}{T_b^m}) s'_{ij} \Big) \tag{1}$$

where $G$ is the sum of head and tail GBs for a burst and is measured in bits.

Let $F_{\max}$ be the maximum over all $|H_i|$ (where $|H_i|$ denotes the number of sessions in $H_i$). Assume that the average amount of multicast traffic for session $j$, $1 \leq j \leq |H_i|$, originating from source $i$ be represented by matrix $R_{|N_e| \times F_{\max}} = [r_{ij}]$. In addition, let $S_{|N_e| \times F_{\max}} = [s_{ij}]$ be the multicast cost matrix for a single multicast session, where $s_{ij}$ is the number of links on the multicast tree rooted at source $i$ for session $j$. For each $MSC_{ij}$, we define the bandwidth gain, $\alpha_{ij}$, due to tree sharing as the ratio of the average amount of multicast traffic carried per link without tree sharing to that with tree sharing, which is equal to

$$\alpha_{ij} = \frac{\sum_{k \in MSC_{ij}} (r_{ik} + \frac{G}{T_b^m}) s_{ik}}{(r'_{ij} + \frac{G}{T_b^m}) s'_{ij}} \tag{2}$$

where it is assumed that the burst assembly time is the same for a multicast session and a MSC.

### 7.4.1 Tree Sharing Strategies

In this subsection, we describe how to decompose a set $H_i$ of multicast sessions into a number of MSCs, each of which uses a ST. Let $N_c$ be the set of all core routers in the network (and $N_e$ and $N_l$ be defined as before). In the following discussion, we model a multicast tree (or session) in the network using a tuple $\mathbf{T} = (C_T, E_T, L_T)$ where $C_T \subseteq N_c$ is the set of core routers, $E_T \subseteq N_e$ is the set of edge routers, and $L_T \subseteq N_l$ is the set of links on the multicast tree **T**, respectively.

We consider three tree sharing strategies, namely Equal Coverage (EC), Super Coverage (SC) and Overlapping Coverage (OC), for deciding which subset of multicast sessions rooted at edge router *i* should become a MSC. For simplicity, we use $MSC_j$ rather than $MSC_{ij}$ to denote the *j*th MSC at edge router *i* hereafter.

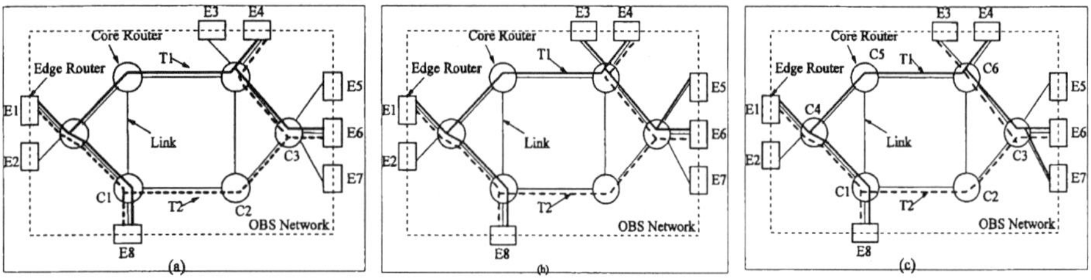


*Figure 7.7.* Tree sharing strategies: (a) Equal coverage (b) Super coverage and (c) Overlapping coverage.

In EC, multicast sessions with the same membership are grouped into one MSC. In other words, the $s \geq 2$ multicast sessions (**T**1 through **T**s) in $MSC_j$ have the same set of member edge routers, i.e., $E_{T_1} = E_{T_2} = \cdots = E_{T_s}$ although each multicast session in $MSC_j$ may have a different multicast tree (or path to each member). Figure 7.7(a) shows an example of EC (where $s = 2$) in which multicast trees **T**1 (solid line) and **T**2 (dashed line) have the same set of edge routers (i.e., E4, E6 and E8) as their members. In such a case, one of the existing multicast trees, **T**1 or **T**2, is selected to be the new ST.

A less restricted tree sharing strategy is SC where the multicast sessions in $MSC_j$ do not necessarily have the same set of edge routers. Specifically, if two multicast sessions, **T**1 and **T**2, are such that $E_{T_2} \subseteq E_{T_1}$ (or $E_{T_1} \subseteq E_{T_2}$), then **T**1 and **T**2 are grouped into one MSC if tree sharing gain for the MSC is above a threshold $\eta$ (see Figure 7.7(b)). Note that in Figure 7.7 (b), IP packets belonging to **T**2 will also be delivered to E3 and E5 via **T**1 since **T**2 $\subseteq$ **T**1, but subsequently discarded by E3 and E5.

The third scheme OC is a more general tree sharing strategy in which it allows a number of multicast sessions having a sufficient degree of overlap in the set of edge routers $E_T$, core routers $C_T$, or links $L_T$ or a subset of multicast sessions resulting in a sufficient tree sharing gain $\alpha$ (from (2)) to be grouped into one MSC. More specifically, we define the degree of overlap as follows. Consider s multicast trees (sessions), $\mathbf{T}k = (C_{T_k}, E_{T_k}, L_{T_k})$ for $k =$

1, 2 ... s. Then, the degree of overlap among these *s* multicast sessions can be defined in terms of edge routers as follows:

$$\gamma^{E} = \frac{\sum_{k=1}^{s} | E_{T_k} | - | \bigcup_{k} E_{T_k} |}{\bigcup_{k} E_{T_k} | s - 1 |} . \qquad (3)$$

The degree of overlap in the above formula is basically the ratio of the number of non-distinct edge routers in the *s* multicast trees to the number of distinct edge routers. Similarly, the degree of overlap is defined in terms of core routers and links [16]. To decompose the set $H_i$ using the OC tree sharing strategy, basically a subset that has a highest degree of overlap using one of the three criteria (i.e., edge router, core router, or link) or a highest tree sharing gain is selected. If its tree sharing gain $\alpha$ is over the threshold, it becomes a MSC. A detailed heuristic algorithm as to how to apply the OC tree sharing strategy using one of the three criteria above and $\alpha$ is referred to [16]. An example of OC is shown in Figure 7.7(c) where **T**1 has member edge routers that do not belong to **T**2 and vice versa, and they are grouped into one MSC if their tree sharing gain exceeds the threshold.

## 7.4.2 Construction of Shared Trees

In TS-MCAST, the multicast source ingress edge router requires the information on the network topology and membership for all multicast sessions originating from it in order to perform the tree sharing strategies described. For some possible approaches to provide the information, readers may refer to [3].

In EC and SC, the construction of a shared tree for a MSC is trivial since in EC any multicast tree among the multicast trees of a MSC can be used as the shared tree, and the super tree of a MSC can be used as the shared tree in SC. On the other hand, in OC, a new shared tree for a MSC needs to be constructed unless the MSC meets the criterion of EC and SC. We consider two approaches for the shared tree construction: centralized approach and distributed approach. In the centralized approach, a centralized algorithm is used to construct a shared tree for a MSC at the source ingress edge router. The shared tree can be Steiner minimum tree using an exact algorithm (e.g., [17]) or can be a Steiner tree using efficient heuristic algorithms (e.g., [18]). After constructing a shared tree for a MSC, the source ingress edge router can send a control packet to each core router in the shared tree so that it can set up a forwarding table for the MSC. Or the source ingress edge router in the shared tree can send the information to each egress edge router in the

MSC for the entire path from the egress edge router to the source edge router so that the egress edge router can perform the join process following the explicit path.

On the other hand, in the distributed approach, to construct a shared tree for a MSC, the source ingress edge router needs to send a packet, say *initiate* packet, in order to initiate the shared tree construction for the MSC. The source edge router sends the initiate packet(s) to all downstreams belonging to the MSC where the initiate packet carries the information of multicast group identities (e.g., multicast labels in the framework of GMPLS) in the MSC. Each core router (which is on one of the multicast trees belonging to the MSC) receiving the initiate packet forwards it to its all downstreams belonging to the MSC until each member egress edge router in the MSC receives it (recall that each core router in a multicast tree has a forwarding table for the multicast tree). If a core router receives the initiate packet more than once for the same MSC, the core router may send a prune message upstream to prevent a cycle. To construct a more efficient shared tree in terms of the number of links (or cost), during the forwarding process of the initiate packet, a Steiner tree can be constructed in a distributed fashion. Another simple way is that the source ingress router sends one initiate packet to each member edge router in the MSC so that the member edge router initiates a join process toward the source ingress edge router via unicasting (which is similar to the join process of PIM-SM).

### 7.4.3 Performance Evaluation

To show the effectiveness of TS-MCAST, we consider two other multicasting approaches, called Separate Multicasting (S-MCAST) and Multiple Unicasting (M-UCAST) and show relative performance between them.

#### Separate Multicasting and Multiple Unicasting

In S-MCAST, multicast traffic of a multicast session is assembled at its ingress edge router into a burst if they arrive within a given assembly time $T_b^m$. Then the bursts containing the multicast traffic of the multicast group are delivered through a multicast tree of the group. That is, multicast traffic is transmitted independently from unicast traffic.

On the other hand, in M-UCAST, multicast traffic is treated as unicast traffic. During the burstification (i.e., burst assembly) process, the source ingress edge router makes multiple copies of a multicast packet belonging to a multicast group, one for each member (egress edge router) of the multicast group, assembles them along with unicast data destined to the same egress edge router into bursts, and delivers the bursts to each member egress edge router via unicasting. By doing so, the control overheads can be reduced since the GB can be shared by both traffic types in a longer burst, and in addition, the number of control packets generated for multicast IP traffic also decreases. In other words, there are no dedicated control packets and GBs to multicast traffic (or bursts).

## Numerical Results

For performance comparison, we use a random network and assume that each link has unlimited bandwidth (i.e., no blocking). Unicast traffic at each ingress edge router is modeled as constant bit rate traffic with uniform distribution and the amount of multicast traffic is given as a percentage of total traffic (unicast plus multicast traffic) of the network. It is assumed that a number of multicast sessions is active at each edge router which acts as their source ingress edge router and each multicast session has the same amount of multicast traffic to transmit. In addition, each multicast session has a pre-determined membership with the creation of multicast session and the membership does not change. The default values for some parameters are as follows. The number of core routers is 40 and the GB size is 2 percent of average unicast traffic. The total amount of multicast traffic in the network is 5 percent and the total number of multicast sessions is 600. The membership size of multicast sessions is 70 percent. To construct a ST for each MSC, we use a simple heuristic, called ST-MEMBER [16]. In the ST-MEMBER algorithm, we start with an existing multicast tree with the largest number of members as the base of the new ST. Then all other members perform an operation similar to a ``join" in CBT [19] (or ``graft" in DVMRP [20]), and by doing so, augments the base tree with additional nodes and links. Interested reader may refer to [16] for detailed network model, simulation set-ups, and additional results not presented here.

Here we present relative performance on the bandwidth savings and the number of control packets generated per link using a parameter, namely the GB size. Note that the bandwidth consumption by TS-MCAST takes into account bandwidth waste by delivery of unwanted packets to some edge routers. Figure 7.8(a) shows the effect of the GB size and the membership size on the bandwidth savings. From Figure 7.8(a), we observe that M-UCAST scheme performs the worst with a small GB size, and becomes better as the GB size increases. It is not difficult to envision that in a mesh network M-UCAST may outperform TS-MCAST under some network conditions since no overhead (due to GBs) exists for multicast traffic. This is because as the GB size increases, the benefit of amortizing GB with both multicast and unicast traffic increases. This is also why the performance of TS-MCAST (OC) improves when the GB size increases. In addition, Figure 7.8(a) also shows the relative performance of TS-MCAST (OC) when applying different criteria. We observe that with a small GB size, the performance difference is not much, but as the GB size increases, the performance difference becomes larger with the tree sharing gain being the best criterion.

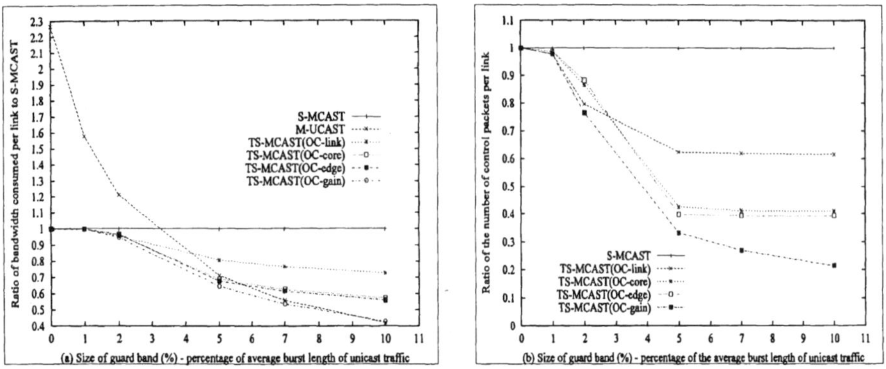


*Figure 7.8.* Effect of the GB on the bandwidth savings and the number of control packets per link.

On the other hand, since one control packet is sent for each burst, a multicast scheme allowing a higher degree of tree sharing will result in a smaller number of control packets (and with longer bursts). The smaller the number of control packets generated, (and thus the longer the bursts,) the lower the processing load at each node, and the better the bandwidth efficiency. Figure 7.8(b) shows relative efficiency (i.e., ratio of the number of control packet per link to that of S-MCAST) of the three multicast schemes in terms of the number of control packets per link in the network as a function of each of the three different parameters. The number of control packets generated at the source edge router per burst assembly time $T_b^m$ is counted. Specifically, since each multicast session generates one control packet per burst assembly time, the number of control packets per link in the network can be calculated by dividing the number of links on its multicast tree by the total number of links in the network. In Figure 7.8(b), the OC tree sharing strategy with traffic sharing gain $\alpha$ as the criterion shows the best performance (i.e., results in the fewest control packets generated) in most cases as the GB size and membership size increase. Note that the TS-MCAST (EC/SC) schemes and the M-UCAST scheme are not shown in Figure 7.8(a)(b) and Figure 7.8(b), respectively. This is because the performance of EC/SC is very close to that of S-MCAST, since very few

MSCs are formed using EC/SC schemes due to their tight requirements in the given simulation conditions, thus also resulting in almost the same number of control packets generated to the S-MCAST scheme. In the case of the M-UCAST scheme, it does not generate control packets dedicated to the multicast traffic (i.e., zero control packets for the multicast traffic), since it shares the GB with the unicast traffic.

## 7.5 Operation and Maintenance for OBS Networks

Operation and Maintenance (OAM) is one of the fundamental issues of any network and OBS network is not any exception. OAM is basically responsible for providing satisfactory level of network operation assuming the inevitable occurrence of failures and defects. It may use either prevention and/or correction in dealing such events. Although we believe this issue has received relatively less attention than other issues in the OBS field, we studied this topic in Alcatel [21] [22] because this issue is important when deploying OBS networks. The rest of the section summarizes the outcome of this study.

Even though previous studies of OAM [23] [24] [25] in different technologies/networks (e.g. ATM, SONET/SDH, WDM, etc.) do share some common ground with OAM study in OBS, there are some unique OBS features, such as spatial and temporal de-couplings of data and control, that need to be separately studied and included to the OAM framework of OBS. These unique features obviously require extra monitoring and synchronization procedures and extensions/modifications on component level. As a result of the unique interactions between data and control, OAM functions are studied in both data and control plane of the OBS network architecture. These OAM functions are performed by OAM units (OAMU), which are integrated into both edge and core nodes, as shown in Figure 7.9.

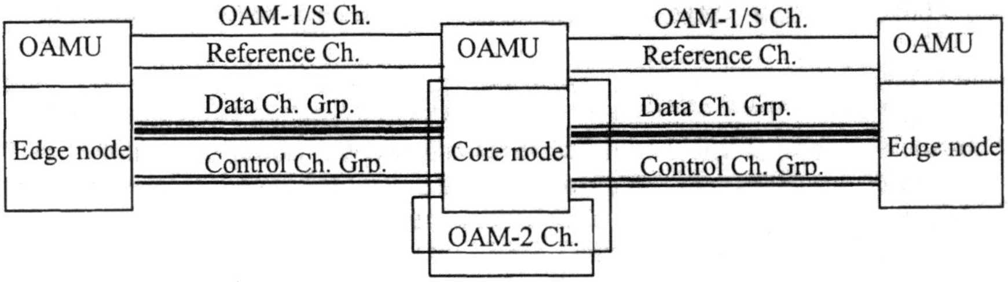


*Figure 7.9.* OAM channels.

In addition to data and control channel groups, there are two separate channels (assigned wavelengths) proposed to be used for OAM activities in control plane: (1) OAM-1/Supervisory (OAM-1/S) channel; and (2) Reference channel. The OAM-1/S channel is used for transmitting OAM-1 packets and supervisory packets. The OAM-1 packets monitor transmission quality on fiber between any two nodes including edge and core. These OAM-1 packets are inserted at the output of an upstream node and extracted at the input of the downstream node. For evaluation of the status, any required optical measurement criteria may be used, (such as attenuation, signal-to-noise (S/N) ratio, etc.). Supervisory packets carry state and control information of network elements in order to invoke any OAM correction function when needed. Reference channel is used to distribute wavelength reference and clock reference in the entire OBS network. A dedicated edge node is responsible for generating the information for this distribution.

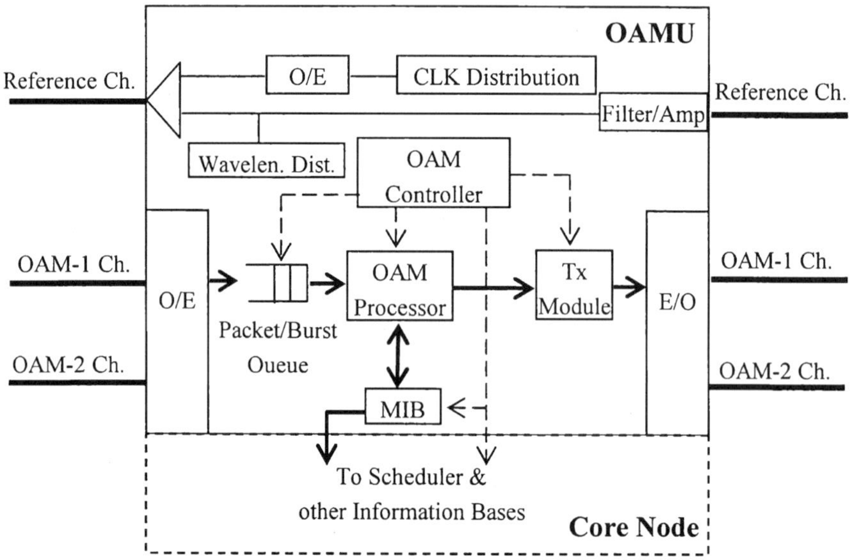


*Figure 7.10.* OAMU.

To check the status of data and/or control channel groups and the integrity of the optical switching in core node on data plane, a specific type of data burst, called OAM-2 DB and its corresponding BHP, called OAM-2 BHP, are used. OAM-2 DBs and OAM-2 BHPs are created during an OAM-2 session which is not different from a user session in terms of DB/BHP transmission. OAM2 DB may carry a set of OAM-related information required for a specific OAM activity. One OAM activity, for instance, that an OAM-2 session is used is to check the core node optical switching integrity. In this activity, a pair of OAM-2 DB and a corresponding OAM-2 BHP is created by the OAM controller of local core node's OAMU, given in

Figure 7.10. OAM-2 BHP is passed to the scheduler and its OAM-2 DB is sent to the Tx Module to be inserted into the OAM-2 Channel. On this channel the OAM-2 DB is looped back to one of the ingress ports of the same core node's optical switch, as shown in Figure 7.9. The OAM-2 DB/BHP pair is processed by the core node as any other user DB/BHP pair would. The OAM-2 DB is switched to a dedicated egress port which is also looped back to the ingress OAM-2 channel of the OAMU. Then, the OAM-2 DB that internally traveled through the optical switch is processed by the OAMU. Another example of OAM-2 session is used to probe the transmission between any two nodes (inserting and extracting nodes) including edge and/or core nodes. The creation and termination of this OAM-2 session occurs similar to the one used for core node optical switch check. Instead of inserting and extracting in the same core node, we inject the OAM-2 DB/BHP pair into the data/control channel groups in the OAM-2 session inserting node and we extract the OAM-2 DB and send it to OAMU in the session extracting node. There are also some passive OAM activities where no OAM specific entity is created. This includes monitoring and evaluating the user traffic and some nodal components by the local OAMU.

After detection of any (unexpected) event, such as unwanted changes, degradations, failures etc, a required OAM correction/response is invoked and/or a warning is released. Supervisory channel is used for communicating related information among OAMUs of different nodes in order not to interfere with user traffic. OAMU of a node also receives all the OAM-1 packets, OAM-2 DBs and supervisory packets into its packet/burst queue, and processes (OAM processor) them according to their specific OAM IDs which reveals the specifics of the particular OAM activity. As a result, Management Information Base (MIB) is updated and a required response and/or correction activity is invoked by using supervisory packets (e.g. Alarm Indication Packets, etc.). MIB internally communicates with other information bases in the routing unit of the conventional OBS node under the coordination of the OAM controller. OAM controller is responsible for initiating, controlling, and coordinating all the OAM activities and tests. It is also in communication with the scheduler for OAM-2 session initiation/termination activity. Reference wavelength signal is split and converted to electrical domain for local clock distribution. Its optical characteristics are also used to provide reference to local light-generating devices. Then, the optical signal is amplified, filtered, and passed to the downstream nodes for further network distribution.

## 7.6 Summary and Concluding Remarks

In this chapter, we presented some of our experiences on OBS related projects in Alcatel research and network strategy group. Among many important issues in OBS networks, the solutions that we described here are on scheduling with QoS support, tree-shared multicasting, and OAM, by taking into account the OBS characteristics such as GBs, BHPs, and temporal separation between DB and BHP. These solutions can be further refined and elaborated to improve their performance. We also described very briefly about a prototype TIPOR employing burst switching technology, which is implemented and demonstrated by Alcatel.

There are many other important issues that need to be studied further in OBS networks, for example, which include efficient burst assembly mechanisms taking into account QoS requirements and traffic conditions in the network, traffic engineering in the combination of GMPLS and OBS, a QoS framework, network survivability, etc. We believe that when considering lack of the current optical technology (especially, optical memory) to realize an optical Internet using optical packet switching, OBS is a promising switching technology to build the optical Internet.

## Acknowledgements

We would like to thank Dr. Saravut Charcranoon for his valuable comments.

## References

[1]C.Qiao and M. Yoo, "Optical Burst Switching (OBS) – A New Paradigm for an Optical Internet", *Journal of High-Speed networks*, pp. 69-84, 1999.

[2]J. Turner, "Terabit Burst Switching'', *Journal of High-Speed networks*, pp. 3-16, 1999.

[3]M. Jeong, H.C. Cankaya, and C. Qiao, "On a New Multicasting Approach in Optical Burst Switched Networks", *IEEE Communications Magazine,* vol. 40, no. 11, pp.96-103, Nov. 2002.

[4] Y. Xiong, M. Vandenhoute, and H. C. Cankaya, "Control Architecture in Optical Burst Switched WDM Networks'', *IEEE JSAC, vol. 18, no. 10,* pp.1838-1851, Oct. 2000.

[5] F. Masseti et al., "Design and implementation of a multi-terabit optical burst/packet router prototype," in Proc. IEEE OFC, Anaheim, CA, Mar. 2002, pp.FD1.1-FD1.3.

[6] D. Chiaroni et al. "First demonstration of an asynchronous optical packet switching matrix prototype for MultiTerabit class routers/switches", in Proc.

ECOC'01, 27th European Conference on Optical Communication, vol.6, 2001, pp.60-61.

[7] J.Y. Wei and R.I. McFarland, "Just-In-Time signaling for WDM Optical Burst Switching Networks," IEEE/OSA Journal of Lightwave Technology, vol.18, no.12, pp. 2019-2037, Dec. 2000.

[8]L. Tancevski, A. Ge, G. Castanon, and L. Tamil, "A New Scheduling Algorithm for Asynchronous Variable Length IP Traffic Incorporating Void Filling", in Proc. IEEE OFC'99, 1999, pp. ThM7-ThM7-3.

[9] V.M. Vokkarane, J.P. Jue, and S. Sitaraman, "Burst Segmentation: An Approach for Reducing Packet Loss in Optical Burst Switched Networks,", in Proc. IEEE ICCC 2002, pp. 2673-2677.

[10] A. Detti, V. Eramo, and M. Listanti, "Performance Evaluation of a new Technique for IP support in a WDM optical Network: Optical Composite Burst Switching (OCBS), IEEE/OSA Journal of Lightwave Technology, Vol. 20, no. 2, pp. 154 –165, Feb. 2002.

[11] H. Cankaya, S. Charcranoon, T. El-Bawab, "A preemptive scheduling technique for OBS networks with service differentiation," To appear in Proc. IEEE Globecom 2003.

[12] C. Dovrolis, D. Stiliadis, and P. Ramanathan, "Proportional Differentiated Services: Delay Differentiation and Packet Scheduling," ACM Computer Communication Review, vol. 29, no. 4, pp.109-20, Oct. 1999.

[13] C. Dovrolis and P. Ramanathan, "Proportional Differentiated Service, Part II: Loss Rate Differentiation and Packet dropping," in Proc. IEEE IWQoS, pp.52-61, June 2000.

[14] Y. Chen, M. Hamdi, and D.H.K. Tsang, "Proportional QoS over OBS Networks," in Proc. IEEE Globecom 2001.

[15] C. Dovrolis, D. Stiliadis, P. Ramanathan, "Proportional Differentiated Services: Delay Differentiation and Packet Scheduling," IEEE/ACM Transactions on Networking, vol.10, no.1, Feb. 2002.

[16] M. Jeong et al., "Tree-Shared Multicast in Optical Burst-Switched WDM Networks," IEEE/OSA Journal of Lightwave Technology, vol. 21, no.1, pp.13-24, Jan. 2003.

[17] S. E. Dreyfus and R. A. Wagner, "The Steiner's Problem in Graphs," Networks, vol. 1, pp.195-207, 1972.

[18] L. Kou, G. Markowsky, and L. Berman, "A Fast Algorithm for Steiner Trees," Acta Informatica, vol. 15, pp. 141-145, 1981.

[19] T. Ballardie, P. Francis, and J. Crowcroft, "Core Based Tree (CBT): An Architecture for Scalable Inter-domain Multicast Routing," in Proc. ACM SIGCOMM, Oct. 1993, pp. 85-95.

[20] T. Pusateri, "DVMRP Version 3," Internet draft, work in progress, draft-ietf-idmr-v3-07.txt, Aug. 1998.

[21] J-D Shin, S. Charcranoon, H. Cankaya and T. El-Bawab, "Procedures and Functions for Operation and Maintenance in Optical Burst-Switching Networks", in Proc. IEEE 2002 Workshop on IP Operations and Management (IPOM 2002), pp. 149-53, Oct. 2002, Dallas, Texas, USA.

[22] J-D Shin, S. Charcranoon, H. Cankaya and T. El-Bawab, "Operation and Maintenance Issues in Optical-Burst Switched Networks," in Proc. SPIE ITCom, Vol. 4872, pp. 230-238, July 2002, Boston, Massachusetts, USA.

[23] Y. Tada, Y. Kobayashi, Y. Yamabayashi, S. Matsuoka, and K. Hagimoto, "OA&M framework for multiwavelength photonic transport networks," IEEE JSAC, vol. 14, pp. 914-922, 1996.

[24] Y. -H. Choi, K. -H. Lee, J. -Y. Lee, and S. -B. Lee, "OAM MIB: an end-to-end performance management solution for ATM," IEEE JSAC, vol. 18, pp. 767-778, 2000.

[25] N. Harrison, et al., "Requirement for OAM in MPLS networks," Internet draft, work in progress, draft-harrison-mpls-oam-req-01.txt, Dec. 2001.

III

# SIGNALING PROTOCOLS AND NETWORK OPERATION

Chapter 8

# GMPLS-BASED EXCHANGE POINTS: ARCHITECTURE AND FUNCTIONALITY

Slobodanka Tomic[1] and Admela Jukan[1,2]

[1] *Vienna University of Technology, Institute of Communication Networks, Favoritenstr. 9/388, A-1040, Vienna, Austria*

[2] *On sabbatical leave at Georgia Institute of Technology, School of Electronic and Computing Engineering, Atlanta, GA*

Email: { slobodanka.tomic,admela.jukan } @tuwien.ac.at

**Abstract** In this paper, we address the architecture and the functionality of the GMPLS-enabled exchange point (GMPLS-XP), which is the transport network equivalent of the Internet Exchange Point (IXP). Consisting of a centralized multi-service switching node or a distributed multi-service switching network, the GMPLS-XP is involved in the topology discovery, routing and automatic connection control in the global multi-provider automatic switched transport network. For this purpose, a GMPLS-XP is modeled as a subnetwork with the multi-layer switching capability, and this topological representation is advertised to the interconnected domains and used to make flexible dynamic inter-connection decisions. In addition to providing flexibility in inter-domain connectivity, by operating over the proposed multi-provider interconnection architecture and supporting policy-based routing, the GMPLS-XP can naturally embed the routing and SLA mediation functionality. As the IXP is critical in supporting today's global Internet operation, the GMPLS-XP will be a fundamental building block for future multi-provider heterogeneous transport network architectures.

**Keywords:** GMPLS, multi-layer routing, multi-domain routing, simulation, interconnection architecture

## 8.1 Introduction

The concepts developed in the Generalized Multi Protocol Label Switching (GMPLS) framework [1], standardized within the IETF, provide for the unified control and operation of the multi-layer transport networks, including both packet and circuit switching technologies. With unified routing and signaling, seamlessly inter-working over different technology layers, the transport networks can evolve towards automatically switched multi-service platforms, with decidedly improved flexibility, operational cost, and manageability of the services offered.

Two major flavors of multi-granular provider-provisioned Virtual Private Network (VPN) services have been identified and proposed for further standardization within the IETF: (i) the Generalized MPLS/BGP VPN (GVPN) defined in [2] reuses proven concept of MPLS/BGP that utilizes the Border Gateway Protocol (BGP) for the distribution of the VPN information and MPLS tunneling; (ii) the Virtual Optical Cross-Connect Service (VOXC) defined in [3] reuses the concept of the "virtual router"-based VPN service. Defined as a generic "network-in-network" service, GVPN/VOXC is applicable for different provider-customer relationships. For example in a carrier-carrier scenario a customer using VPOX can control and manage this provisioned infrastructure in a "virtual provider" role. By means of the proposed GMPLS-based mechanisms, the resources of the provider network, including the switching capability, are either (i) be dynamically used by the customer initiated end-to-end virtual links, i.e., generalized label switched paths (LSP) or (ii) are partitioned and assigned to a number of VPN customers for the purpose of interconnecting each customer's physically divers locations (sites).

Although extensive, the current service definitions [2,3] do not explicitly consider the scenario where several interconnected providers are involved in the service provisioning. Therefore, it is necessary to expand the administrative scope of the service model from one-provider to a multi-provider environment is obvious. The fact that the proliferation of Internet services was possible because Internet has mechanisms to function as a "network of networks" clearly indicates that the multi-provider inter-working is essential for service availability.

Some aspects of multi-provider networking have been already addressed for optical [4] and GMPLS networks: possible scenarios for inter-area routing supporting traffic engineering (TE) within one provider domain are studied in [5] and BGP extensions for routing over multiple concatenated provider-domains are proposed in [6]. However, the considered scenarios do not include all the critical building blocks, as the GMPLS framework does not address possible equivalents of "peering" and "transit" in the multi-ISP Internet. This is partially due to the fact that the business-models and requirements related to inter-domain inter-working in the new heterogeneous networks are still not fully understood. Therefore the potential capabilities of the unified automatic switching approach could still not be fully explored.

In this chapter, we extend the concept of the Internet Exchange Point (IXP) or the Internet Business Exchange (IBX) [7] into the transport context.

Considering the importance of the role that the IXPs today have in supporting the global Internet operation, a GMPLS-enabled exchange point, here referred to as GMPLS-XP, may be an important "missing link" in the GMPLS "big picture". In future multi-provider heterogeneous transport networks, GMPLS-XP may enable the architecture where provider domains are not simply statically connected but can be arbitrarily and on-demand interconnected over exchange points.

Introducing the concept of GMPLS multi-layer automatic switching into the traditional Exchange Point, could add a new degree of flexibility and dynamic in the network operation. In the network architecture deploying GMPLS-XPs, traditionally off-line "bandwidth engineering" concepts can be newly positioned for dynamic, on-line, adoption. Thus, the architecture and functionality of GMPLS-XP could be of a considerable importance for the flexibility of the generalized virtual network service.

In this chapter, we focus on the logical representation of GMPLS-based exchange point necessary to provide the flexibility of dynamic interconnections. In addition, we address the functionalities supporting inter-domain routing over exchange points which may initially maintain only static connections,

## 8.2 Interconnection Architectures

In the global Internet today, the exchange points and the mechanisms for inter-domain inter-working are crucial for operating and expanding the global network, and for the provisioning of services with global reach.

The necessity for inter-connection clearly exists in both the wireline and the wireless Internet. In the conventional wireline IP networking, at the Internet Exchange (IX) or Internet Business Exchange (IBX) [7], the Autonomous Systems (AS) are statically interconnected either directly or through a layer 2 switch. What enables inter-domain operation is the process in which the routes, i.e., the reachability information, of the domains are exchanged. In the multi-domain Internet today, this process is governed by the Border Gateway Protocol (BGP) [8], by which each of the autonomous systems advertises its local routes to other autonomous systems. The routes are distributed and installed according to the specific policies. For each of the imported routes the end systems learn which autonomous systems will be traversed on this route. The route within each autonomous system is determined internally to that domain. In this way, domains effectively hide their internal topology and still take part in the global routing at the higher hierarchical level, i.e., the AS level.

In the architecture of all-IP wireless (or mobile) networks, the concept of the exchange has its representation in GRX Exchange Points where the providers of the GPRS Roaming Exchange service (GRX) [9] are interconnected according to GRX peering agreements. GRX plays the crucial role for users' roaming, enabling not only the global connectivity but also new mobile VPN services.

In the optical domain, by introducing the "distributed exchange" concept based on the optical BGP [10], the CA*net4 research network has proposed an important approach for the global optical fiber network. In the CA*net4, the institutions interested in direct interconnecting acquire dark fibers from different network carriers and connect to the optical cross-connects of the optical core of the CA*net4 network. The optical BGP distributes the reachability over the optical core. When one institution wants to establish a direct peering session for high-bit-rate applications to another reachable institution an IP BGP session is initiated and the lightpath is established between these two institutions. The Lightpath Route Arbiter is the component responsible for the lightpath establishment. When a direct lightpath is established, the interconnected institutions can establish BGP peering sessions between them. In this way, the optical core acts as a re-configurable distributed exchange point.

In the MPLS community, the application of exchange points is also gaining attention. In [11] the exchange architecture based on MPLS technology called MPLS-IX was proposed. MPLS-IX is data-link independent and can unify two IX architectures prevailing today, being (i) Local Area Networks (LANs), such as FDDI, Ethernet or Gigabit Ethernet, and (ii) Permanent Virtual Circuits (PVC) ATM. The MPLS mechanisms are used to configure virtual back-to-back links, or back-to-back LSPs between interconnecting domains. Over those virtual interconnections, traditional bi-lateral peering models can be deployed. MPLS-IX is an important architectural advance and, by substituting MPLS with GMPLS, the same concept can be applied to the IX with optical technology.

The GMPLS exchange architecture may be even more flexible than that of the MPLS-IX model, if the process of interconnection set-up is integrated in the on-line network operation supported by the mechanisms of the *control plane*. This is the key feature of the GMPLS-XP application we propose and describe in the chapter.

To highlight the control-plane-centric nature of GMPLS-XP solution, a comparison to the bandwidth trading interconnection architecture based on the Pooling Points [12] can be made. The pooling points are infrastructure facilities where bandwidth sellers and bandwidth buyers are connected and placed physically and legally in the market. The pooling point solution is basically a *management* solution that provides an interface through which resources can be offered for trading and acquired. When the request for the resource provisioning is completed, the management scripts are produced and service provisioning can take place either manually or through the management system. The important feature of the system is that the pooling point operator is responsible for the whole provisioning process, which requires special operational agreements. Obviously there is a strong business case for deploying control plane provisioning of connections within each of the domains. What GMPLS-XP concept adds to this architecture is flexible interconnection point controlled by the control plane mechanisms.

## 8.3 Introducing GMPLS-XP in the Multi-Domain Multi-Provider Architecture

In this section we will show how the flexibility of the network can be increased with the introduction of the GMPLS-XP. We will start with the network architecture based on the traditional bi-lateral mechanisms (as inherent to BGP) and the static XP, and extend it with a trusted routing mediation service, added at the exchange points, that can operate over statically interconnected network and deal with the complexity of the multi-domain and multi-layer routing. We will complete the architecture with the flexible interconnection point for which we propose the topological modeling of the internal architecture. This topological information complements the topological representation of the global network and may be disseminated to the interconnecting domains. Therefore, the interconnections between domains (type, bandwidth) could be dynamically selected, configured and reconfigured. In this way, the static XP with no specific routing intelligence may be transformed into the proposed GMPLS-XP.

We start with the example network depicted in Figure 8.1, where four GMPLS-enabled provider networks are interconnected over one Exchange Point (XP) supporting only static interconnections. Two of the interconnections are established through the switch and one is a direct back-to-back link. Over this global network the GVPN service is provided between three geographically diverse sites (C1, C2, C3) of one client virtual network with possibly specific dynamic topology constraints. We assume that the standard mechanisms and relationships are deployed: the clients get the routing information from the provider, by means of the routing protocols (i.e., OSPF, IS-IS or BGP) extended for GMPLS, and use it for the LSP route calculation and for LSP set-up signaling by means of signaling protocols (i.e., RSVP-TE or CR-LDP) extended for GMPLS. Between two connected parties, e.g., between a site C1 and a domain D1, or between D1 and D3, GMPLS can support different interconnection models, overlay augmented or peer. On this interface one of the parties has a client role and the other provider role. The interconnections themselves, such as the one between D2 and D4, are provisioned through the management system. The inter-connection decision depends on the anticipated bandwidth requirements and on business objectives, and may be an outcome of an off-line optimization process. In this static case, once the interconnections are established, GMPLS-enabled domains can exchange their routing information, e.g., by using extension of BGP as proposed in [6].

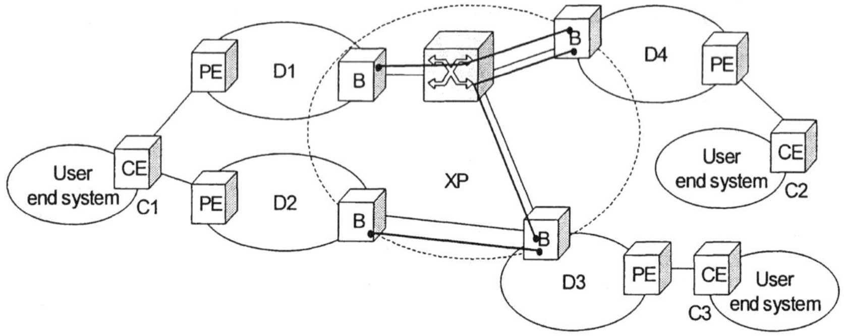


*Figure 8.1.* Multi-provider, multi-domain architecture thee static inter-domain links.

In the example of the Figure 8.1, service member C1, a multi-homed user, invokes a set-up of a generalized label switched path (LSP) to C2. This LSP may span different domains, e.g., D1 and D4, or D2, D3 and D4, and possibly different levels of GMPLS hierarchy. Hence, C1 needs to acquire the topology information (reachability) over the interfaces to D1 and D2 and to determine the end-to-end path. The other possibility is to delegate to the domain D1 (and possibly also to D2) the task to calculate the path to C2, as discussed in [5]. In general, the access to functionality such as routing delegation required in multi-domain TE scenarios, similar to those defined in [5], between interconnected parties (domains, sites) would be part of a contract between those parties, i.e., a part of a control plane service level agreement (SLA). The important aspect of this control plane SLA could be related to the supported interconnection model (peer or overlay). The control plane interconnections between routing services of clients and domains for the previous example are illustrated in Figure 8.2.

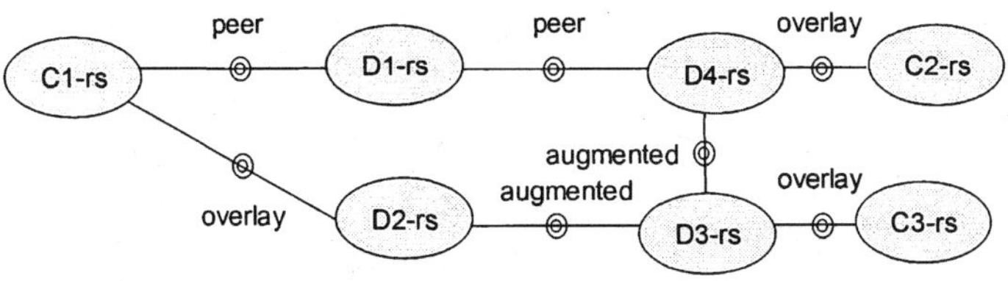


*Figure 8. 2.* Control plane interfaces between routing services (rs) of clients and domains.

In our example, a client C1 is multi-homed with static links to D1 and D2, and its routing service maintains two control plane SLAs. For example a component of an SLA with D1 and D2, related to a negotiated interconnection model, may state that the peer is deployed on the interface to D1 and the overlay model on the interface to D2. The routing information (global rechability or link state information) that C1 receives strongly depends on those SLAs as well as the SLAs existing between domains over the exchange point. Since this information is necessarily inaccurate due to the state changes and information hiding between domains, sub-optimal routing may occur. Whilst more frequent updates can account for the frequent state changes in the network, introducing new routing mediation functionality between the interconnecting parties may approach the inaccuracy due to the information hiding.

## 8.3.1 Multi Provider Edge Routing Service

In routing over the global network, the availability of the global information will determine the quality of the path. Because the inter-domain control interface is not considered to be a trusted interface, only aggregated routing information (reachability) will be transported over it. On the other hand, this problem may be approached by introducing the trusted distributed routing service. In Figure 8.3, such routing service functionality, referred to as Multi Provider Edge Routing Service (MPE-rs), is added to the statically interconnected network, i.e., a network with static XPs.

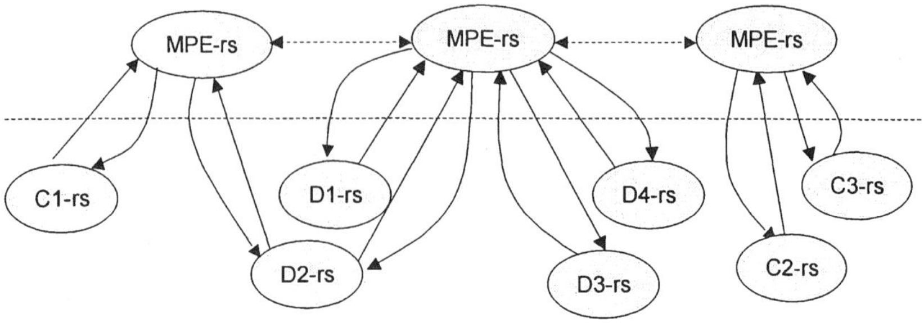


*Figure 8. 3.* Multi-provider, multi-domain architecture thee static inter-domain links.

MPE-rs is conceptually a distributed trusted layer in the hierarchical global routing toward which all the domains may send the full routing information. In this architecture, the local policies and preferences can be also imported into the MPE, which could than export customized information to its clients so that they can invoke LSP set-up. Each single SLA, now migrated to MPE, can be modified in more scalable way. We introduced and studied MPE in our previous work [13, 14].

While the MPE routing service illustrated in Figure 8.3 can improve the inter-domain routing, we still assume that interconnections between domains are static. In Figure 8.4, we show one multi-layer XP with static interconnections. Domains connect to the XP over links of different types; the interconnection can be established (through the management system or manually) between the end-points of the links of the same type. So domain D1 connecting to an XP over one wavelength LSC link can connect to the domain D4. Similarly, D6 can connect to D5 and D2 to D3.

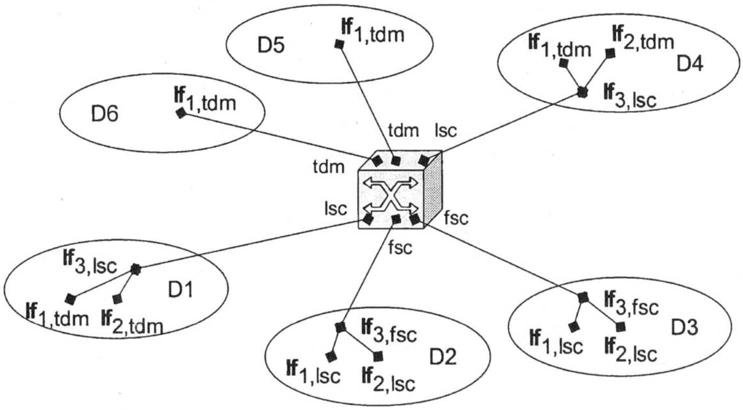


*Figure 8.4.* Multi-provider, multi-domain architecture static inter-domain link; Link are lambda switching capable (LSC,) fiber switching capable (FSC) and TDM switching capable (TDM).

On the other hand, however, it would be beneficial to enhance the XP so that, at one instance in time, the domain D1 could connect over one TDM component to the domain D6 and over the other TDM component to the domain D4. Alternatively, at another time, the domain D1 could connect to the domain D2 over the LSC component. To achieve such flexibility, we introduce the GMPLS-XP multi-layer concept in the model of the exchange point.

## 8.3.2 Modeling of GMPLS-XP Internal Architecture

The multi-layer switching core based on the optical cross-connect is what makes the GMPLS-XP architecture flexible. The optical links and the wavelengths on those links can be flexibly configured for de-multiplexing and switching on the different layers in the digital hierarchy and in the different time instances according to the configuration of the switch fabric. To access and leverage this equipment flexibility for establishing dynamic interconnections, an appropriate model is necessary.

We adopt the concept of GMPLS Regions [1] to model the internal architecture of GMPLS-XP. This results in the topological representation of different internal switching and multiplexing capabilities. In other words, just as conventional Internet Exchange Point can be configured as one Autonomous System, a GMPLS-XP can be represented as one GMPLS domain composed of a number of LSP Regions. Here, the multiplexing and switching capability of both access and internal interfaces of the GMPLS-XP can be logically mapped to the LSP Regions according to the type of the interface. In fact, the domains connecting to a GMPLS-XP, can be considered as logically connected to the appropriate LSP region, depending on the type of the access link and the configuration of its ingress and egress interfaces. By adopting this internal representation and using the proposed extended routing architecture, the GMPLS-XP can act as a routing peer for the connecting domains and advertise its own internal "topology" together with the reachability of the connected domains. Consequently, the choice of which GMPLS-XP internal links (i.e., domain inter-connections) are used can become a fully dynamic decision.

To better highlight this key proposition for the GMPLS-XP let us compare static XP with GMPLS-XP.

When two providers decide to interconnect over static XP, the interconnection link must first be provisioned. Once this link is configured, the providers may exchange routes. The decision to peer is not an on-line decision. On the other hand, if the decision as to which interconnection to establish on the XP is to be made on-line, the information needed to make this decision must be also available on-line. Since GMPLS-XP is represented as GMPLS switching sub-network, the LSPs established over this sub-network are in-fact on-demand inter-domain links. Due to this representation, the routing service (MPE-rs) is also involved in GMPLS-XP routing control.

Figure 8.5 depicts one GMPLS-XP comprising multiple LPS Regions and interconnecting several domains. D1 is connected to the GMPLS-XP over a link, with wavelength switching capable interfaces ($if_{1,lsc}$ on the D1 side and $if_{1,lsc}$ on the XP side). In this example we focus on one wavelength component of this link. At one point in time this wavelength may be switched to the interface $if_{4,lsc}$ towards the domain D4. At the other point in time D1 may require interconnections to D5 and D6 at the TDM level. In this case the XP interface $if_{1,lsc}$ will connect to $if_{3,lsc}$, and the TDM components will be accordingly extracted and switched at the TDM layer.

In some other point in time, interface $if_{1,lsc}$ may be switched to $if_{2,lsc}$, being a component of the fiber switching capable interface $if_{1,fsc}$ towards the domain D6.

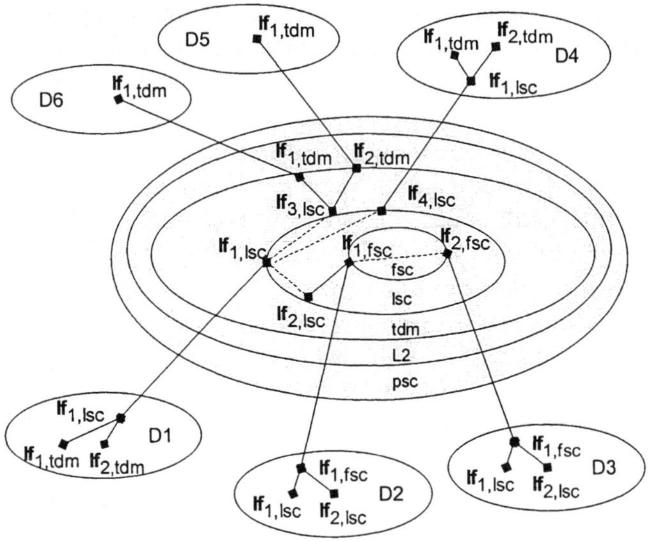


*Figure 8.5.* GMPLS-XP internal topology model.

The most important enabler of this concept is the discovery and the self-configuration process in which this topological representation should be build.

### 8.3.3 GMPLS-XP with the MPE Routing Service

The functional flexibility of the GMPLS-XP can be exploited even further by integrating it with the MPE routing service. Before the interconnections between domains are established, all the connecting domains can be engaged in the exchange of the routing information. This exchange is performed through the MPE, and therefore the different domains do not engage in bi-lateral exchange of routes. The MPE also acquires the topology information of the GMPLS-XP as illustrated in Figure 8.6 and can act as a routing peer of the connecting domains or some kind of a trusted routing mediator over the links connecting to the GMPLS-XP. Via the MPE, the GMPLS-XP imports the routing information and policies from the connecting domains. As a result, it may collect the complete routing and policy information and, therefore, act as a routing proxy for the connected domains. For example, the policy can specify the metric for which the least-cost is to be achieved, the preference list of domains, etc.

Networks can connect to the GMPLS-XP (with integrated MPE) in a client role, provider role, or both client and provider roles. In this case, the *client* is the party that initiates the request for a generalized LSP set-up spanning several domains. Conversely/Similarly, the *provider* is the party that, on the one hand, provides required routing information to the routing service or is otherwise involved in the routing service, and, on the other hand, takes part in the request accommodations during request signaling.

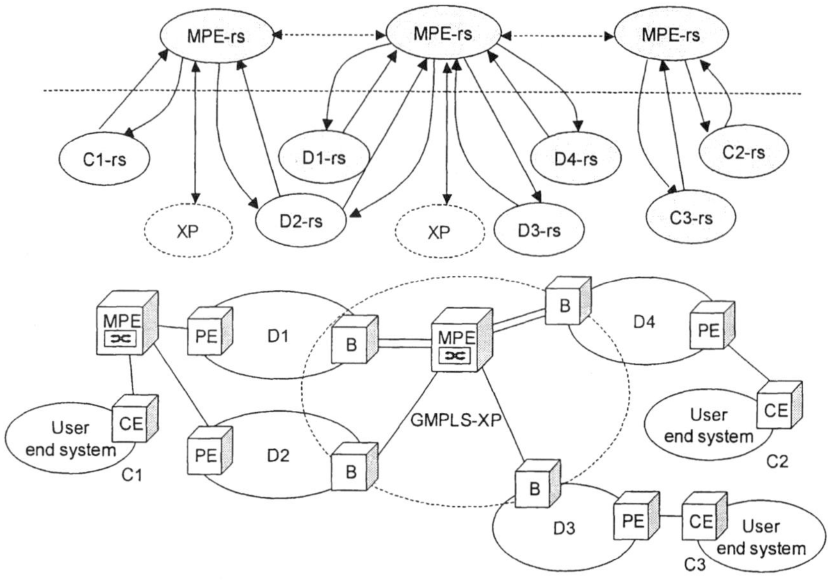


*Figure 8.6.* GMPLS-XP with the embedded MPE.

Figure 8.6 shows the interconnected architecture based on GMPLS-XPs with the embedded MPE routing service. Being connected to GMPLS-XP with MPE, a client network C1 can choose between two provider domains, i.e., D1 or D2, via one physical connection only. Which domain would be used to establish a connection can be completely resolved in the MPE.

## 8.4 Final Remarks

As Internet control concepts are being introduced into the transport network, the latter is gaining the look-and-feel of a self-configurable network in which innovative dynamic inter-layer "bandwidth engineering" approaches can be deployed. While the focus of the standardization community is on the seamless control of the multiple-layer hierarchy at the intra-domain interfaces, the *inter-domain* multi-layer interworking has attracted limited interest so far. Nevertheless, inter-domain multi-layer interworking will be critical for automatically switched transport network to provide global services. In this context, one of the Internet concepts that have not been thoroughly considered is that of the Internet Exchange Point (IXP). IXP is the invaluable enabler of the global Internet. It is the point where the business relationships between domains are defined and the inter-domain routing mechanisms are deployed. In this chapter, we addressed this "missing link" and defined the GMPLS-XP as an architectural element that can dynamically interconnect domains on different GMPLS hierarchy levels.

We suggest modeling of a GMPLS-XP as a sub-network with the multi-layer switching capability, i.e., comprising several different GMPLS LSP Regions. In this scenario, domains interconnecting at the GMPLS-XP will use this topological representation when making flexible on-line inter-connection decisions. The basic enabler for this architecture is the discovery process in which the architectural representation of an XP is built. In addition, we envisage the routing and SLA mediation functionality, the MPE, embedded in the GMPLS-XP, supporting this flexible multi-provider interconnected architecture. We believe that the proposed architecture will be key to enable networking scenarios in which on-line interconnections will be established with the operational goal to achieve more optimal global routing and resource utilization and support network self-organization.

## References

[1]Eric Mannie (Editor): "Generalized Multi-Protocol Label Switching (GMPLS) Architecture" work in progress draft-ietf-ccamp-gmpls-architecture-03.txt, Expiration date: Feb. 2003

[2]H. Ould-Brahim, Y. Rekhter, D. Fedyk, P. Ashwood-Smith, E.C. Rosen, E. Mannie, L. Fang, J. Drake, Y. Xue, R. Hartani, D. Papadimitrio, L. Berger, "GVPN: Generalized Provider-provisioned Port-based VPNs using BGP and GMPLS" work in progress draft-ouldbrahim-ppvpn-gvpn-bgpgmpls-01.txt Expiration Date: December 2002

[3]H. Ould-Brahim, Y. Rekhter, M. Carugi, Y. Xue, R. Hartani, D. Papadimitrio "VPOXC Provider Provisioned Virtual Private Optical Cross-Connect Service", work in progress draft-ouldbrahim-ppvpn-vpoxc-01.txt, Expiration Date: December 2002

[4]G.M. Bernstein, V. Sharma, L. Ong, "Interdomain optical routing", Journal Of Optical Networking (JON), Vol. 1, No. 2 - February 2002 Page(s):80-92

[5]K. Kompella, Y. Rekhter, J.P. Vasseur, T. W. Chung, "Multi-area MPLS Traffic Engineering", work in progress draft-kompella-mpls-multiarea-te-03.txt Expiration Date: December 2002

[6]Y. Xu, A. Basu, Y. Xue, "A BGP/GMPLS Solution for Inter-Domain Optical Networking", work in progress draft-xu-bgp-gmpls-02.txt Expiration Date: Dec 2002

[7]Metz, C. “Interconnecting ISP networks” IEEE Internet Computing , Volume: 5 Issue: 2 , March-April 2001 Page(s): 74 –801

[8]A Border Gateway Protocol 4 (BGP-4), IETF RFC 1771

[9]Blyth, K.J.; Cook, A.R.J. “Designing a GPRS roaming exchange service”, Second International Conference on 3G Mobile Communication Technologies, 2001., 2001 Page(s): 201 –205

[10] Bill St. Arnaud, Michael Weir, John Coulter “BGP Optical Switches and Lightpath Route Arbiter” Optical Networks Magazine March/April 2001

[11] Nakagawa, I.; Esaki, H.; Nagami, K., “A design of a next generation IX using MPLS technology” Applications and the Internet, 2002. (SAINT 2002). Proceedings. 2002 Symposium on , 2002 Page(s): 238 -245

[12] G. Cheliotis, B. Liver “Brokerage of Bandwidth Commodities”, 7th International Conference on Intelligence in Services and Networks (IS&N), 23-25 Feb. 2000, Athens

[13] S.Tomic, A.Jukan, "MPFI: The Multi-provider Network Federation Interface for Interconnected Optical Networks"; IEEE Global Telecommunications Conference (GLOBECOM 2002), Taipei, Taiwan; 17.11.2002 - 21.11.2002; in: "Proceedings Globecom 2002".

[14] S. Tomic, A. Jukan: "*GMPLS-based Exchange Points: Architecture and Functionality*";, presented at the Workshop on High Performance Switching and Routing (HPSR 2003), Torino, Italy - June 24-28, 2003; in: "*Proceedings HPSR 2003*".

Chapter 9

# THE GMPLS CONTROL PLANE ARCHITECTURE FOR OPTICAL NETWORKS

## *Routing and Signaling for Agile High Speed Switching*

David Griffith *

*National Institute of Standards and Technology (NIST), 100 Bureau Drive, Stop 8920, Gaithersburg, MD 20879*

Email: david.griffith@nist.gov

**Abstract** Optical networking technology underwent a major revolution in the 1990s as the old paradigm of SONET/SDH systems to support circuit-switched connections began to give way to a new mesh network of transparent, high-speed optical switches with the capability to support a variety of higher-layer traffic types and services. In order for such a network to operate efficiently, it is essential that it employ a control plane that is capable of managing both legacy framed traffic and new traffic types. Also, the optical control plane must be able to operate across network boundaries, both at the edge interface to the customer's equipment and at administrative domain boundaries in the core network. The Internet Engineering Task Force (IETF) has tasked several working groups to develop the architecture for such a control plane as well as protocols to support its functioning. These groups' work has built on previous work in the IETF on Multi-Protocol Label Switching (MPLS), which was developed to allow packet routers to operate more efficiently. In this chapter, we describe the GMPLS architecture and related protocols, specifically RSVP-TE, OSPF-TE, and LMP.

**Keywords:** GMPLS, WDM, CR-LDP, RSVP-TE, ISIS-TE, OSPF-TE, LMP, protection and restoration, traffic engineering

*The identification of any commercial firm, product, or trade name does not imply endorsement or recommendation by the National Institute of Standards and Technology.

## 9.1 Introduction

Optical communications technology has experienced tremendous growth in the nearly 30 years since the first telephone calls were transmitted commercially over fiber by AT&T. Within a few decades, improvements in semiconductor lasers, optical detectors, tunable gratings, transparent optical switch fabrics, optical fibers, and other physical layer technologies have produced a vast network that can transmit information on multiple wavelengths at rates of up to 40 Gbps on each wavelength. Much of this expansion was created by a burst of demand during the 1990s that was caused by the public's enthusiastic adoption of computer networking services in the wake of the introduction of the World Wide Web.

Optical networks were originally conceived as high-bandwidth systems that could multiplex large numbers of voice calls while meeting the strict grade-of-service requirements demanded by the traffic that they were carrying. The steady growth of data traffic, which now constitutes the majority of global information flow, has produced a need for a network control plane that is capable of supporting both circuit-switched and packet-switched connections. In addition, the steady rollout of services that require ever-greater bandwidth, such as video on demand (VOD), is generating a need on the part of major carriers for the ability to create, modify, and tear down connections rapidly and with a minimum amount of human intervention.

The first set of optical standards, which emerged in the early 1990s, described mechanisms to multiplex TDM digital signals using hierarchical framing. Two similar sets of standards were developed: a version used in North America, known as Synchronous Optical Network (SONET), the other version, used virtually everywhere else in the world, is called Synchronous Digital Hierarchy (SDH). Optical communications systems using SONET/SDH were widely deployed during the last decade but they were intended primarily to support circuit-switched voice traffic. As the quantity of packet-switched traffic has increased, the SONET/SDH optical networking model has come to be viewed as adding too much complexity and expense to data network implementations. It is more efficient to run Internet Protocol (IP) traffic directly over optics, eliminating the intervening layers, but this requires an optical control plane that can support all of the administrative functions that such a network requires.

For the past several years, the Internet Engineering Task Force (IETF) and other standards development organizations (SDOs) have been working to define the architecture and protocols that can be used to compose a flexible control plane for optical networks. The IETF's work has built on the development of MultiProtocol Label Switching (MPLS), extensions to existing routing and signaling protocols, and the creation of new protocols where needed. The resulting generalized MPLS (GMPLS) architecture offers the possibility of efficient, automated control of optical networks supporting a variety of services. This chapter examines the structure of the GMPLS control plane and describes the basic functions of its parts.

This chapter is organized as follows. In Section 9.2, we describe the essential features of the GMPLS architecture. We discuss how the label switching paradigm was extended to encompass a large range of interface types, and the types of optical networking architectures that are possible under GMPLS. In Section 9.3, we discuss the signaling protocols that are used for connection establishment and maintenance, particularly RSVP-TE. In Section 9.4, we examine routing in GMPLS networks. We focus on OSPF-TE, which is an extension of link-state routing for packet-switched networks to support constrained routing in networks that contain a variety of interface types and whose elements do not necessarily share traditional types of routing adjacencies. In Section 9.5, we discuss the Link Management Protocol (LMP), which was developed to support the creation of control channels between nodes and to verify interface mappings between connected switches. We conclude in Section 9.6 by comparing the work on GMPLS in the IETF to related work taking place in other SDOs.

## 9.2 The GMPLS Architecture

Generalized Multiprotocol Label Switching (GMPLS) [1, 2] grew out of the development of MPLS and was prompted by the need for an IP-oriented control plane that could incorporate traffic engineering capabilities along with support for a multitude of traffic types, particularly legacy systems such as ATM and SONET/SDH. GMPLS builds on the label switching model while adding features that enable it to work with optical networks.

### 9.2.1 MultiProtocol Label Switching (MPLS)

MPLS evolved from several parallel development efforts. The story of its creation has been recorded in detail elsewhere [3], but we give the main details here. In the early 1990s, Asynchronous Transfer Mode (ATM) was a strong contender for dominance in large data networks, mostly because ATM switches at that time were faster than IP routers. Because this situation created a need to tunnel IP traffic over ATM networks, the IETF developed a set of Requests for Comments (RFCs), which are the set of freely available documents that define Internet protocols, to describe how this could be done. Several router vendors began working on an IP/ATM architecture that did not use ATM's User-Network Interface (UNI) signaling (described in the ATM Forum's UNI 1.0 document [4] and by the ITU [5]), which was considered cumbersome, in favor of a more lightweight approach. The need for a standardized version of label-based switching motivated the creation of a working group in the IETF.

The version of label switching that ultimately emerged from the MPLS working group incorporates elements of all the major vendor schemes but is most closely related to the approach, proposed by engineers from Cisco, known as tag switching. Tag switching incorporated mechanisms for forwarding IP datagrams over multiple link layer protocols such as ATM but also the Point-to-Point Protocol (PPP) and IEEE 802.3. Tag switching was novel in that it used the RSVP protocol to set up connections with specific traffic characteristics in order to support some degree of Quality of Service (QoS).

The idea behind all the approaches that the IETF harmonized to create MPLS is that of labels that can be applied to IP packets to forward them without having to examine the IP header's contents. The labels are encoded using a thin encapsulation header (known as a shim header) to tunnel IP flows through various types of layer 2 clouds [6]. The shim header is located between the IP header and the link layer protocol header.

MPLS uses forwarding adjacencies (FAs) to set up label switched paths (LSPs) across networks of label switching routers (LSRs). All IP packets that have some set of parameters in common are considered to belong to a particular FA. In practice, membership in a FA is based solely on the IP packet's destination address field. Once a complete set of label mappings has been installed in each LSR on the LSP for an FA, packets belonging to that FA can be forwarded over the LSP from the ingress LSR to the egress LSR without having the contents of their IP headers examined or modified. For this reason, the MPLS architecture describes mechanisms that can be used to decrement the IP header's Time To Live (TTL) field by the appropriate amount when the packet leaves the MPLS cloud [7].

MPLS supports hierarchical LSPs by allowing multiple shim headers to be stacked between the layer 2 and layer 3 headers, so that LSPs can be tunneled through other LSPs. The label stack is processed using the last-in-first-out (LIFO) discipline, so that the label associated with the highest LSP in the hierarchy is located at the bottom of the stack. When a packet bearing a label stack arrives at the input port of an intermediate LSR (i.e., an LSR that lies along the packet's path but is neither the source nor the destination), the LSR removes (or "pops") the topmost label from the stack and uses it to locate the appropriate Next Hop Label Forwarding Entry (NHLFE), which indicates to which output port the packet should be sent and which value should be assigned to the outgoing label that will be applied to the packet. A new label with the appropriate value is then created and it is pushed onto the top of the stack before the packet is forwarded to the next LSR.

An illustration of how labels are used to forward packets is given in Figure 9.1. The router shown in the figure is an intermediate LSR for the LSP that is carrying Packet A. The LSR examines Packet A's label and uses the NHLFE table to determine the output port to which it should send Packet A and the outgoing label that it should apply to the departing packet. Packet B is being tunneled through a lower-layer LSP, and the LSR in the figure is the egress point for that LSP. In this case, the LSR removes the topmost label and forwards the packet, whose remaining label is meaningful to the LSR that is directly downstream (i.e., towards the destination) from the LSR in the figure.

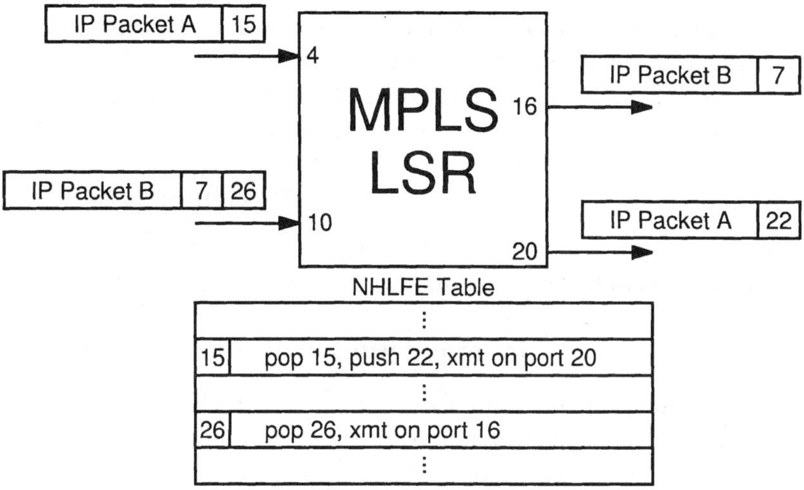


*Figure 9.1.* An example of packet forwarding in an MPLS router.

## 9.2.2 Generalizing MPLS

As the development of the MPLS architecture proceeded, people realized that there were several interesting parallels between the tunneling of IP traffic along LSPs consisting of a sequence of LSRs and the proposed tunneling of IP traffic along lightpaths consisting of a sequence of optical cross-connects (OXCs) with wavelength conversion capabilities. If an optical network could be made to support a kind of MPLS architecture, wavelengths could play the same role as numerical labels in an electrical MPLS network of routers. Instead of a table of next hop label forwarding entries, each OXC would maintain a table of port and wavelength mappings, so that data associated with the lightpath arriving on any particular input port on a particular wavelength would automatically be mapped to a predetermined wavelength on a desired output port. The set of port and wavelength mappings would be implemented by configuring the switching fabric in the switch's core.

Using this basic realization, the IETF began work in early 2000 on the GMPLS architecture [1]. Originally known as MP$\lambda$S or MPLambdaS, it is an extension of the MPLS architecture that includes mechanisms for encoding labels that are associated with interfaces that are not packet-switch capable. The GMPLS architecture supports the following five interface types.

1 Packet Switch Capable (PSC)
An interface that is uses the information contained in the packet header to forward individual packets.

2 Layer-2 Switch Capable (L2SC)
An interface that can read layer 2 frame headers and use them to delineate individual frames and forward them.

3 Time-Division Multiplex Capable (TDM)
An interface that switches data based on the position of the slot that it occupies in a TDM frame.

4 Lambda Switch Capable (LSC)
An interface that switches traffic carried on an incoming wavelength onto a different outgoing wavelength.

5 Fiber Switch Capable (FSC)
An interface that switches information streams or groups of streams based on the physical resource that they occupy.

GMPLS is supported by separate protocols that perform routing (OSPF-TE and ISIS-TE), signaling (RSVP-TE and CR-LDP), and link management and correlation (LMP). Their places in the IP-based protocol stack are shown in Figure 9.2. Each protocol name is connected to the layer that it uses to encapsulate and transport its messages. In only two cases (LMP and CR-LDP) are layer 4 segments used to carry GMPLS messages. The stack shown in the figure is a "heavy stack," in that a large number of layers reside between the IP layer and the optical layer. Future architectures use protocol stacks with fewer layers, or even let IP traffic run directly over optics, without any framing technologies in between.

While MPLS supports a variety of label distribution modes, GMPLS protocols work solely with the downstream-on-demand mode, in which label mappings are distributed by LSRs to their upstream neighbors only in response to a label mapping request that comes downstream. This enables GMPLS to support circuit-switching operations. Labels in GMPLS, known as generalized labels, are carried in signaling messages and are typically encoded in fields whose internal structure depends on the type of link that is being used to support the LSP. For example, Port and Wavelength labels are jointly encoded in a 32-bit field and have meaning only to the two nodes exchanging label

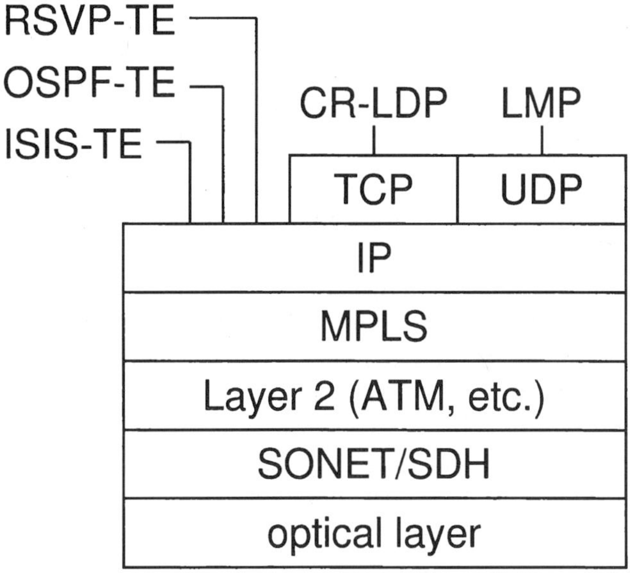


*Figure 9.2.* The GMPLS protocol stack, showing relative positioning of essential routing, signaling, and link management protocols.

mapping information. MPLS shim labels and Frame Relay labels are encoded right-justified in the 32-bit general label space. For ATM labels, the Virtual Path Identifiers (VPIs) are encoded right-justified in bits 0-15, and the Virtual Channel Identifiers (VCIs) are encoded right-justified in bits 16-31. Encodings also exist for SONET/SDH time slots, as shown in Figure 9.3. They are as follows:

```
 0                   1                   2                   3
 0 1 2 3 4 5 6 7 8 9 0 1 2 3 4 5 6 7 8 9 0 1 2 3 4 5 6 7 8 9 0 1
+-+-+-+-+-+-+-+-+-+-+-+-+-+-+-+-+-+-+-+-+-+-+-+-+-+-+-+-+-+-+-+-+
|               S               |   U   |   K   |   L   |   M   |
+-+-+-+-+-+-+-+-+-+-+-+-+-+-+-+-+-+-+-+-+-+-+-+-+-+-+-+-+-+-+-+-+
```

S: Index of a particular STS-1 SPE in a STS-N multiplex.
U: (SDH only) Index of a particular virtual container (analogous to virtual tributary) in an STM-1
K: (SDH only) Index of a particular branch in a VC-4.
L: Index of a branch in a TUG-3, VC-3, or STS-1 SPE. If L = 1 the SPE is not subdivided; L = 2,3,...,8 indicates one of 7 possible VT groups in the SPE (if SONET is used).
M: Index of a branch in a VT group. M = 0 indicates an unstructured STS-1 SPE. Otherwise it refers to a particular VT-1.5, VT-2, or VT-3.

*Figure 9.3.* Encodings for SONET/SDH indices in the generalized label.

## 9.3 GMPLS Signaling

The MPLS architecture does not state explicitly how labels are to be requested and distributed. The mechanism for doing this is left up to the network operator, although the IETF has created a signaling architecture for GMPLS [8]. Tag switching, the MPLS ancestor developed by Cisco, used RSVP to request and distribute labels. It was therefore natural that RSVP would be proposed as the standard protocol for signaling LSP setup and teardown in GMPLS. A small group of equipment vendors led by Nortel networks proposed a competing signaling scheme, known as the Label Distribution Protocol (LDP), which they had created explicitly for MPLS. LDP was later extended to support constrained routing (CR) [9]. CR-LDP's proponents pointed out several shortcomings of conventional RSVP such as its lack of scalability due to its need to periodically refresh connection state information by retransmitting signaling messages. This issue and others were addressed in the modified version of RSVP described in RFC 3209 [10]. Additional modifications to RSVP, specifically to support traffic engineering and GMPLS, are described in RFC 3473 [11].

The struggle for dominance between the two protocols lasted for several years and affected discussions in other SDOs, such as the OIF, as well as those in the MPLS working group and other working groups in the sub-IP area in the IETF. For approximately two years, the issue was addressed by "agreeing to disagree" and supporting both proposed signaling protocols. This led to considerable document bloat in the affected working groups as each proposed signaling protocol extension required defining new objects for RSVP-TE and for CR-LDP. The issue was resolved after the IETF's meeting in July, 2002, when RSVP-TE was designated as standards track, while CR-LDP was relegated to informational status [12]. This decision was influenced by market realities, which were reflected in an implementation survey conducted by the MPLS working group [13] that showed that while RSVP-TE was supported by nearly all of the respondents, CR-LDP was implemented by a small minority, and those vendors who supported CR-LDP also tended to support RSVP-TE. For this reason, in this section we describe only RSVP-TE and how it is used to manage connections in optical networks.

### 9.3.1 RSVP-TE

The RSVP protocol and its extension, which are described in detail in [10], [11], and [14], was originally designed to support Integrated Services (IntServ) in IP networks. It does this by reserving resources in routers to achieve a desired QoS. RSVP is concerned only with signaling, and it is functionally separate from other networking functions such as routing, admission control, or policy control.

**RSVP Operations.** RSVP in its classical form supports receiver-initiated reservations for multicast sessions. Applications with data to transmit, known as senders, advertise their status by transmitting Path messages downstream to one or more receivers. As Path messages traverse the network, they establish state information in the RSVP-capable routers that they pass through. This information generally consists of a traffic specification that includes information necessary to support QoS functions (e.g., peak data rate, peak burst size, etc.). It also contains information that identifies the sender that created the Path message and the router immediately upstream (i.e., toward the sender) from the router that received the message. Once a Path message reaches its destination, that node can start sending Reservation (Resv) messages upstream to the originating source node. As the Resv message propagates toward the sender, it causes RSVP-capable routers along the route to reserve resources to support the traffic characteristics (or flowspec) that are advertised in the Resv message. When the session's sender receives a Resv message, it can begin sending data to the receiver. Because the exchange of Path and Resv messages supports only unidirectional flows, a separate set of Path and Resv messages must be exchanged to support a bidirectional session. The structure of Path and Resv messages in RSVP-TE for GMPLS is shown in Figure 9.4 and Figure 9.5, respectively. In both figures, gray fields indicate optional objects. In Figure 9.5, the structure of the flow descriptor list depends on the filter being used by the source that generated the Path message. A list can include any number of label blocks, but only one label can be associated with each FILTER_SPEC in the list.

When the Multiprotocol Label Switching protocol was being developed by the MPLS working group, RSVP was extended to allow it to support traffic engineering (TE) by requesting and distributing label bindings [11]. They are used to support the creation of LSP tunnels, i.e., LSPs that are used to tunnel below standard IP-based routing. The modified version of RSVP, known as RSVP-TE, builds support for MPLS into RSVP by defining new objects for transporting label requests and mappings using Path and Resv messages. Details about the Forwarding Equivalence Class (FEC) that is to receive the mapping are encoded in new versions of the SESSION, SENDER_TEMPLATE, and FILTER_SPEC objects that identify the ingress of the LSP tunnel. In addition, LABEL_REQUEST and LABEL objects have been added to the Path and Resv messages, respectively. RSVP-TE supports only downstream-on-demand label distribution mode with ordered control. The LABEL object allows the Resv message to carry a label stack of arbitrary depth. The LABEL_REQUEST and LABEL objects must be stored in the Path and Reservation state blocks, respectively, of each node in the LSP tunnel, and they must be used in refresh messages, even if there has been no change in the tunnel's state. This will tend to increase the signaling overhead. If an intermediate node doesn't sup-

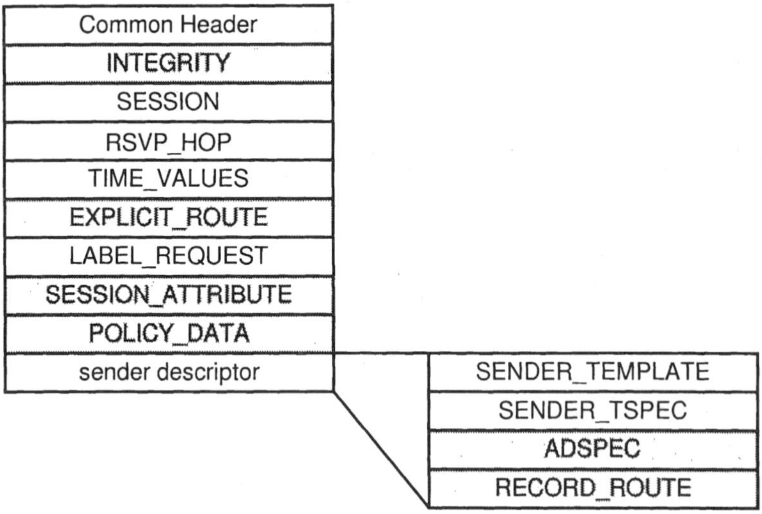


*Figure 9.4.* Structure of the Path message in RSVP-TE.

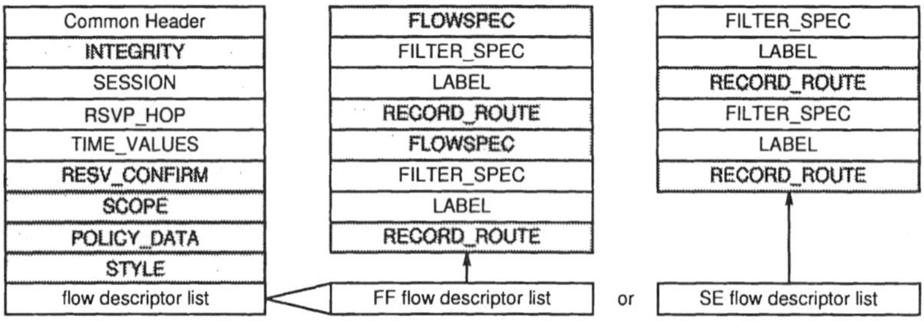


*Figure 9.5.* Structure of the Resv message in RSVP-TE.

port LABEL_REQUEST objects or has no label bindings available (e.g., all its resources are in use), it sends a PathErr message back to the source node.

If traffic engineering is being used to route LSP tunnels, there are a number of situations where an active tunnel can be rerouted, such as when there is a change of next hop at a particular point in the tunnel or when a node or link failure occurs. When tunnels are rerouted, the preferred course of action is to set up the alternate route before tearing down the existing route so that there is no disruption of the data traffic (this process is known as "make before break"). In order to support this functionality, RSVP-TE must use the shared explicit (SE) reservation style, in which an explicitly-defined set of senders is associated with a given session. This allows the tunnel ingress to specify a tunnel detour associated with a new label descriptor at the same time that it maintains the old tunnel, over which data is still flowing. After the detour has been established, the defunct portion of the tunnel can be torn down or allowed to time out.

Once a receiver is finished receiving data or a sender no longer has data to send, they can delete the state that they created by respectively using ResvTear and PathTear teardown messages. Originally RSVP used soft state, so that it was possible to avoid using teardown messages and simply allow the state to time out on its own; however, this is discouraged in [14] and circuit-switched applications require explicit teardown messages. Intermediate routers along a path can also send teardown messages if either the Path or the Reservation state associated with a particular session times out. The messages are forwarded immediately by each router that receives them until they reach a node where they do not cause that node's state to change. Because these messages do not run over a transport layer protocol that can request retransmission of missed segments, they can be lost. In old RSVP networks this was not a fatal problem because the state would time out on its own eventually; when this happened the node whose state had timed out would transmit the appropriate teardown messages and the ordered teardown process would continue. RSVP-TE uses message acknowledgment to prevent the loss of teardown messages.

**Explicit Routing and Route Recording.** RSVP-TE also includes support for explicit routing by incorporating an EXPLICIT_ROUTE object (ERO) into the Path message and a RECORD_ROUTE object (RRO) into both the Path and the Resv message. These objects cannot be used in multicast sessions. The RECORD_ROUTE object is used to do loop detection, to collect path information and report it to both ends of the tunnel, or to report the path to the tunnel ingress so that it can send an EXPLICIT_ROUTE object in its next Path refresh to pin the route. The EXPLICIT_ROUTE object contains a sequence of abstract nodes through which the LSP tunnel must pass. The ERO can be modified by nodes that forward it. RSVP-TE can also specify nodes that should be avoided [15].

**RSVP-TE Hello Messages.** Unlike RSVP, RSVP-TE allows directly connected neighbors to exchange Hello messages to detect node failures. The Hello feature is optional. Hello messages carry either HELLO_REQUEST or HELLO_ACK objects, both of which contain 32 bit instance numbers for the nodes at both ends of the connection. A node that uses the Hello option sends Hello_Request messages to its neighbor at regular intervals (the default interval is 5 msec); a participating recipient replies with a Hello_Ack message. If the sender of the Hello_Request hears nothing from the receiver after a fixed period of time (usually 3.5 Hello intervals), it assumes that the link between them is broken. If either node resets or experiences a failover, it uses a new instance number in the Hello messages it transmits. This allows an RSVP-TE node to indirectly alert its neighbor that it has reset. If there are multiple numbered links between neighboring nodes (i.e., each interface has its own IP address), then Hello messages must be exchanged on all the links.

**Label Suggestion and Upstream Labels.** Optical lightpaths are typically bidirectional and follow the same route. While it is possible to create a bidirectional lightpath that comprises two disjoint unidirectional paths, this creates management problems for the network operator and is avoided whenever possible. Thus, when a bidirectional lightpath is created, there will be two exchanges of signaling information, one for each unidirectional component.

For MPLS networks operating in downstream-on-demand mode, labels are not installed in a LSR until a Resv message is received. This mode of operation poses problems when it is applied to photonic switches that route light directly from input ports to output ports without converting the received signal into electrical form. Switch fabrics based on Micro Electro-Mechanical Systems (MEMS) (e.g., tiltable mirrors) require tens of milliseconds to respond to reconfiguration commands. This adds considerably to the LSP establishment time.

To reduce the time required to set up a new connection, RSVP-TE can include suggested labels in Path messages. The suggested label allows the upstream OXC to tell its downstream neighbor from which interface it would like to receive traffic. When an OXC transmits a Path message bearing a suggested label, it begins configuring its own switch fabric in anticipation of being able to receive data on the suggested interface. When the OXC receives a Resv message from its downstream neighbor, it checks the label contained in the message. If this label matches the suggested label, the OXC continues configuring itself; otherwise it aborts the configuration changes associated with the suggested label and begins configuring its switch fabric in accordance with the label mapping contained in the Resv message.

To further expedite the lightpath creation process, OXCs can include an upstream label object in the Path message. This object informs the downstream neighbor which interface it should use for the reverse direction counterpart to the lightpath that is being actively signaled. An example of label suggestion and upstream labels appears in Figure 9.6. In the figure, each ordered pair $(p, \lambda)$ denotes a particular wavelength $\lambda$ on a particular fiber port $p$. OXC A begins by sending a Path message to OXC B that suggests transmitting on $\lambda_2$ on OXC B's port 1 for the upstream path and assigns $\lambda_1$ on OXC's port 6 for the upstream path. OXC A begins configuring its switch fabric while OXC B sends a Path message to OXC C that assigns $\lambda_1$ on port 7 of OXC B for the upstream path and suggests $\lambda_2$ on OXC C's port 1 for the downstream path.

OXC C is unable to honor the suggestion and instead assigns $\lambda_3$ on its port 2, forcing OXC B to restart its configuration process. OXC B confirms OXC A's suggestion in a Resv message, and both lightpaths are established.

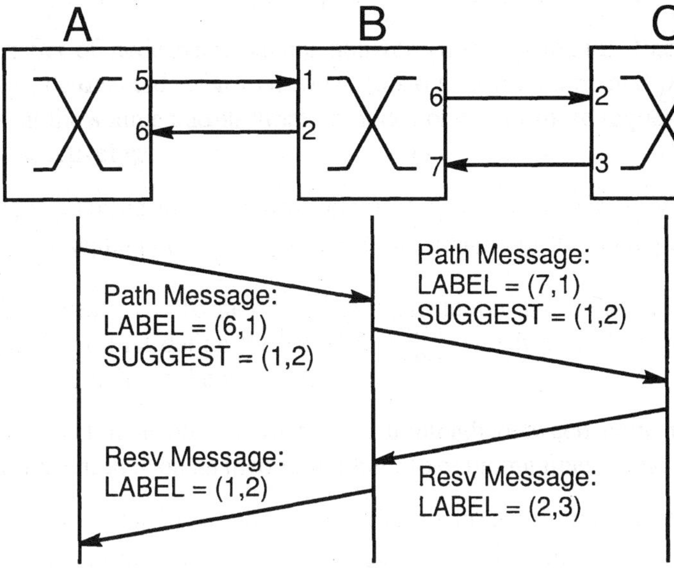


*Figure 9.6.* Label suggestion and upstream labels for path establishment.

An important issue associated with lightpath setup is the problem of contention for resources by multiple LSPs. For an example of this consider Figure 9.6, and suppose that OXC B sends a Path message to OXC A suggesting that $\lambda_2$ on port 1 be used for the reverse path, which collides with OXC A's upstream label assignment. GMPLS uses a simple mechanism to resolve contention issues; the node whose node_ID is greater wins the contention. Assuming OXC A's node_ID is greater, OXC B is forced to use a different label.

## 9.4 GMPLS Routing

One of the most important functions of an optical network management system is determining which resources should be assigned to support a new traffic flow. GMPLS uses IP routing with extensions for traffic engineering to carry out this function, using constrained routing to account for restrictions that the physical layer or network operator may impose (e.g., routing diversity for paths that share a common protection resource). Both the Open Shortest Path First (OSPF) [16] and Intermediate System-to-Intermediate System (IS-

IS) routing protocols were extended to support GMPLS routing features [17]; we focus on OSPF-TE in this section.

### 9.4.1 OSPF-TE

The OSPF-TE extensions use the concept of opaque Link State Advertisements (LSAs), which are defined in RFC 2370 [18]. Three types are defined, and in each case the Type Value field determines the flooding scope. Opaque LSAs with a Type Value of 9 are flooded only on a local link or subnet. Opaque LSAs whose Type Value is 10 are flooded within a single area in an Autonomous System (AS), while those whose Type Value is 11 are flooded throughout an entire AS. Because the Type Value field is used to indicate flooding scope, the Link State ID field is used to differentiate the various types of opaque LSAs. The first eight bits of the Link State ID contain the Opaque Type field, while the other 24 bits contain the Opaque ID, which serves the same function as the Link State ID Field in non-opaque LSAs. Opaque LSAs are flooded only to routers that are capable of understanding them; each router's knowledge of the ability of its neighbors to understand opaque LSAs is created during the database exchange process.

OSPF-TE [19] includes a new Traffic Engineering LSA for carrying attributes of routers (and switches), point-to-point links, and connections to multi-access networks. The TE LSA has an LS Type value of 10, so it is flooded only within areas, not over an entire AS. Network LSAs (Type 2 LSAs) are also used in TE routing calculations. Instead of the Link State ID that typically appears in the LSA header, the TE LSA uses an LSA ID that consists of 8 bits of type information and a 24-bit Instance field that is used to allow a single system to maintain multiple TE LSAs. In all other respects, the TE LSA header is identical to the standard LSA header. The body of the TE LSA consists of a single object that conforms to the Type/Length/Value (TLV) format. The OSPF-TE draft defines two such objects, which are the Router Address TLV and the Link TLV. Each TLV contains one or more nested sub-TLVs. The Router Address TLV specifies a stable IPv4 address associated with the originating node that is reachable if the node has at least one active interface to the rest of the network. This address must not become unreachable if a set of interfaces goes down.

Nine Link TLV sub-TLVs are defined in the OSPF-TE extensions draft. They are Link Type, Link ID, Local Interface IP Address, Remote Interface IP Address, TE Metric, Maximum Bandwidth, Maximum Reservable Bandwidth, Unreserved Bandwidth, and Administrative Group. The Link Type and Link ID sub-TLVs must appear exactly once in the Link TLV. The Link Type sub-TLV currently identifies the link as point-to-point or multi-access, but it can support up to 255 link types. The Link ID sub-TLV carries a 32-bit field that identifies the entity to which the sourcing node is connected via the link in question; if the connection is a point-to-point link, it is the Router ID of the neighboring node.

The other sub-TLVs carried by the Link TLV are optional but none can appear more than once. Of particular interest are those related to GMPLS TE operations. The TE Metric sub-TLV carries a 32-bit TE link metric, which is not necessarily the same as the OSPF link metric. The TE metric could be the link distance in meters, for instance, which is an important consideration when determining reach for connections over transparent optical domains. The Maximum Bandwidth sub-TLV specifies the link capacity in units of bytes/sec using the 32-bit IEEE floating point format; the Maximum Reservable Bandwidth sub-TLV likewise uses a floating-point metric in units of bytes/sec to specify the maximum bandwidth that can be reserved on the link in the direction leading from the source router; this can be larger than the Maximum Bandwidth in certain cases (such as when the link is oversubscribed). The Unreserved Bandwidth sub-TLV consists of eight floating point values that describe the free bandwidth (in units of bytes/sec) that is available at each of eight priority levels, with level 0 appearing first in the sub-TLV and level 7 appearing last.

The GMPLS extensions draft [20] defines the following four additional sub-TLVs for the Link TLV: Link Local/Remote Identifiers, Link Protection Type, Interface Switching Capability Descriptor, and Shared Risk Link Group. We describe two of these sub-TLVs here. The Link Protection Type sub-TLV is used to indicate the type of recovery scheme that is being used to protect traffic on the advertised link, or to indicate that the advertised link is serving as a backup resource for protected traffic on another TE link. Protection information is encoded in a set of flags that are contained in the first eight bits of the sub-TLV. The Shared Risk Link Group (SRLG) sub-TLV is used to identify the set of SRLGs to which the advertised link belongs. A given TE link may belong to multiple SRLGs.

### 9.4.2 Link Bundling

The designers of the GMPLS architecture faced a significant scaling issue in the case of core optical networks, which consist of large switches that have hundreds of fiber ports, and whose fiber links may carry hundreds of wavelengths on each fiber. A pair of switches may be connected by dozens of fibers; it is impractical to transmit a separate LSA for each one of these links. In a large network, the control plane traffic overhead would be considerable.

Since the desire is to minimize message overhead, the concept of link bundling was introduced in the MPLS working group [21]. Link bundling extends the TE link concept by addressing a problem that occurs in the creation of some types of TE links. In some cases, the elements of a TE link can be unambiguously identified using a ⟨Link_ID, label⟩ ordered pair. An example of this would be a TE link associated with a fiber port of a FSC switch whose labels correspond to the wavelengths on the fiber. In other cases, the TE construct does not produce unambiguous resource identifiers. Consider the case where the FSC switch we just described has bundled all its fiber links into a single TE link, while using the same wavelength identifiers for labels at each port. In this case the link bundling construct is required to resolve the ambiguity. Elements of bundled links are identified by an ordered triple of the form ⟨Bundle_ID, Component_ID, label⟩. The set of component links that compose a bundle are a minimal partition of the bundle that guarantees an unambiguous meaning for each ⟨Component_ID, label⟩ ordered pair. Examples of link bundling are shown in Figure 9.7. In Figure 9.7(a), two FSC nodes connected by a large number of bidirectional parallel links aggregate all of them into a single TE link that can be advertised using OSPF-TE. In Figure 9.7(b), we show two nodes in a BLSR/2 network connected by two unidirectional fibers, where half of the bandwidth on each fiber is available for normal traffic and the other half of the bandwidth is reserved for protection. Since both nodes are LSC the network operator chooses to aggregate all the working wavelengths on each fiber into a single TE link and to do the same for all the protection wavelengths on each fiber.

All component links in a link bundle must have the same OSPF link type, TE metric, and resource class. There are 10 link bundle traffic parameters that can be advertised by OSPF-TE. In addition to describing the link type and metric, they also describe the current bandwidth usage and the interfaces at each end of the link bundle.

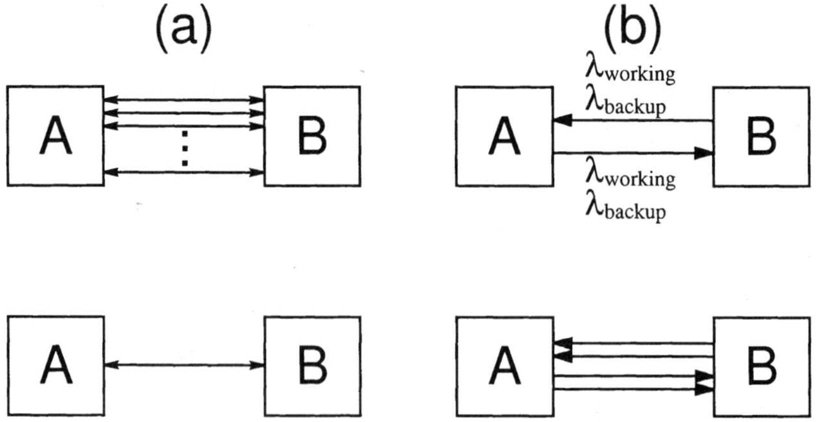


*Figure 9.7.* Two examples of link bundling.

### 9.4.3 Advertising Physical Layer Information in Routing Updates

Optical networks, because of the constraints imposed by the physical layer, are fundamentally different from the packet-switched networks for which routing protocols like OSPF were designed. The limitations on lightpath reach that result from linear and nonlinear fiber impairments, as well as coupling losses in transparent network elements, demand a more sophisticated approach to routing than the minimum hop calculations that are typically used to route IP packets. For this reason the IETF's IP/Optical (IPO) working group developed a set of input guidelines for extending link-state routing to optical networks [22]. This document, which at the time of this writing is in the RFC Editor's queue, does not specify how the relevant physical layer parameters are to be encoded in OSPF LSAs. Rather, it describes how a network operator may choose to encode information that the control plane can use to establish optical paths and lists some of the constraints that must be taken into consideration when doing so.

Physical impairments become an issue when we consider optical networks in which it is possible to create a light path where the distance between successive regeneration points is great enough to allow the optical signal quality to degrade below the minimum acceptable level. While this is less of a concern in networks that feature opaque optical switches, the transition to higher data rates and the introduction of transparent switching technologies (e.g., MEMS) means that this problem will become more acute in the future.

The draft [22] defines a domain of transparency to be an optical subnetwork in which amplification of optical signals may take place, but not regeneration. When a lightpath is created across the domain, it must be routed so that physical layer impairments are kept within acceptable limits. Transparent optical systems introduce both linear and nonlinear impairments to the signals that move through them. The draft considers how to encode linear impairments and does not recommend using LSAs to explicitly advertise information about nonlinear impairments, because they arise from complex interactions between different signals on a fiber. The best way to deal with nonlinear impairments may be to quantify the penalty that they impose and factor this into the optical link budget when computing lightpath routes.

Examples of linear impairments are polarization mode dispersion (PMD) and amplifier spontaneous emission (ASE). While PMD is not an issue in networks with widely deployed dispersion compensation devices and fibers, it can present a significant problem in legacy networks with inadequate compensation. In order for the PMD to fall within acceptable limits, the time averaged differential group delay (DGD) between two orthogonal polarizations must be less than the quantity $a/B$, where $B$ is the bit rate and $a$ is a parameter that depends on the margin associated with PMD, the outage probability, and the sensitivity of the receiver. As transmission rates increase, PMD will become more of an issue, particularly on legacy systems which may not have the compensating devices that are present in greenfield deployments. DGD is measured in units of psec/$\sqrt{\text{km}}$; the PMD on a fiber span of length $L$ kilometers, measured in psec, is found by taking $L \cdot \text{DGD}^2$. This parameter can be encoded and distributed by the routing protocol and used by the control plane to ensure that the PMD associated with particular lightpath is below the maximum acceptable value. The other impairment, ASE, increases the probability of bit error by introducing noise into the optical signal whenever it passes through an optical amplifier. It can be encoded for each link as the sum of the noise on the link's component fiber spans. Using this information, the control plane can choose a route such that the optical signal-to-noise ratio is greater than the minimum acceptable value as defined by the operator. Because PMD and ASE do not change rapidly over time, they could be stored in a central database. However, impairment values often are not known to the degree of precision that would be required to make accurate routing decisions. Hence, improved measurement and record-keeping capabilities are needed in order to properly use this information.

Alternatively, the network operator can use a maximum distance routing constraint, which depends on the data rate. Many carriers do this now and may choose to extend the practice to the transparent optical networks that they will deploy in the future. It can be used as the sole constraint if PMD is not an issue (either because the operator has deployed compensation devices or low-PMD fiber) and if each fiber span has the same length or the control plane handles variable-length spans by adjusting the maximum distance upper bound so that it reflects the loss associated with the worst span in the network. If this approach is used, the only information that needs to be disseminated by the routing protocol is the length of each link.

An additional issue involves routing of lightpaths to guarantee diversity, so that damage to the network's physical resources, whether caused by nature or humans (inadvertently or maliciously), has a minimal impact on the network operator's customers. To support this capability, standards development groups such as the IETF have employed the concept of the Shared Risk Link Group (SRLG), introduced in [23]. Two network elements belong to the same SRLG if they are physically co-located (e.g., two fibers that lie within a common fiber bundle would probably be assigned to the same SRLG). The use of SRLGs imposes an additional constraint when routing multiple lightpaths for a customer, namely that no pair of lightpaths should share any SRLGs.

The approach proposed in [22] is to advertise two parameters for each SRLG that capture the relationship between the links that compose the SRLG. Those two parameters are Type of Compromise and Extent of Compromise. The Type of Compromise parameter would capture the degree of closeness of the SRLG elements (e.g. shared cable, shared right of way (ROW), or shared office). The Extent of Compromise parameter would give the distance over which the closeness encoded by the Type of Compromise parameter exists.

## 9.5 Link Management Protocol

In order for the optical network to propagate signaling and routing messages, a set of control channels must be set up to link the various optical switches together and to facilitate the management of the TE data links that connect them. A key feature of GMPLS optical networks is that the control plane can be physically decoupled from the data plane by creating control channels that do not share any physical resources with the data channels that they supervise. In GMPLS, control channels can be created only when it is possible to terminate them electrically and when their endpoints are mutually reachable at the IP layer.

The Link Management Protocol (LMP), defined in [24], provides a mechanism for creating and managing multiple control channels between pairs of GMPLS nodes. It supports neighbor discovery by allowing connected nodes to verify the proper connection of their data links and to correlate those links' properties. LMP can also be used to support fault management, and thus plays a role in supporting protection and restoration capabilities in GMPLS optical networks.

### 9.5.1 LMP Operations

LMP operates by transmitting control messages that are encapsulated in UDP packets. Like many other protocols, LDP uses a common header that indicates the type of message being transmitted as well as the total length of the message. The message itself comprises multiple LMP objects, each of which has a standard TLV format. The first bit in each object is used as a flag to indicate whether the object is negotiable or non-negotiable.

Because LMP uses an unreliable transport layer protocol, it must incorporate a mechanism that guarantees that messages are properly received. LMP messages that initiate a process (e.g., the Config message that initiates control channel setup) carry a Message_ID object that contains a 32-bit sequence number that is incremented and wraps when it reaches its maximum value. When the recipient of a message that carries a Message_ID object sends a reply, the reply must carry a Message_ID_Ack object that contains the same 32-bit sequence number that the original Message_ID object carried.

**Control Channel Management.** In order for an LMP adjacency to exist between two nodes, there must be at least one functioning control channel established between the nodes. A control channel is bidirectional and consists of two unidirectional connections that are logically associated by the channel's endpoints. LMP creates control channels by exchanging messages that allow the nodes at the control channel's termination points to discover the IP addresses of the destination for each unidirectional control traffic flow. So that two nodes that share an LMP adjacency can distinguish between multiple control channels, each control channel destination point is assigned a unique 32-bit identifier by the node that owns it.

Four message types are used by LMP to create and maintain control channels. They are Config, ConfigAck, ConfigNack, and Hello. The first three messages are used to advertise and negotiate control channel parameters. The Hello message is used to support a fast keep-alive function that enables LMP to respond to control channel failures within the keep-alive decision cycle of the link-state routing protocol that the network is using. Hello messages do not need to be used if other mechanisms for detecting control channel failures (e.g., link layer detection) are available.

Control channel setup begins with the transmission of a Config message by one of the channel end points. In addition to identifying the control channel ID at the transmitting node, the Config message carries a set of suggested parameter values for the fast keep-alive mechanism. If the suggested values are acceptable, the receiving node will respond with a ConfigAck message that carries the 32-bit ID number for its control channel. Unacceptable parameter values are handled by transmitting a ConfigNack message with alternative parameter values that can be inserted into a new Config message to be sent by the original transmitting node as part of a new attempt. Once a ConfigAck message has been successfully received by the node that transmitted a Config message, the control channel is considered established, and both endpoints can begin transmitting Hello messages.

LMP also supports moving control channels to an inactive, or "down," state in a graceful manner. The common header for LMP messages contains a flag that can be set to indicate that the control channel is going down; a node that receives an LMP message with this flag set is expected to move the unidirectional control channel to the down state and may stop sending Hellos.

**Link Property Correlation.** The link property correlation function is used to guarantee consistency in TE link assignments between nodes that have an LMP adjacency. Link property correlation is initiated by transmitting a LinkSummary message over the control channel associated with the data links of interest. TE links can be identified using IPv4 or IPv6 addresses, or they can be unnumbered. Similar addressing options are used for data links. Data links are also identified by the interface switching type that they support, which we listed in Section 9.2.2, and by the wavelength that they carry. LinkSummary messages are also used to aggregate multiple data links into TE links or to modify, exchange, or correlate TE link or data link parameters. If the recipient of a Link Summary message, upon examining its database, finds that its Interface_ID mappings and link property definitions agree with the contents of the Link Summary message, it signals this by transmitting a Link Summary Ack message. If there is disagreement, a Link Summary Nack message is sent instead. If a node that sends a Link Summary message receives a Link Summary Nack message in reply, it is recommended that link verification, described the next subsection, should be carried out on those mappings that have been flagged as incorrect.

**Link Connectivity Verification.** The verification process begins when a BeginVerify message is transmitted over the control channel. The BeginVerify message establishes parameters for the verification session. The parameters include the time between successive Test messages, the number of data links that will be tested, the transport mechanism for the Test messages, the line rate of the data link that will carry the Test messages, and, in the case of data links that carry multiple wavelengths, the particular wavelength over which the Test messages will be transmitted.

The Test messages themselves are transmitted over the data link that is being tested, not any of the control channels. Test messages are transmitted repeatedly at the rate specified in the BeginVerify message until either a TestStatusSuccess message or a TestStatusFailure message is received from the destination node. The Test Status reply messages are repeated periodically until they are either acknowledged or the maximum number of retransmissions has been reached.

The IETF has also developed a draft that describes how LMP Test messages should be encoded for SONET/SDH systems [25]. In such an environment, Test messages can be sent over the Data Communications Channel (DCC) overhead bytes or sent over the control channel and correlated with a test pattern in the J0/J1/J2 Section/Path overhead bytes.

**Fault Management.** LMP can also be used to support protection and restoration operations through rapid dissemination of fault information over control channels. The actual detection of failures occurs at a lower layers; in a network of transparent optical switches, for instance, failures are detected by a loss of light (LOL) indication. LMP's fault management mechanism is decoupled from the network's fault detection mechanisms, and so will work over any opaque or transparent network.

LMP propagates fault information by using the ChannelStatus message. Each node that detects a fault transmits a ChannelStatus message to its upstream neighbor. Each upstream node responds by transmitting a ChannelStatusAck message to the sending node. Nodes that receive ChannelStatus messages correlate the failure information contained within. If the upstream node is able to isolate the failure, then signaling can be used to initiate recovery mechanisms.

### 9.5.2 LMP-WDM

In large-scale WDM systems it is sometimes desirable to manage the connections between an optical switch and the optical line system (OLS), which is responsible for transmitting and receiving photonic signals. The Common Control and Measurement Plane (CCAMP) working group has devised a set of extensions to allow LMP to run between the switch and the OLS [26]. In Figure 9.8 we show an example of LMP and LMP-WDM adjacencies.

The emphasis in the draft is on opaque OLSs (i.e., SONET/SDH and Ethernet ports); LMP-WDM can be extended to transparent switching but requirements for this are not yet clear. LMP-WDM supports the same four management functions as LMP. The draft defines a new LMP-WDM_CONFIG object that carries two flags indicating support for LMP-WDM extensions and whether the sender is an OLS. It also defines additional Data Link sub-objects for use in the Link Summary message that convey additional link information, such as the SRLGs that are associated with a particular data link, estimates of the bit error rate (BER) on a data link, and the length of the physical fiber span of the link. The draft also defines a new object to be used in the Channel Status message for fault localization.

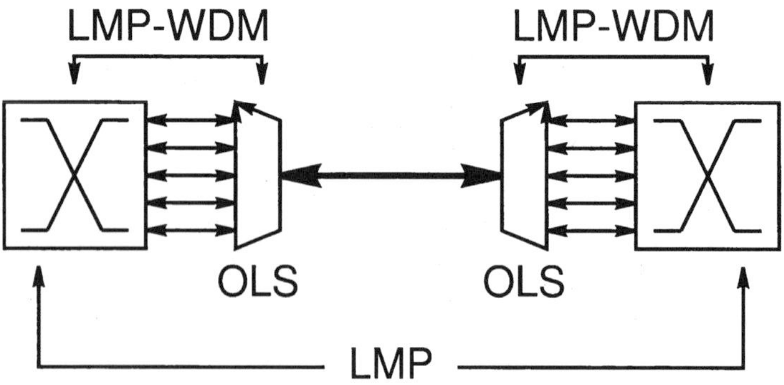


*Figure 9.8.* LMP and LMP-WDM adjacencies.

### 9.5.3 LMP for Protection and Restoration

Recently, the CCAMP working group received a proposal [27] to use LMP to propagate fault information using the same type of flooding mechanism that OSPF uses to propagate link state advertisements. The proposal defines two new LMP messages, FaultNotify and FaultNotifyAck. When a node detects a failure, it sends FaultNotify messages to each neighbor with which it shares in LMP adjacency. The FaultNotify message carries either the identifier of the failed link or a set of SRLG identifiers as well as the ID of the node that is sending the message.

The node that receives a FaultNotify message checks to see whether the failed entities identified in the message already exist in a database of failed network entities that each node would be required to maintain. If all of the failed elements are already represented in the database, the receiving node does not forward the message; otherwise it transmits a copy of the FaultNotify message to each of its neighbors. In either event, the receiving node sends a FaultNotifyAck message to the sender.

If a node that receives a FaultNotify message determines from the information that it contains that it lies on the recovery path for the traffic affected by the reported failures, it independently reconfigures its switching matrix to support the recovery path. A potential problem with this approach is the misdirection of extra traffic to unintended destinations when a switch in the middle of a protection path reconfigures itself. It is also not clear how this proposed approach will perform relative to the canonical failure reporting mechanism for GMPLS, which is targeted signaling that is confined to the working and recovery paths. These issues are being debated in the CCAMP working group at the time of this writing.

## 9.6 Relationship to OIF and ITU-T Work

It must be kept in mind that the primary focus of the IETF with respect to optical networks is on developing extensions to IP protocols to enable IP to work well over those new types of networks. The network architecture itself is the province of the International Telecommunications Union (ITU), which has, over the past several years, generated framework recommendations that define the Automatically Switched Optical Network (ASON) [28] and Automatically Switched Transport Network (ASTN) [29]. The ITU has also developed recommendations that define the signaling network [30], routing, [31], distributed connection management [32], and automatic discovery [33].

In order to harmonize the work in these two groups, the ITU and IETF have a formal liaison relationship in which elected representatives from each group attend the others meetings and provide updates and status reports that are used, along with formal written communications, to coordinate work and request information or the undertaking of certain actions.

The other major SDO that is working in this area is the Optical Internetworking Forum (OIF). The OIF has already produced an implementation agreement that defines signaling, addressing, and other aspects of the User-Network Interface (UNI) [34]. Rather than defining new protocols, the OIF has always sought to use the machinery developed by groups such as the IETF and ITU and adapt it, where necessary, to meet the requirements enumerated by major carriers and service providers. The OIF is currently engaged in two major implementation agreement development efforts. The first, UNI 2.0, aims to define new services that will be supported across the UNI. The second effort is to develop signaling and routing for intra-carrier Network-Network Interfaces (NNIs). Both of these projects will draw extensively from the output of the IETF and the ITU.

## 9.7 Summary

In this chapter we described the overall structure of the GMPLS control plane and the major protocol building blocks that are required to implement it. We examined the many extensions to existing IP-related protocols such as RSVP and OSPF that were required to enable traffic engineering and path management. We examined the Link Management Protocol, which was created specifically for managing control channels and verifying link connectivity between optical network elements. We also considered how the IETF's work in this area has affected the work of other standards bodies and how various conflicts and issues are being addressed.

## References

[1] Mannie, E., Ed., "Generalized Multi-Protocol Label Switching Architecture," IETF Internet Draft.

[2] Banerjee, A., Drake, J., Lang, J. P., Turner, B., Kompella, K., and Rekhter, Y., "Generalized Multiprotocol Label Switching: an Overview of Routing and Management Enhancements," *IEEE Communications Magazine*, vol. 39, no. 1, pp. 144–150, January 2001.

[3] Davie, B. and Rekhter, Y., "MPLS: Technology and Applications," Morgan Kaufmann Publishers, San Francisco, 2000.

[4] The Optical Internetworking Forum, "User Network Interface (UNI) 1.0 Signaling Specification," http://www.oiforum.com/public/documents/OIF-UNI-01.0.pdf, October 2001.

[5] "Digital Subscriber Signalling System No. 2 - User-Network Interface (UNI) layer 3 specification for basic call/connection control," ITU-T Rec. Q.2931, 1995.

[6] E. Rosen, D. Tappan, G. Fedorkow, Y. Rekhter, D. Farinacci, T. Li, and A. Conta, "MPLS Label Stack Encoding," IETF RFC 3032, January 2001.

[7] P. Agarwal and B. Akyol, "Time To Live (TTL) Processing in Multi-Protocol Label Switching (MPLS) Networks," IETF RFC 3443, January 2003.

[8] "Generalized Multi-Protocol Label Switching (GMPLS) Signaling Functional Description," L. Berger, Ed., IETF RFC 3471, January 2003.

[9] P. Ashwood-Smith, Ed., "Generalized Multi-Protocol Label Switching (GMPLS) Signaling: Constraint-based Routed Label Distribution Protocol (CR-LDP) Extensions," IETF RFC 3472, January 2003.

[10] Awduche, D., Berger, L., Gan, D., Li, T., Srinivasan, V., and Swallow, G., "RSVP-TE: Extensions to RSVP for LSP Tunnels," RFC 3209, December 2001.

[11] L. Berger, Ed., "Generalized Multi-Protocol Label Switching (GMPLS) Signaling: Resource ReserVation Protocol-Traffic Engineering (RSVP-TE) Extensions," IETF RFC 3473, January 2003.

[12] L. Andersson and G. Swallow, "The Multiprotocol Label Switching (MPLS) Working Group decision on MPLS signaling protocols," IETF RFC 3468, February 2003.

[13] Berger, L. and Rekhter, Y., Eds., "Generalized MPLS Signaling–Implementation Survey," IETF Internet Draft.

[14] R. Braden, Ed., "Resource ReSerVation Protocol (RSVP) – Version 1 Functional Specification," IETF RFC 2205, September 1997.

[15] Lee, C-Y., Farrel, A., and De Cnodder, S., "Exclude Routes–Extension to RSVP-TE," IETF Internet Draft.

[16] Moy, J.T., "OSPF: Anatomy of an Internet Routing Protocol," Addison-Wesley, 1998.

[17] Kompella, K., and Rekhter, Y., Eds., "Routing Extensions in Support of Generalized MPLS," IETF Internet Draft.

[18] Colt, R., "The OSPF Opaque LSA Option," IETF RFC 2370, July 1998.

[19] Katz, D., and Kompella, K., Eds., "Traffic Engineering Extensions to OSPF Version 2," IETF Internet Draft.

[20] K. Kompella, Ed., "OSPF Extensions in Support of Generalized MPLS," IETF Internet Draft.

[21] Kompella, K., Rekhter, Y., and Berger, L., Eds., "Link Bundling in MPLS Traffic Engineering," IETF Internet Draft,

[22] J. Strand and A. Chiu, "Impairments and Other Constraints on Optical Layer Routing," IETF Internet Draft.

[23] "Smart Routers–Simple Optics: An Architecture for the Optical Internet," Hjálmtýsson, G., Yates, J., Chaudhuri, S., and Greenberg, A., *IEEE/OSA Journal of Lightwave Technology*, vol. 18, no. 12, pp. 1880–1891, December 2000.

[24] J. Lang, Ed., "Link Management Protocol (LMP)," IETF Internet Draft.

[25] J. P. Lang and D. Papadimitriou, Eds., "SONET/SDH Encoding for Link Management Protocol (LMP) Test Messages," IETF Internet Draft.

[26] A. Fredette, Ed., "Link Management Protocol (LMP) for Dense Wavelength Division Multiplexing (DWDM) Optical Line Systems," IETF Internet Draft.

[27] Soumiya, T. and Rabbat, R., Eds., "Extensions to LMP for Flooding-Based Fault Notification," IETF Internet Draft.

[28] "Architecture for the Automatically Switched Optical Network (ASON)," ITU-T Rec. G.8080/Y.1304, November 2001.

[29] "Requirements For Automatic Switched Transport Networks (ASTN)," ITU-T Rec. G.807/Y.1301, July 2001.

[30] "Architecture and specification of data communication network," ITU-T Rec. G.7712/Y.1703, March 2003.

[31] "Architecture and Requirements for Routing in the Automatic Switched Optical Networks," ITU-T Rec. G.7715/Y.1706, June 2002.

[32] "Distributed Call and Connection Management (DCM)," ITU-T Rec. G.7713/Y.1704, November 2001.

[33] "Generalized automatic discovery techniques," ITU-T Rec. G.7714/Y.1705, November 2001.

[34] Jones, J., "User-Network Interface (UNI) 1.0," *Optical Networks Magazine*, vol. 4, no. 2, pp. 85–93, March/April, 2003.

Chapter 10

# OPERATIONAL ASPECTS OF MESH NETWORKING IN WDM OPTICAL NETWORKS

Jean-Francois Labourdette, Eric Bouillet and Chris Olszewski
*Tellium, Inc.*
Email: labourdette@ieee.org, ebouillet@ieee.org, cjo496@yahoo.com

**Abstract** Networks that transport optical connections using Wavelength Division Multiplexing (WDM) systems and route these connections using intelligent optical cross-connects (OXCs) are being established as the core constituent of next generation long-haul networks [2, 3, 4, 5, 6, 7, 8, 9]. Actual national US networks implement shared mesh restoration using intelligent optical switches that protect against single link and node failures [17]. As mesh networks are replacing ring-based networks, operators are faced with changes in the way they operate their networks. The increased efficiency and robustness of mesh networks, compared to ring-based networks, have to be managed appropriately to actually achieve those benefits. In this work, we discuss four key economic and operational benefits from mesh optical networking. They are (1) end-to-end routing and provisioning, (2) fast and capacity-efficient restoration, (3) re-provisioning, and (4) re-optimization. They contribute in various ways to network cost savings and higher network and service reliability in mesh networks compared to traditional ring-based networks.

**Keywords:** Mesh networking, optical networking, fast restoration, survivable networks.

## 10.1 Introduction

Optical transport networks have been traditionally based on ring architectures, relying on standardized ring restoration protocols such as SONET BLSR and UPSR to achieve fast restoration (50 msec). However, many carrier networks are indeed meshed. Core optical networks consist of backbone nodes interconnected by point-to-point WDM fiber links in a mesh interconnection pattern. Each WDM fiber link carries multiple wavelength channels (e.g. 160 OC-192 channels). Transmission rates of wavelength channels on long-haul WDM systems are currently evolving from OC-48 to OC-192, and are expected to evolve to OC-768 in the future. Multiple conduits (each containing multiple fibers) usually terminate at the backbone nodes from adjacent nodes. Figure 10.1 illustrates a core optical network. Diverse edge equipments, such as IP routers, FR/ATM switches, and Multi Service Provisioning Platforms (MSPPs) are connected to an optical switch. For discussion of the value of connecting an IP backbone network over a reconfigurable optical layer, see for example [30, 31]. Given this realization, and the many operational drawbacks of ring-based networks (such as provisioning across multiple rings, stranded capacity, and inflexibility to handle traffic forecast uncertainties), it is therefore not a surprise that carriers have evolved, or are considering evolving, their core network architecture from ring to mesh. Such mesh networks are based on next-generation Optical Cross-Connects (OXCs) that support fast, capacity-efficient shared path restoration [1]. There has also been interest in intermediate architectures (for example, those improving on traditional span-based restoration [20] or based on "p-cycles" [44]) that also try to achieve the speed of ring restoration in ways more closely aligned with traditional ring architectures, such as shared span restoration.

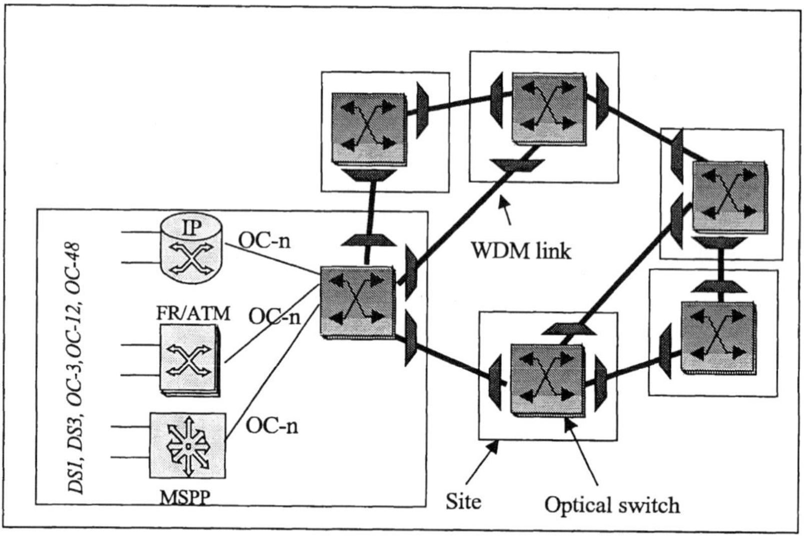


*Figure 10.1.* Optical mesh network.

As mesh networks are replacing ring-based networks, operators are faced with changes in the way they operate their networks. The increased efficiency and robustness of mesh networks, compared to ring-based networks, have to be managed appropriately to actually achieve those benefits. In this paper, we discuss four key economic and operational benefits from mesh optical networking. They are (1) end-to-end routing and provisioning, (2) fast and capacity-efficient restoration, (3) re-provisioning, and (4) re-optimization. They contribute in various ways to network cost savings and higher network and service reliability in mesh networks compared to traditional ring-based networks.

## 10.2 Capacity-Efficient End-to-End Routing and Provisioning

In mesh networks, routing and provisioning of connections, or lightpaths, are carried out on an end-to-end basis by computing and establishing a primary path, and a back-up path in case of protected service [8, 9]. In the case of dedicated mesh protection, the back-up path is dedicated to protecting a unique primary path. With shared mesh restoration (detailed in Figure 10.2), backup paths can share capacity according to certain rules. In both cases, the back-up paths are pre-established, and are unique, or failure independent. Restoration is guaranteed in case of a single failure by ensuring that primary and back-up paths are mutually diverse, and in the case of shared back-up paths, by ensuring that sharing is only allowed if the corresponding primary paths are mutually diverse. Then, the primary paths cannot fail simultaneously in case of a single failure, preventing any contention for the back-up capacity. This is different from dynamic re-provisioning where a back-up path is computed and established after the failure occurs, and after the failure has been localized, using any remaining available capacity. Dynamic re-provisioning can still complement dedicated mesh protection and shared mesh restoration by providing a means, albeit slower, to recover in case of multiple failure scenarios (discussed in Section 3).

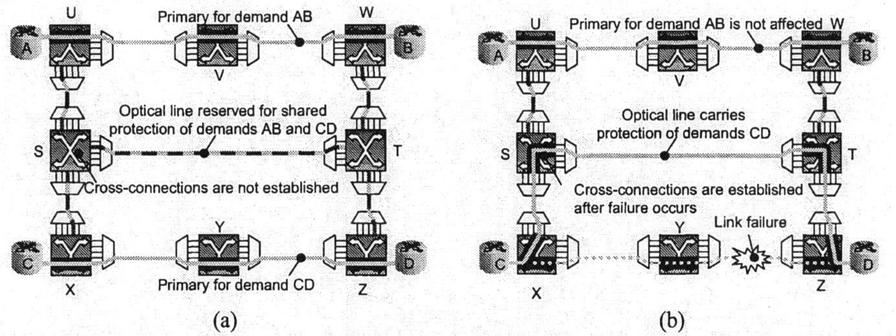


*Figure 10.2.* Shared mesh restoration: (a) Network connections before a failure occurs; (b) Network connections after recovery from the failure.

Diversity of routes in a mesh optical network is defined using the notion of Shared Risk Link Groups [8, 26]. A set of optical channels that have the same risk of failure is called a Shared Risk Link Group (SRLG). SRLGs are configured by network operators with the knowledge of the physical fiber plant of the optical network. Most SRLGs can be expressed as one or a combination of three possible primary types [8]. We describe them in Figure 10.3(b). The default, and most conventional type, is type a) in the figure, which associates an optical channel risk failure with a fiber cut. Another type of SRLG very likely to be encountered is type b). This type is typical of fibers terminating at a switch and sharing the same conduit into the office; that is, a conduit cut would affect all the optical channels terminating at the switch. Types a) and b) can be characterized in a graph representation as pictured by a') and b'). For instance, the removal of the edge in a'), or the middle node in b'), disconnects all the nodes which is tantamount to an SRLG failure in each case. Using these elementary transformations it is possible to model the network as a graph onto which established shortest-path operations can be applied. Other types of SRLG, such as c) in the example, are the most difficult kind to model and to provide diverse routing for. They occur in a few instances, such as fibers from many origins and destinations routed into a single submarine conduit, or dense metropolitan areas. Contrary to type a) and b), there is no convenient way to graphically represent type c) SRLGs and their presence can increase dramatically the complexity of the SRLG-diverse routing problem. A naive representation of the type c) SRLG would be to present it as graph c'). Such a representation, however, is erroneous, as it introduces additional paths not present in the original network topology, which could lead us to routing computations that are not physically feasible.

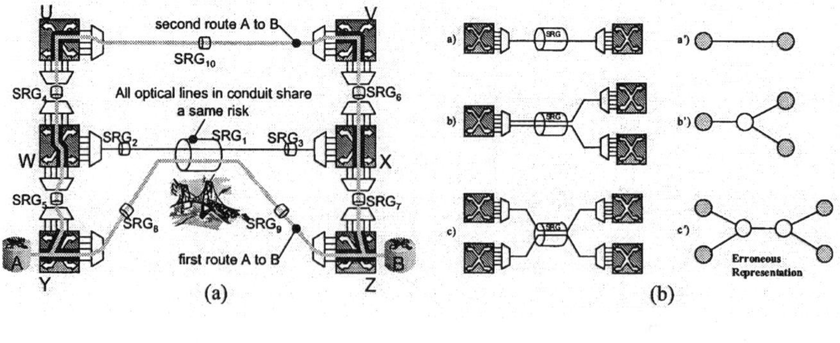


*Figure 10.3.* (a) Shared Risk Link Groups (SRLGs); (b) SRLG classification.

Appropriate diversity of primary and back-up paths, as well as sharing rules of back-up capacity, guarantee successful restoration in case of single link/SRLG failures, or single link/SRLG-and-node failures (with more restrictive rules). Compared to traditional ring-based protection or dedicated mesh protection, shared mesh restoration allows considerable saving in terms of capacity required [8, 9, 21]. Figure 10.4 shows some illustrative results for a 75 node network taken from [26]. In addition, the backup resources can be utilized for lower priority pre-emptable traffic in normal network operating mode. Note that the restoration of shared mesh protected lightpaths may be slower than with dedicated protection, essentially because it involves signaling and path-setup procedures to establish the backup path, yet is still within the realm of SONET 50 msec restoration times [14, 32]. In particular, the restoration time will be generally proportional to the length of the backup path and the number of hops, and if recovery latency is an issue this length must be kept under acceptable limits. However this constraint may increase the cost of the solutions, as it is sometime more cost-effective to use longer paths with available shareable capacity than shorter paths where shareable capacity must be reserved, as shown in Section 2.1. An appropriate cost model in the route computation algorithm [10, 11, 12] can handle this tradeoff.

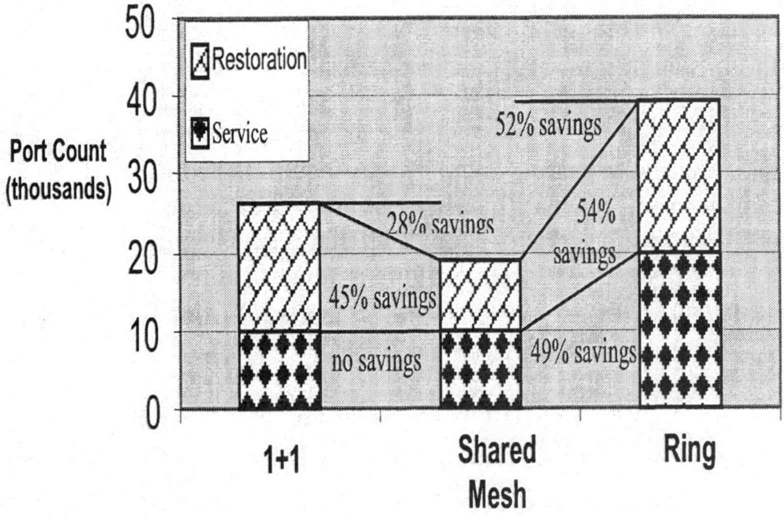


*Figure 10.4.* Dedicated mesh (1+1), shared mesh, and ring configuration comparison.

### 10.2.1 On-line routing algorithm

On-line routing handles requests as they are received, and as a result, lightpaths are provisioned based on the current state of the network, without considering any future requests. In this discussion of on-line routing, assume that a $k$-shortest path approach is used for computing both the primary and back-up paths (see [8] for discussion on the complexity of the on-line routing problem). For routing purposes, the algorithms utilized in an intelligent optical mesh network use cost models that assign weights to links in the network. The policy used for assigning weights to the links is different for primary and backup path computation. The weight assigned to a link for computing a primary path is typically a user-defined cost that reflects the real cost of using a channel on that fiber. The weight assigned to a link $e$ for computing a backup path is a function of the primary path [10, 11, 12] as follows:

- infinite weight if link $e$ intersects with an SRLG used by the primary path,
- weight $w_e$ if new capacity is required on link $e$ to provision the path, and
- weight $\epsilon w_e$, $0 \leq \epsilon \leq 1$, if the path can share existing capacity reserved on link $e$ for other pre-established backup paths.

The routing of each lightpath attempts to minimize the total cost of all channels in the lightpath route, i.e., the goal is to share the existing capacity amongst multiple backup paths while keeping their length under control. Setting $\epsilon$ to zero favors solutions that take the most advantage of sharing, but may lead to longer paths. In contrast, setting $\epsilon$ to one gives priority to shorter backup paths, and thus solutions that are capable of faster restoration times, but it does not attempt to maximize sharing and may thus require more backup capacity. Experiments have shown that a value of $\epsilon$ in the range $0.2$ to $0.4$ achieves a good tradeoff in terms of average backup path lengths versus capacity requirements over a broad range of networks, varying in number of nodes, fiber connectivity, and demand load. This observation is illustrated in Figure 10.5 on a sample network.

## 10.2.2 Capacity Efficiency of Mesh Routing

Simulation experiments were run on two networks. $N17$ is a 17-node, 24-edge network that has a degree distribution of (8,6,1,2) nodes with respective degrees (2,3,4,5). $N100$ is a 100-node, 137-edge network that has a degree distribution of (50,28,20,2) nodes with respective degrees (2,3,4,5). These are realistic topologies representative of existing networks. It is assumed that these networks have infinite link capacity, and that SRLGs comprise exclusively all the optical channels in individual links. In network $N17$, demand is uniform, and consists of two bi-directional lightpaths between every pair of nodes. That amounts to 272 lightpaths. In network $N100$, 3278 node-pairs out of 4950 possible node pairs are connected by one bi-directional lightpath. For each network, five protection/restoration architectures are considered: no

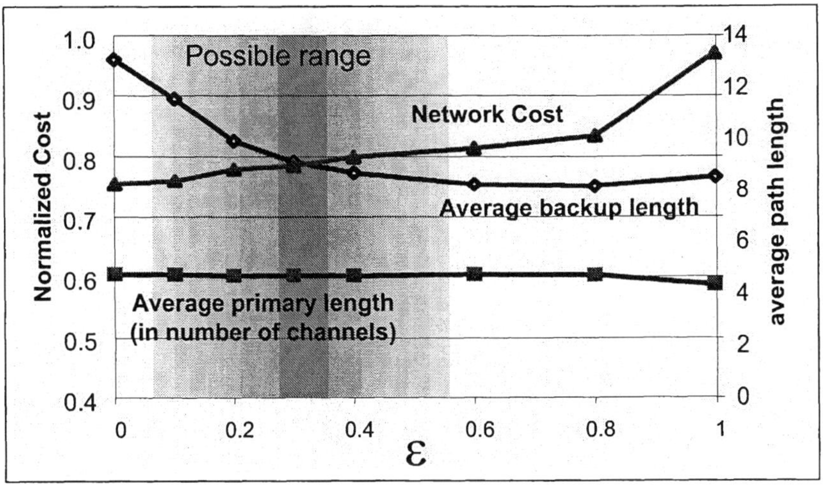


*Figure 10.5.* Impact of cost model ($\epsilon$) on average path length and capacity requirement.

protection; dedicated mesh protection with guaranteed recovery from single link failures; dedicated mesh protection with guaranteed recovery from single link-and-node failures; shared mesh restoration with guaranteed recovery from single link failures; shared mesh restoration with guaranteed recovery from single link-and-node failures. Shared mesh back-up paths are computed using a weight of 0 ($\epsilon = 0$) in order to make the maximum use of sharing. Requests for lightpaths arrive one at the time (on-line routing) in a finite sequence. The order is arbitrary but common to all architectures within each network to ensure a fair comparison. Figures of merit are capacity requirements separated into their primary and protection/restoration parts, and expressed in units of bi-directional OC-48 channels.

Results are presented in Figure 10.6 and Figure 10.7. The quantities shown on the charts are averaged over series of 10 experiments using various demand arrival orders. These results indicate that link-disjoint and node-disjoint dedicated mesh protection approaches (which ensure guaranteed recovery against single link, respectively node and link, failures) consume approximately the same total amount of capacity. This is expected since dedicated protection against link failures protects against node failures as well for nodes up to and including degree three, and these nodes constitutes a majority of the nodes in these two networks. This property does not apply to shared mesh restoration and to lightpaths that must be protected against single node failures. Mainly because of this, shared mesh restorable lightpaths that recover from single link-and-node failures use more resources (relatively to the dedicated schemes) than those that recover from single link failures. The difference however should be

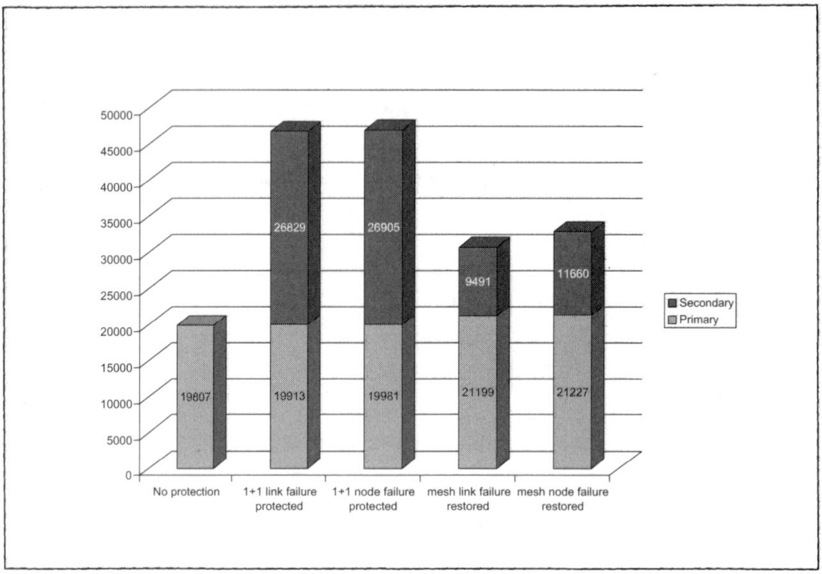


*Figure 10.6.* Comparison of capacity usage for different protection and restoration architectures (17 nodes).

considered in light of the benefits of protecting the network against single node failur

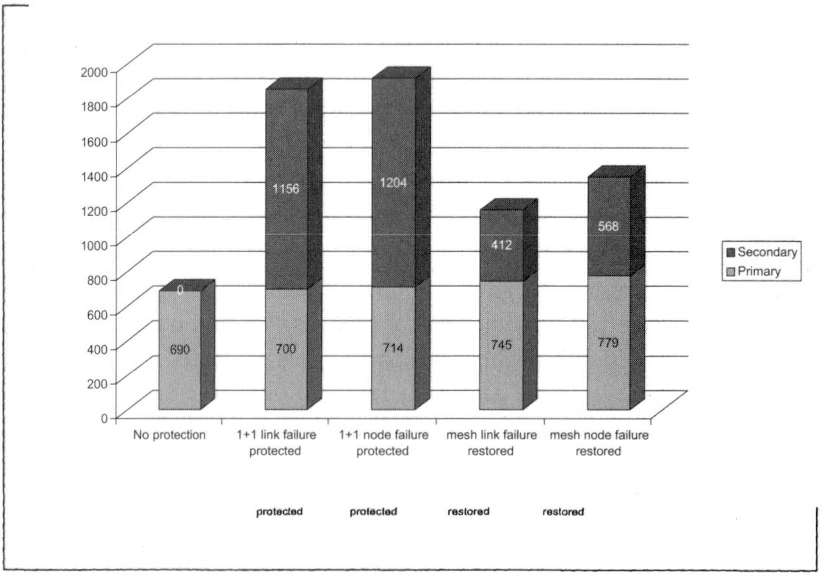


*Figure 10.7.* Comparison of capacity usage for different protection and restoration architectures (100 nodes).

| # of lightpaths using a given shared channel | 1 | 2 | 3 | 4 | 5 | 6 | 7 | 8 | 9 |
|---|---|---|---|---|---|---|---|---|---|
| # of occurrences | 10 | 12 | 6 | 6 | 18 | 7 | 1 | 8 | 4 |
| # of lightpaths using a given shared channel | 10 | 11 | 12 | 13 | 14 | 15 | 16 | 17 | 18 |
| # of occurrences | 4 | 2 | 0 | 1 | 4 | 2 | 2 | 0 | 1 |

*Table 10.1.* Distribution of number of lightpaths protected by shared channels (average number of lightpaths per shared channel is about six).

### 10.2.3 Controlling the Amount of Sharing

If the routing algorithm does not discriminate among shared channels while routing a lightpath, some shared channels may protect a large number of lightpaths (although on average the number of lightpaths protected by a shared channel is small). Table 10.1 illustrates the distribution of the number of lightpaths protected by shared channels for a typical mesh optical network with 45 nodes and a demand of 80 lightpaths. One observes that there is a shared channel that protects 18 lightpaths, although, on the average, a shared channel protects about six lightpaths.

If a shared channel protecting a large number of lightpaths fails, then those lightpaths are at risk upon a subsequent failure on their primary paths. Similarly, in the single instance of a shared channel protecting 18 lightpaths in Table 10.1, if one of those lightpaths needs to be restored, the remaining 17 lightpaths are unprotected upon any subsequent failure affecting their primary path. They would have to be re-provisioned as discussed in Section 4. By limiting the sharing on protection channels, we limit the number of lightpaths that would experience re-provisioning (order of seconds) as opposed to restoration (order of 10's to 100's of msec) in cases of double failures. In this work, we analyze a scheme that limits the number of lightpaths protected by a shared channel. The goal is to eliminate the extreme cases of shared channels protecting a large number of lightpaths, while at the same time ensuring that the protection capacity does not increase significantly. In an approach that we refer to as capping, we set a hard limit on the number of lightpaths that can use a given shared channel as part of their back-up path. The routing algorithm considers only those shared channels that have not exceeded the limit (which we call $Cap$) of the number of lightpaths that they protect. We first show that a well-chosen limit can be robust to the network topology and demand pattern. Let

$$R = \frac{\textit{number of protection channels}}{\textit{number of working channels}}.$$

$L_{av}$ = *average number of lightpaths using a shared channel.*

$L_{max}$ = *maximum number of lightpaths using a shared channel.*

$h_w$ = *average number of working hops among all lightpaths.*

$h_p$ = *average number of backup hops among all lightpaths.*

Then, by definition [18], $R = (h_p/h_w)(1/L_{av})$ is inversely proportional to the average number of lightpaths using a shared channel. Then $R \geq (h_p/h_w)(1/L_{max})$. A limit of $L_{max} = 1$ in which each protection channel can protect at most one lightpath, is equivalent to dedicated mesh protection. Since the lower bound on $R$ is inversely proportional to $L_{max}$, we expect that with a sufficiently large choice of $L_{max}$, small changes in the value of $L_{max}$ will cause small changes in $R$, and therefore that the specific choice of the sharing limit will not impact the ratio $R$ of protection channels to working channels.

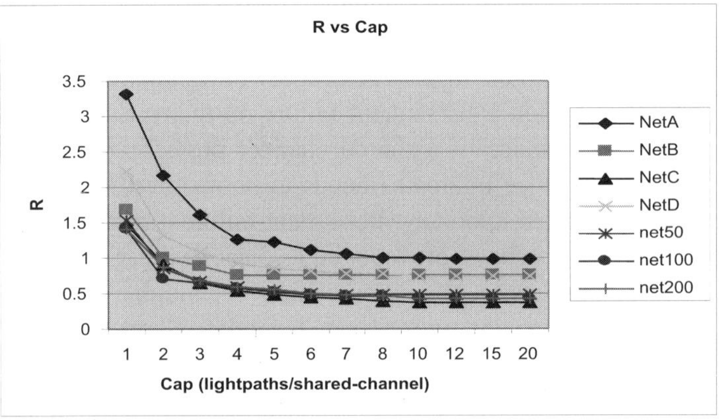


*Figure 10.8.* Ratio $R$ of protection to working capacity versus the sharing limit ($Cap$).

We conducted a set of experiments to study the impact and sensitivity of the sharing limit value $Cap$ on a variety of network parameters. The set of networks and demands in our experiments is a mix of representative real networks and demands, and randomly generated networks and demands. We considered four representative real networks: $netA$ (45 nodes), $netB$ (17 nodes), $netC$ (50 nodes), and $netD$ (100 nodes), and three randomly generated networks: $net50$ (50 nodes), $net100$ (100 nodes), and $net200$ (200 nodes). Figure 10.8

illustrates the ratio $R$ of protection channels to working channels for the different networks for different values of the sharing limit ($Cap = L_{max}$). We observe that as the sharing limit $L_{max}$ increases, $R$ decreases sharply at first, and then decreases gradually, and then remains flat. Most of the sharing gains are obtained when the sharing limit is below about five lightpaths. In many networks, beyond a sharing limit of 10 lightpaths, there are no incremental gains in protection capacity. Figure 10.8 also illustrates that $R$ is inversely proportional to the sharing limit. We observe that for larger networks, sharing saturates at larger values of the sharing limit. Based on these samples, a sharing limit of around six lightpaths appears to be robust across a variety of networks.

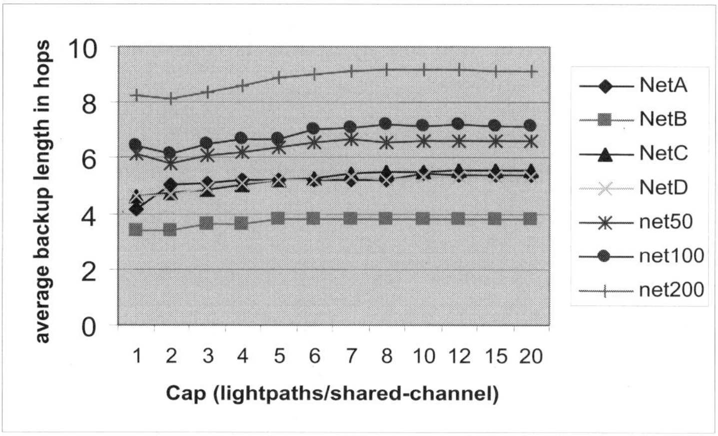


*Figure 10.9.* Average number of backup path hops versus the sharing limit ($Cap$).

Figure 10.9 illustrates the average number of hops on the backup paths as the sharing limit varies. Backup path hops directly influence the restoration time upon a failure, with a larger number of backup hops generally yielding a longer restoration time. We observe that as the sharing limit increases, the average number of backup hops, and thus the restoration times, increase marginally. This is because, as the sharing limit increases, there are more opportunities for sharing, and the backup path may traverse a longer distance trying to use links with sharable channels. Figure 10.10 plots the impact of the sharing limit on the percentage of lightpaths that are restored upon double failures for a typical network. The values are averaged over all double failures. Also plotted is the capacity requirement for each value of the sharing limit. We observe that as the sharing limit increases, the required capacity decreases (due to a decrease in protection capacity because of better sharing). However, due to better sharing, there are more contentions for capacity upon double failures, and as a result, the percentage of lightpaths whose restoration fails under double failure scenarios increases. The figure indicates that to achieve better resiliency against double failures, sharing limit needs to be decreased below five, consequently the capacity requirement increases.

Based on the experiments we conclude that capping the amount of sharing on back-up channels achieves the following (this is also consistent with more recent work [46]):

- Imposing a sharing limit on shared channels eliminates the cases of shared channels protecting a large number of lightpaths. When the sharing limit is well chosen, there is no capacity penalty.
- A sharing limit of around six lightpaths is robust across a variety of topologies and demand sets.
- Load-balancing among feasible shared channels on the same link is another way of limiting sharing and one that has a quantifiable, but non-guaranteed, effect [18].

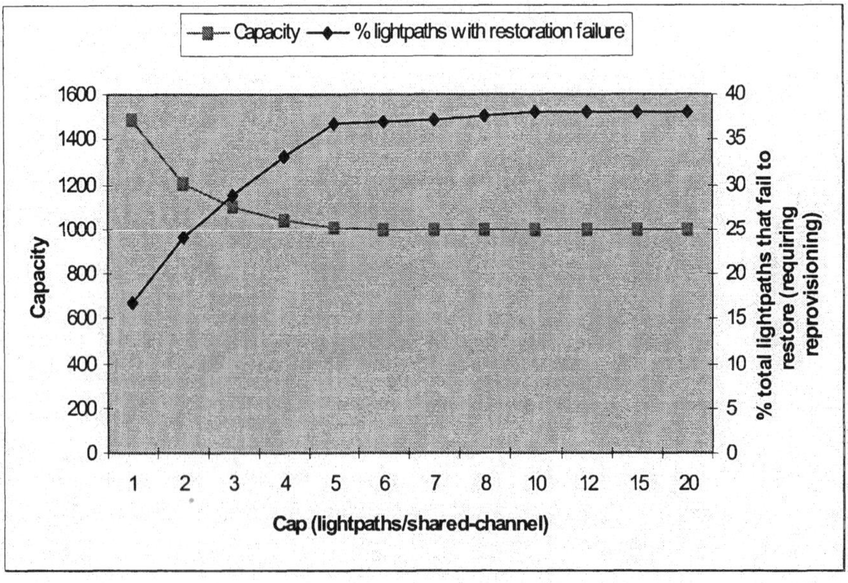


*Figure 10.10.* Impact of the sharing limit $Cap$ on restoration upon double failures.

### 10.2.4 Distributed Routing

Recently, distributed control architectures have been proposed for optical transport networks as a means to automate operations, enhance interoperability and scalability as well as facilitate the deployment of new applications, such as unified traffic engineering [33, 37]. In particular, the ability to provision connections/services dynamically and automatically through a distributed control plane has attracted a lot of interest. Efforts to standardize such a distributed control plane have reached various stages in several bodies such as the Internet Engineering Task Force (IETF) [34], International Telecommunications Union (ITU) [35] and Optical Internetworking Forum (OIF) [36]. The IETF is defining Generalized MPLS (GMPLS) which describes the generalization of MPLS protocols to control not only IP router networks but also various circuit switching networks including OXCs, photonic switches, ATM switches, and SONET/SDH systems. GMPLS extends MPLS signaling and Internet routing protocols to facilitate automatic service provisioning and dynamic neighbor and topology discovery across multi-vendor intelligent transport networks, as well as their clients. The OIF has assumed an overlay model of integration between optical transport networks and their clients [33] and defined an implementation agreement for the optical User-to-Network Interface (UNI) based on the GMPLS protocols. The ITU has also assumed the overlay model and is defining a distributed control plane for the Automatic Switched Optical Network (ASON) [34], and specifically the Network-to-Network Interface (NNI). This work is progressing in coordination with the OIF.

With the application of a distributed control plane, each network node participates in routing protocols that disseminate topology and link state information and is thus able to perform path computation for connection provisioning. In general, connections are signaled using explicit routes, i.e., the connection path is available at the ingress node and is carried in the connection establishment message. Connection requests may be initiated either by the management plane (EMS/NMS) or from a client directly connected to the optical network (via a UNI). In the former case, it is possible that the path is computed by the management plane and provided to the ingress node of the connection. Alternatively, a fully distributed path computation model could be supported where the ingress node is responsible for computing the explicit route.

Path computation must take into account various requirements and constraints, including bandwidth and delay constraints, recovery and survivability, optimal utilization of network resources. This requires dissemination of information about the network topology and various link attributes to every node in the network using routing protocols. GMPLS has extended traditional IP routing protocols such as OSPF and IS-IS, to support explicit path computation and traffic engineering in transport networks. In order to reduce overhead and improve routing scalability, however, typically only aggregated link information is disseminated via these routing protocols, leading to loss of information. For example, in the case of core optical switches with hundreds of ports, there may be multiple links between a pair of nodes. Links between the same pair of nodes, with similar characteristics, can be bundled together and advertised as a single link bundle or a traffic engineering (TE) link into the routing protocol. Therefore, unlike the (centralized) management plane, each network element does not have complete information about the network topology and link state. Clearly, path computation with complete information is expected to achieve higher efficiency compared to the distributed case. The challenge of distributed path computation is, therefore, to disseminate appropriately aggregated information so that the computed explicit paths meet all the requirements and constraints and incur minimal penalty in network utilization. While not addressed in the paper, distributed algorithms have been proposed and analyzed [38, 28, 40, 41] and shown to incur a relatively small efficiency penalty, while offering the benefits of distributed path computation.

## 10.3 Fast Mesh Restoration

Contrary to ring-based architectures where restoration speed on a given ring is relatively well-known, restoration speed in mesh networks will vary, as a function of the network size, routing of lightpaths (both primary and back-up paths), loading of the network, and failure location. It is therefore critical to possess tools that allow network operators to estimate the restoration times that would be experienced in a live network. Such modeling tools are used in conjunction with network dimensioning, planning and optimization algorithms and tools [22, 24, 29] in an iterative way to validate the restoration performance of a network design, or modify the design to insure the desired restoration performance will be achieved (see Figure 10.11).

The restoration performance modeling tool models the restoration protocols and signaling implemented in the actual optical switch, and can use a discrete event simulation engine to determine restoration times after a failure. Furthermore, the tool can be calibrated against lab results obtained from failures of a few tens of lightpaths, or against simulated failures in the field. The restoration simulation studies involve failing single conduits, which result in the simultaneous failures of the multiple primary lightpaths that traverse these conduits. The maximum restoration times correspond to the last lightpath restored as a result of a conduit failure (indicating the end of the restoration process) [14]. As expected, restoration latency generally increases as more lightpaths are failed. Results presented in [14] are representative of a what-if type study to determine the range of restoration latencies that can be expected from a network upon single link or node failures. For illustration purpose, we show

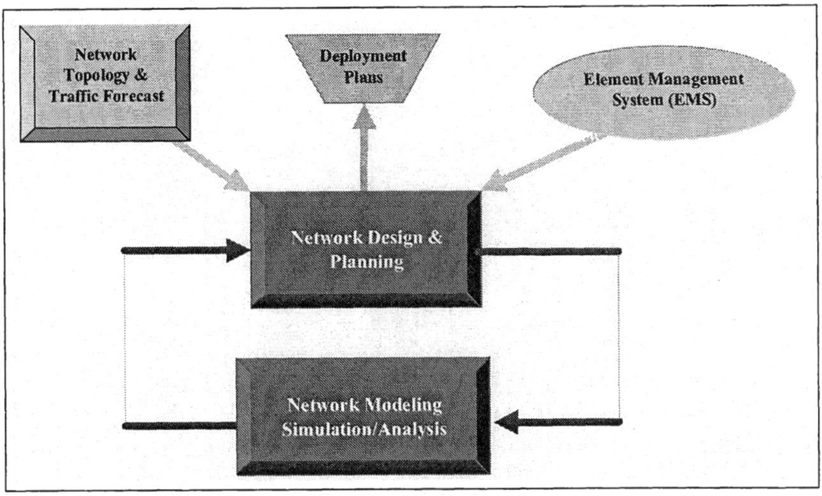


*Figure 10.11.* Interaction between network design and planning, and network restoration modeling.

in Figure 10.12 the results from the modeling tool for a 50 node hypothetical network, with 910 lightpaths routed over it. Five different single failure scenarios are shown, involving the restoration of 51, 56, 61, 69, and 72 lightpaths, respectively.

We also conducted experiments and obtained results from an actual network. Provisioning of shared mesh restored lightpaths in an actual mesh network was carried out over time, relying on the EMS/NMS routing engine to calculate the working and backup paths using the weight assignment as described above. The actual network was on the order of 45 nodes and 75 trunks, and was carrying shared mesh restored demands amounting to several hundred gigabits of service [17]. Upon a single link or single node failure, restoration times ranging from a few tens to a couple of hundred msecs were observed. The maximum restoration times observed were less than 200 msecs in the worst case (when a large number of lightpaths have to be restored simultaneously as a result of a single failure). This was consistent with the restoration times predicted by a modeling tool designed to conduct restoration simulation studies.

## 10.4 Re-Provisioning for Improved Reliability

While fast restoration ($\sim$ 100 msec) is an important aspect of service availability, also critical is the ability to recover from multiple failures quickly, as opposed to the several hours it takes in practice to replace a damaged equipment in a central office. Service availability in mesh network has been studied [15] and [19]. One of the key factors that improve service availability is the ability to recover from multiple failures, with, for example, a combi-

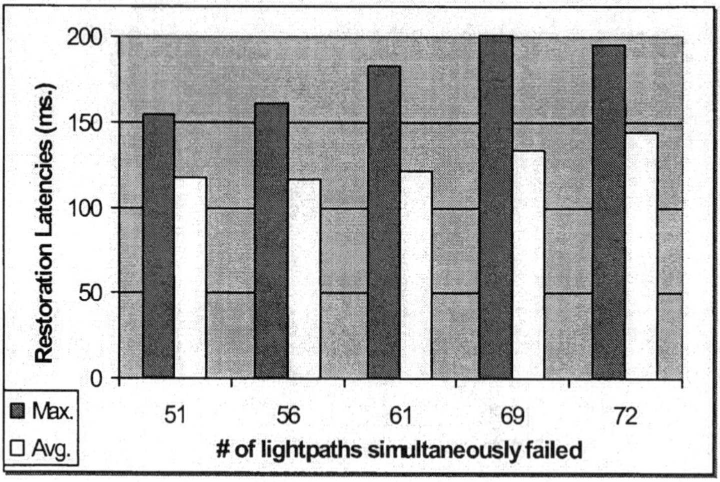


*Figure 10.12.* Restoration latencies for different sets of failed lightpaths.

nation of restoration for the first failure, and re-provisioning for the subsequent failure(s). In the case of multiple failures, a mesh network, utilizing intelligent OXCs, can also support lightpath re-provisioning. Lightpath re-provisioning tries to establish a new backup path when the restoration on the original backup path does not succeed. Re-provisioning uses existing spare capacity and unused shared capacity to find a new backup path to immediately restore the failed lightpath. There are three conditions that result in lightpath re-provisioning:

- failure of the primary path followed by a failure of the backup path prior to the repair of the primary path,
- failure of the backup path followed by a failure of the primary path prior to the repair of the backup path, and
- failure of the primary path of a lightpath $L_1$ sharing backup capacity with a lightpath $L_2$, followed by a failure of the primary path of lightpath $L_2$.

This last case would cause a contention situation where more than one lightpath need to use the shared backup capacity. In this case, lightpath $L_1$ is restored onto its backup path after the failure, thus occupying the shared backup resources. When lightpath $L_2$ fails, it cannot restore onto its backup (because resources are being used), resulting in a re-provisioning attempt, although the likelihood that such a scenario occurs can be mitigated using real-time channel assignment strategies [39]. Note that re-provisioning may fail if there is not enough capacity available. However, the presence of lightpath re-provisioning increases the service availability of a mesh network. Service unavailability occurs as a result of multiple concurrent failure scenarios and the time it takes to fix the failure (e.g., hours if a fiber cut in a remote area needs to be repaired). Re-provisioning a lightpath that becomes unavailable after a double failure will improve the service availability of the network, by reducing the time that the service is unavailable from hours to tens of seconds. This is particularly significant in trans-oceanic networks where the time to repair a damaged undersea cable could be as much as 48 hours. Simulation studies showed that, compared to traditional protection schemes, mesh restoration provides higher reliability due to the implementation of re-provisioning after a second failure, resulting in up to a two digit percentage decrease in unavailability [15].

## 10.5 Network Re-optimization

During the network operation, requests for services are received and provisioned by the EMS/NMS using an online routing algorithm with the cost model defined above. Both the primary and backup paths of each new demand request are computed according to the current state of the network, i.e., the routing algorithm takes into account all the information available at the time of the connection request to make the appropriate routing decision. As the network changes with the addition or deletion of fiber links and capacity and traffic patterns evolve, the routing of the existing demands is very likely to become sub-optimal. Re-optimization by re-routing the backup and/or primary paths gives the network operators the opportunity to regain some of the network bandwidth that is currently being misused. In particular, re-routing only the backup paths is an attractive way to regain some of the protection bandwidth and reduce backup path length while avoiding any service interruption. Figure 10.13 illustrates how re-optimization temporarily eliminates the difference between the current solution and the best-known off-line solution[1] that is achievable under the same conditions. Re-optimization is also a good approach to bring the network to a desired state after a multiple failure event and re-provisioning of lightpaths around failed network components. Rather than going back to the routes before the multiple failure event, it seems better to move forward to the new best routes.

Figure 10.14 and Table 10.2 show the gains achieved in an actual live network [17] as the backup paths were twice re-optimized over a period of one year. The original network consisted of 45 nodes, 75 links, and 70 shared mesh restorable demands with their routes determined by the network operator through on-line routing. In both instances, all backup paths were re-optimized and tested during a maintenance window. The entire re-optimization procedure thus took just a few hours. Note that Table 10.2 refers only to ports used

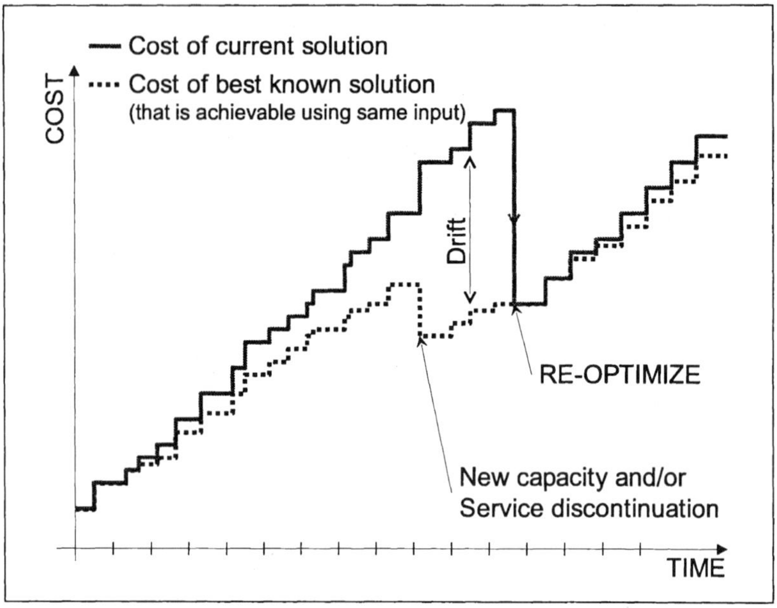


*Figure 10.13.* Current cost versus best possible cost with cost-benefit of re-optimization.

for the protection channels and shows the port counts and numbers of backup hops measured before and after re-optimization. Clearly, re-optimization is beneficial both in terms of the number of protection ports used, and of the length of the backup paths. Specifically, as shown in the table, the first backup path re-optimization saved 31% of the protection ports, which in turn translated to 20% savings in the total number of ports. Also, the average length of the backup paths decreased from 5.87 to 4.88 hops [16, 17]. The results of a second re-optimization carried out several months later show comparable improvements.

The importance of re-optimization to the network is threefold. Firstly, the reduced number of protection ports translates into freed capacity, which can then be used to carry new traffic and postpone new capacity deployment. Secondly, the reduction in backup path length translates into a reduction of restoration latency. In particular, in the first re-optimization of this actual network, the reduction of the average length of the backup path reduced the restoration latency by 26% (calculated using the average length of the backup path in miles before and after the re-optimization) [16, 17]. Finally, re-optimization allows network operators to make efficient (i.e., cost-effective) use of new nodes and links that are deployed over time in the network, and to better adapt to churn in traffic demand.

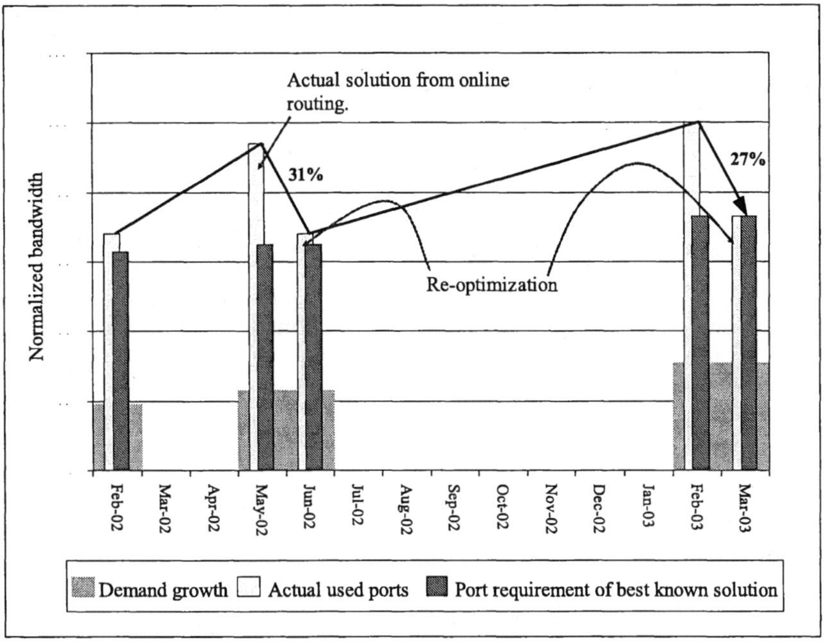


*Figure 10.14.* Actual back-up path re-optimizations.

| | backup port count | | | avg. # of backup hops | | max. # of backup hops | |
|---|---|---|---|---|---|---|---|
| | before | after | % save | before | after | before | after |
| $1^{st}$ re-opt. (06/02) | 310 | 214 | 31% | 5.87 | 4.88 | 15 | 10 |
| $2^{nd}$ re-opt. (03/03) | 332 | 242 | 27% | 4.8 | 3.57 | 13 | 9 |

*Table 10.2.* Backup Path Re-optimization.

## 10.6 Network Planning and Maintenance

Network planning and dimensioning of mesh networks is expected to be much more robust to traffic forecast uncertainties than planning of ring-based networks. The constrained routing imposed by interconnected rings becomes inefficient if the traffic does not match the network configuration, resulting in available but unused capacity on certain rings and lack of available capacity needed on other rings. In a mesh network, the diversity of routes makes the network much more robust to errors in traffic forecast. See [47] for a methodology to model forecasting errors and an analysis of their impact on the design of span-restorable and path-protected mesh networks. Finally, re-optimization in mesh networks allows network operators to correct any temporary non-optimal routing and resource utilization.

In networks built out of interconnected rings, maintenance activities are carried out in different rings concurrently without impacting what is happening in other rings. In mesh networks, back-up capacity in one location may be shared between lightpaths originating and/or terminating in widely separated areas. Network operators can nevertheless use their inventory of routes and corresponding sharing between lightpaths to define a schedule of non-interfering concurrent maintenance activities on network links and other network components[2]. This can be achieved thanks to network self-discovery abilities of intelligent optical switches (such as automatic neighbor and port discovery, automatic topology discovery), which insures complete accuracy of the network database residing in the EMS/NMS.

## 10.7 Conclusions

Long-haul national networks utilizing optical switches are becoming *intelligent mesh networks*. They offer service providers the ability to support many new operational capabilities. They are (1) end-to-end point-and-click routing and provisioning, (2) fast capacity-efficient shared mesh restoration with times comparable to SONET ring restoration times, (3) re-provisioning of connections in the event of double failures and (4) network re-optimization to regain some of the network capacity that is not optimally used. In this work, we have discussed and analyzed these many operational aspects of optical mesh networks. In addition, network planning and design of mesh networks, while only slightly touched on here, is expected to be much more robust to traffic forecast uncertainties than ring-based designs, where stranded capacity is an expensive reality. Finally, maintenance of mesh networks is another area of interest where new operational procedures will be required to make this strategic network evolution a reality.

## Acknowledgments

The authors would like to acknowledge their colleagues Ramu Ramamurthy, George Ellinas, Ahmet Akyamaç, and Sid Chaudhuri.

## Notes

1. See [16] for off-line methods to determine near-optimal routing solutions.
2. Network operators could also temporarily move back-up paths away from a maintenance area.

## References

[1] Ramu Ramamurthy, Zbigniew Bogdanowicz, Shahrokh Samieian, Debanjan Saha, Bala Rajagopalan, Sudipta Sengupta, Sid Chaudhuri, and Krishna Bala, "Capacity Performance of Dynamic Provisioning in Optical networks", IEEE Journal of Lightwave Technology, vol. 19, no. 1, pp. 40-48, January 2001.

[2] T.E. Stern, K. Bala, Multiwavelength Optical Networks: A Layered Approach, Prentice Hall, May 1999.

[3] S. Lumetta, M. Medard, "Towards a Deeper Understanding of Link Restoration Algorithms for Mesh Networks", in Proc. of IEEE Infocom 2001, Anchorage, Alaska, April 2001.

[4] E. Modiano, A. Narula-Tam, "Survivable Routing of Logical Topologies in WDM Networks", in Proc. of IEEE Infocom 2001, Anchorage, Alaska, April 2001.

[5] B. Doshi, et al, "Optical Network Design and Restoration", Bell-Labs Technical Journal, Vol. 4, No. 1, pp. 58-84, January-March 1999.

[6] S. Ramamurthy and B. Mukherjee, "Survivable WDM mesh networks. Part I: Protection", in Proc. IEEE Infocom 1999, Vol. 2, pp. 744-751, 1999.

[7] S. Ramamurthy and B. Mukherjee, "Survivable WDM mesh networks. Part II: Restoration", in Proc. of IEEE ICC 1999, Vol. 3, pp. 2023-2030, 1999.

[8] G. Ellinas, et al, "Routing and Restoration Architectures in Mesh Optical Networks", Optical Networks Magazine, Issue 4:1, Jan/Feb 2003.

[9] J-F. Labourdette, Eric Bouillet, Ramu Ramamurthy, Georgios Ellinas, Sid Chaudhuri, Krishna Bala , "Routing Strategies for Capacity-Efficient and Fast-Restorable Mesh Optical Networks", Photonic Network Communications, vol. 4, pp. 219-235, July/Dec 2002.

[10] E. Bouillet, J. Labourdette, R. Ramamurthy, S. Chaudhuri, "Enhanced Algorithm Cost Model to Control tradeoffs in Provisioning Shared Mesh Restored Lightpaths", in Proc. OFC 2002, Anaheim, CA, March 2002.

[11] J. Doucette, W. Grover, T. Bach, "Bi-Criteria Studies of Mesh Network Restoration: Path Length vs. Capacity Tradeoffs", in Proc. of OFC 2001, Anaheim, CA, March 2001.

[12] C. Qiao, Y. Xiong and D. Xu, "Novel Models for Shared Path Protection", in Proc. OFC 2002, Anaheim, CA, March 2002.

[13] R. Ramamurthy, J. Labourdette, S. Chaudhuri, R. Levy, P. Charalambous, C. Dennis, "Routing Lightpaths in Optical Mesh Networks with Express Links", in Proc. OFC 2002, Anaheim, CA, March 2002.

[14] A. Akyamaç, Jean-Francois Labourdette, Kuolin Hua, Ziggy Bogdanowicz , "Optical Mesh Network Modeling: Simulation and Analysis of Restoration Performance", in Proc. NFOEC 2002, Dallas TX, September 2002.

[15] A. Akyamaç, S. Sengupta, J. Labourdette, S. Chaudhuri, S. French, "Reliability in Single Domain vs. Multi Domain Optical Mesh Networks", in Proc. NFOEC 2002, Dallas TX, September 2002.

[16] E. Bouillet, et al, "Lightpath Re-optimization in Mesh Optical Networks", in Proc. 7th European Conference on Networks & Optical Communications (NOC), Darmstadt, Germany, June 2002.

[17] P. Charalambous Georgios Ellinas, Chad Dennis, Eric Bouillet, Jean-Francois Labourdette, Ahmet A. Akyamç, Sid Chaudhuri, Mikhail Morokhovich, David Shales, "A National Mesh Network Using Optical Cross-Connect Switches, OFC 2003, Atlanta, GA, March 2003.

[18] R. Ramamurthy, J-F. Labourdette, A. Akyamac, S. Chaudhuri, "Limiting Sharing on Protection Channels in Mesh Optical Networks", OFC 2003, Atlanta, GA, March 2003.

[19] C. Janczewski et al., "Restoration Strategies in Mesh Optical Networks: Cost, Performance, and Service Availability", in Proc. NFOEC 2002, Sept 2002, Dallas, TX.

[20] W. Grover and J. Doucette, "Increasing the Efficiency of Span-Restorable Mesh Networks on Low-Connectivity Graphs", 3rd international workshop on design of reliable communication networks (DRCN 2001), Budapest, Hungary, October 2001.

[21] J. Doucette and W. Grover, "Comparison of Mesh Protection and Restoration Schemes and the Dependency on Graph Connectivity", 3rd international workshop on design of reliable communication networks (DRCN 2001), Budapest, Hungary, October 2001.

[22] R. Iraschko, M.H. MacGregor, and W. Grover, "Optimal Capacity Placement for Path Restoration in STM or ATM Mesh-Survivable Networks", IEEE/ACM Transactions on Networking, Vol. 6, No. 3, June 1998.

[23] S. Datta, et al, "Efficient Channel Reservation for Backup Paths in Optical Mesh Networks", IEEE GLOBECOM 2001, San Antonio, TX, Nov. 2001.

[24] Y. Liu, O. Tipper, and P. Siripongwutikom, "Approximating Optimal Spare Capacity Allocation by Successive Survivable Routing", Infocom 2001, Anchorage, AL, April 24-28, 2001.

[25] M. Kodialam and T.V. Lakshman, " Dynamic Routing of Bandwidth Guaranteed Tunnels with Restoration", IEEE INFOCOM 2000, Tel Aviv, Israel, March 26-30, 2000.

[26] R. Ramamurthy ,Zbigniew Bogdanowicz, Shahrokh Samieian, Debanjan Saha, Bala Rajagopalan, Sudipta Sengupta, Sid Chaudhuri, and Krishna Bala, "Capacity Performance of Dynamic Provisioning in Optical Networks", IEEE Journal of Lightwave Technology, vol. 19, issue 1, pp. 40-48, January 2001.

[27] R.D. Doverspike, Gokhan Sahin, John L. Strand, Robert W. Tkach, "Fast Restoration in a Mesh Network of Optical Cross-connects", OFC 1999, San Diego, CA, Feb. 1999.

[28] E. Bouillet, J-F. Labourdette, G. Ellinas, S. Chaudhuri, "Local Optimization of Shared Back-up Channels in Optical Mesh networks", OFC 2003, Atlanta, GA, March 2003.

[29] J. Labourdette, E. Bouillet, R. Ramamurthy, A. Akyamac, "Fast Approximate Dimensioning and Performance Analysis of Mesh Optical Networks", DRCN 2003, 19-22 October 2003, Banff, Alberta, Canada.

[30] J. Labourdette, E. Bouillet, S. Chaudhuri, "Role of Optical Network and Spare Router Strategy in Resilient IP Backbone Architecture", DRCN 2003, 19-22 October 2003, Banff, Alberta, Canada.

[31] S. Sengupta, V. Kumar, and D. Saha, "Switched Optical Backbone for Cost-Effective Scalable IP networks", IEEE Communications Magazine, June 2003.

[32] A. Akyamac,Subir Biswas, Jean Labourdette, Sid Chaudhuri, Krishna Bala, "Ring Speed Restoration and Optical Core Mesh Networks", in Proc. 7th European Conference on Networks & Optical Communications (NOC), Darmstadt, Germany, June 2002.

[33] B. Rajagopalan, D. Pendarakis, D. Saha, R. Ramamurthy and K. Bala, "IP over Optical Networks: Architectural Aspects", IEEE Communications Magazine, September 2000.

[34] E. Mannie, editor, "Generalized Multi-Protocol Label Switching (GMPLS) Architecture," Internet Draft, Work in Progress, draft-ietf-ccamp-gmpls-architecture-03.txt, March 2002.

[35] ITU-T Recommendation G.8080, "Architecture for the ASON".

[36] The Optical Internetworking Forum (OIF), "User Network Interface (UNI) v1.0 Signaling Specification", December 2001.

[37] D. Katz, et.al., "Traffic Engineering Extensions to OSPF Version 2", Internet Draft, Work in Progress, October 2002.

[38] R. Ramamurthy, S. Sengupta, S. Chaudhuri, "Comparison of Centralized and Distributed Provisioning of Lightpaths in Optical networks", OFC 2001, Anaheim, CA, March 2001.

[39] E. Bouillet, J-F. Labourdette, G. Ellinas, R. Ramamurthy, and S. Chaudhuri, "Stochastic Approaches to Route Shared Mesh Restored Lightpaths in Optical Mesh Networks", Proc. of IEEE INFOCOM 2002, New York, NY, June 2002.

[40] C. Qiao, Dahai Xu, "Distributed Partial Information Management (DPIM) Schemes for Survivable Networks - Part I", Proc. of IEEE INFOCOM 2002, New York, NY, June 2002.

[41] G. Li,Dongmei Wang, Charles Kalmanek, Robert Doverspike, "Efficient Distributed Path Selection for Shared Restoration Connections", Proc. of IEEE INFOCOM 2002, New York, NY, June 2002.

[42] Z. Dziong, S. Kasera, R. Nagarajan, "Efficient Capacity Sharing in Path Restoration Schemes for Meshed Optical Networks" in Proc. NFOEC 2002, Dallas TX, September 2002.

[43] H. Liu, Eric Bouillet, Dimitrios Pendarakis, Nooshin Komaee,Jean-Francois Labourdette, Sid Chaudhuri, "Distributed Route Computation Algorithms and Dynamic Provisioning in Intelligent Mesh Optical Networks", ICC 2004, Paris, France, June 20-24 2004.

[44] W. D. Grover and D. Stamatelakis, "Cycle-oriented distributed preconfiguration: Ring-like speed with mesh-like capacity for self-planning network restoration", in Proc. IEEE Int. Conf. Communications (ICC'98), Atlanta, GA, June 1998, pp. 537-543.

[45] B. Doshi, D. Jeske, N. Raman, A. Sampath, "Reliability and Capacity Efficiency of Restoration Strategies for Telecommunication Networks", DRCN 2003, 19-22 October 2003, Banff, Alberta, Canada.

[46] J. Doucette, M. Clouqueur, W. Grover, "On the Availability and Capacity Requirements of Shared Backup Path-Protected Mesh Networks", Optical Networks Magazine, Nov/Dec 2003.

[47] D. Leung and W. Grover, "Comparative Ability of Span-Restorable and Path-Protected Network Designs to Withstand Error in the Demand Forecast", in Proc. NFOEC 2002, Dallas TX, September 2002.

IV

# TRAFFIC GROOMING

Chapter 11

# TRAFFIC GROOMING IN WDM NETWORKS

Jian-Qiang Hu[1] and Eytan Modiano[2]

[1]*Department of Manufacturing Engineering & Center for Information and Systems Engineering, Boston University, Brookline, MA 02446*

[2]*Department of Aeronautics and Astronautics & Laboratory for Information and Decision Systems, Massachusetts Institute of Technology, Cambridge, MA 02139*

Email: hqiang@bu.edu, modiano@mit.edu

**Abstract** In today's WDM networks, the dominant cost component is the cost of electronics, which is largely determined by how traffic is groomed at each node. Therefore, the issue of traffic grooming is extremely important in the design of a WDM network. In this article, our goal is to introduce various aspects of the traffic grooming problem to the reader. We start with the static traffic grooming problem and illustrate how it can be solved based on the Integer Linear Programming formulation and various heuristic approaches. We then discuss variants of the problem including grooming dynamic traffic, grooming with cross-connects, grooming in mesh and IP networks, and grooming with tunable transceivers.

**Keywords:** Wavelength Division Multiplexing (WDM), traffic grooming.

## 11.1 Introduction

Wavelength Division Multiplexing (WDM) is emerging as a dominant technology for use in backbone networks. WDM significantly increases the capacity of a fiber by allowing simultaneous transmission of multiple wavelengths (channels), each operating at rates up to 40Gbps. Systems with over 80 wavelengths are presently being deployed and capacities that approach several Terabits per second can be achieved. While such enormous capacity is very exciting, it also places a tremendous burden on the electronic switches and routers at each node that must somehow process all of this information. Fortunately, it is not necessary to electronically process all of the traffic at each node. For example, much of the traffic passing through a node is neither sourced at that node nor destined to that node. To reduce the amount of traffic that must be electronically processed at intermediate nodes, WDM systems employ Add/Drop multiplexers (ADMs), that allow each wavelength to either be dropped and electronically processed at the node or to optically bypass the node electronics, as shown in Figure 11.1.

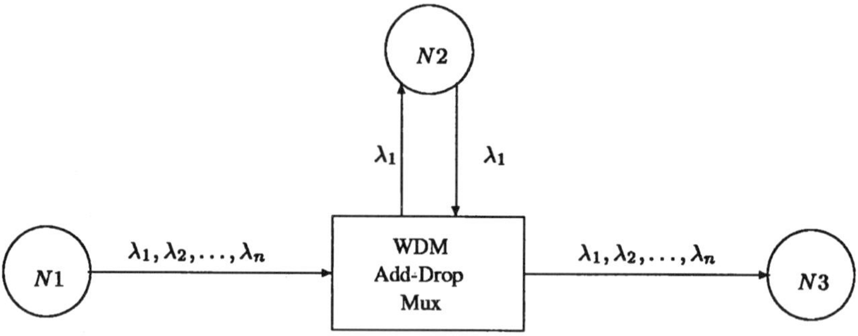


*Figure 11.1.* Using ADM to provide optical bypass.

Much of today's physical layer network infrastructure is built around Synchronous Optical Network (SONET) rings. Typically, a SONET ring is constructed using fiber (one or two fiber pairs are typically used in order to provide protection) to connect SONET ADMs. Each SONET ADM has the ability to aggregate lower rate SONET signals into a single high rate SONET stream. For example, four OC-3 circuits can be multiplexed together into an OC-12 circuit and 16 OC-3's can be multiplexed into an OC-48. The recent emergence of WDM technology has provided the ability to support multiple SONET rings on a single fiber pair. Consider, for example, the SONET ring network shown in Figure 11.2a, where each wavelength is used to form an OC-48 SONET ring. With WDM technology providing dozens of wavelengths on a fiber, dozens of OC-48 rings can be supported per fiber pair using wavelength multiplexers to separate the multiple SONET rings. This tremendous increase in network capacity, of course, comes at the expense of additional electronic multiplexing equipment. With the emergence of WDM technology, the dominant cost component in networks is no longer the cost of optics but rather the cost of electronics.

The SONET/WDM architecture shown in Figure 11.2a is potentially wasteful of ADMs because every wavelength (ring) requires an ADM at every node. As mentioned previously, not all traffic needs to be electronically processed at each node. Consequently, it is not necessary to have an ADM for every wavelength at every node, but rather only for those wavelengths that are used at that node. Therefore, in order to limit the number of ADMs required, the traffic should be groomed in such a way that all of the traffic to and from a given node is carried on the minimum number of wavelengths. As a simple and illustrative example, consider a unidirectional ring network (e.g., Uni-directional Path

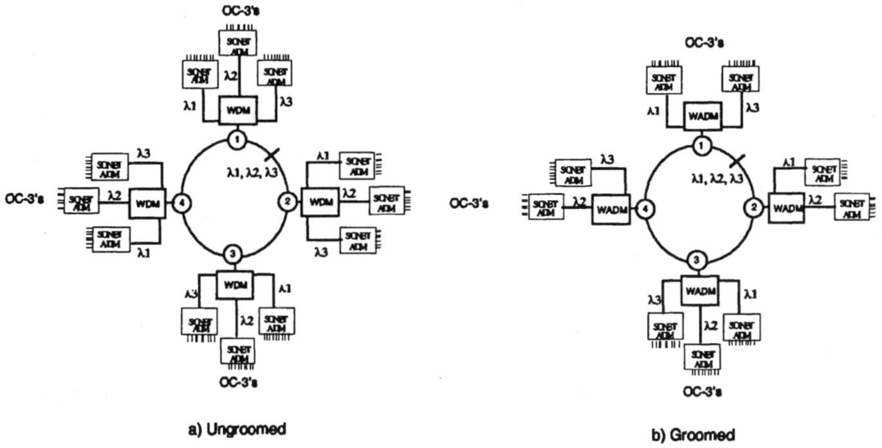


*Figure 11.2.* SONET/WDM rings.

Switched Ring, UPSR) with four nodes. Suppose that each wavelength is used to support an OC-48 ring, and that the traffic requirement is for 8 OC-3 circuits between each pair of nodes. In this example we have 6 node pairs and the total traffic load is equal to 48 OC-3's or equivalently 3 OC-48 rings. The question is how to assign the traffic to these 3 OC-48 rings in a way that minimizes the total number of ADMs required. Consider, for example, the two traffic assignments listed in Tables 11.1 and 11.2. With the first assignment, each node has some traffic on every wavelength. For example wavelength $\lambda_1$ carries the traffic between nodes 1 and 2 and the traffic between nodes 3 and 4. Therefore, each node would require an ADM on every wavelength for a total of 12 ADMs. With the second assignment each wavelength contains traffic from only 3 nodes and hence only 9 ADMs are needed. Notice that both assignments carry the same amount of total traffic (8 OC-3's between each pair of nodes). The corresponding ADM allocations for both assignments are shown in Tables 11.1 and 11.2, respectively.

In a bi-directional ring the amount of electronics is determined not only by how circuits are groomed but also by how circuits are routed (since a circuit in a bi-directional ring can be routed either clockwise or counter-clockwise) and how wavelengths are assigned to grooming lightpaths, i.e., the traffic grooming problem has to be considered in combination with routing and wavelength assignment (RWA) problem. Together, we have a traffic grooming and routing and wavelength assignment (GRWA) problem. A special case of the GRWA is the routing and wavelength assignment problem when all circuits are light-

*Table 11.1.* Assignment #1.

| Circuits | |
|---|---|
| $\lambda_1$ | between nodes 1 and 2<br>between nodes 3 and 4 |
| $\lambda_2$ | between nodes 1 and 3<br>between nodes 2 and 4 |
| $\lambda_3$ | between nodes 1 and 4<br>between nodes 2 and 3 |

*Table 11.2.* Assignment #2.

| Circuits | |
|---|---|
| $\lambda_1$ | between nodes 1 and 2<br>between nodes 1 and 3 |
| $\lambda_2$ | between nodes 2 and 3<br>between nodes 2 and 4 |
| $\lambda_3$ | between nodes 1 and 4<br>between nodes 3 and 4 |

paths (i.e., no grooming is needed). RWA is important to allow end-to-end lightpaths to share common ADMs [11]. In a SONET Bi-directional Line-Switched Ring (BLSR), an ADM is responsible for adding/dropping both the upstream and down stream data. This is done so that the data in one direction can be switched to the opposite direction in case of a failure. Consequently, if an ADM has working traffic in one direction of a lightpath (for example, upstream), and is not supporting traffic in the opposite direction (down stream), then its capability is not fully utilized and the bandwidth in the unused direction is wasted. This is analogous to what is commonly called stranded bandwidth in BLSR except it is occurring at the lightpath level.

To illustrate the importance of RWA of (groomed) lightpaths, compare the two RWAs, listed in Tables 11.3 and 11.4, of the same set of nine lightpaths, $\{1 \Leftrightarrow 2, 1 \Leftrightarrow 3, 2 \Leftrightarrow 3, 4 \Leftrightarrow 5, 4 \Leftrightarrow 6, 5 \Leftrightarrow 6, 7 \Leftrightarrow 8, 7 \Leftrightarrow 9, 8 \Leftrightarrow 9\}$, for a BLSR with 9 nodes ($i \Leftrightarrow j$ indicates a bi-directional lightpath between node i and node j). In these assignments, the circuit from $i$ to $j$ is routed in the direction opposite to the circuit from $j$ to $i$. In RWA #1, all circuits are routed via shortest paths, while in RWA #2, circuits are more cleverly packed to make efficient use of the ADMs in both directions. For example, on $\lambda_1$ circuits $1 \Leftrightarrow 2$ and $2 \Leftrightarrow 3$ are routed via the shortest path, while $1 \Leftrightarrow 3$ is routed along the "longer path" $3 \Leftrightarrow 1$. Both RWAs support the same set of traffic demands. The first RWA uses 15 ADMs and 2 wavelengths, and the second RWA uses more wavelengths, but it only requires 9 ADMs.

The above example also illustrates a few characteristics of the overall problem of network cost minimization. First, the minimum number of ADMs is often not achieved with the minimum capacity usage. In the example, the method that uses the minimum number of ADMs requires an additional wavelength. Standard RWA algorithms that focus on minimizing the number of wavelengths cannot be directly applied to ADM cost minimization. Instead algorithms that attempt to jointly optimize the cost of ADMs and Wavelengths

*Table 11.3.* RWA #1.

| | Lightpaths | ADMs |
|---|---|---|
| $\lambda_1$ | $1 \Leftrightarrow 2, 2 \Leftrightarrow 3, 4 \Leftrightarrow 5$<br>$5 \Leftrightarrow 6, 7 \Leftrightarrow 8, 8 \Leftrightarrow 9$ | 9 |
| $\lambda_2$ | $1 \Leftrightarrow 3, 4 \Leftrightarrow 6, 7 \Leftrightarrow 9$ | 6 |

*Table 11.4.* RWA #2.

| | Lightpaths | ADMs |
|---|---|---|
| $\lambda_1$ | $1 \Leftrightarrow 2, 2 \Leftrightarrow 3, 3 \Leftrightarrow 1$ | 3 |
| $\lambda_2$ | $4 \Leftrightarrow 5, 5 \Leftrightarrow 6, 6 \Leftrightarrow 4$ | 3 |
| $\lambda_3$ | $7 \Leftrightarrow 8, 8 \Leftrightarrow 9, 9 \Leftrightarrow 7$ | 3 |

are more desirable (e.g., see [17, 31] for the joint optimization problem). Second, the minimum number of ADMs is not achieved with shortest path routing. Since shortest path is desired to reduce network latency, a tradeoff exists between network latency and ADM costs. Lastly, the RWA example shows that ADM saving is possible by appropriate RWA without the aid of grooming. This gives us two methods in reducing ADMs: grooming and RWA of groomed lightpaths. It would be tempting for a network planner to design the network in two steps: 1) low level grooming of tributaries onto lightpaths and 2) RWA of the resulting lightpaths. Unfortunately, this two-step process will lead to a sub-optimal solution. In fact, it was shown in [11] that an improvement of up to 20% could be achieved if the two steps are jointly considered in the design process.

Both grooming and RWA have the characteristic of grouping and packing problems. Such problems are often difficult. This intuitively explains why the ADM minimization problem is so complex. In fact, it was shown in [23] that traffic grooming problem is NP-complete by showing that the Bin Packing problem can be transformed into the traffic grooming problem in polynomial time. Since the Bin Packing problem is known to be NP-complete the traffic grooming problem must be NP-complete as well. As a result, many papers on grooming rely on heuristics and simulations to evaluate the heuristics.

As we mentioned earlier, the majority of optical networks in operation today have been built based on the ring architecture, however, carriers have increasingly considered the mesh architecture as an alternative for building their next generation networks, which have a compelling cost advantage over ring networks and are also more resilient to various network failures and more flexible in accommodating changes in traffic demands. On the other hand, in order to capitalize on these advantages, it is even more important to efficiently groom traffic in mesh networks. Similar to bi-directional rings, the traffic grooming problem for mesh networks has to be considered in combination with RWA. But RWA is much more complicated for mesh networks since circuits can be routed more flexibly in mesh networks.

In this article, we attempt to expose the reader to the basics of the traffic grooming problem. Our discussion is in no way meant to be an exhaustive exposition of the vast literature on the topic. Good survey articles on traffic grooming literature can be found in [9], [24]. In the next section we introduce the static grooming problem and discuss both the Integer Linear Programming formulation and heuristic approaches to its solution. In subsequent sections we discuss variants of the problem including grooming dynamic traffic, grooming with cross-connects, grooming in Mesh and IP networks, and grooming with tunable transceivers.

## 11.2 Grooming Static Traffic

The static traffic grooming problem is a special instance of the virtual topology design problem. Given a traffic demand of low rate circuits between pairs of nodes, the problem is to assign traffic to wavelengths in such a way that minimizes the number of ADMs used in the network. Virtual topology design problems can be formulated as a mixed integer programming problem. In the next subsections we discuss the integer linear programming (ILP) formulation for the traffic grooming problem followed by heuristic algorithms for solving the grooming problem.

### 11.2.1 ILP Formulation

We start by introducing the integer linear programming (ILP) formulation for the traffic grooming problem. The ILP formulation has been previously used in [8, 29, 17] for the traffic grooming problem. In [8], the objective function considered is electronic routing and the goal there is to derive bounds based on the ILP formulation. In [29], the authors concluded that the ILP formulation is not computationally feasible for rings with 8 nodes or more. Hence, they propose instead to use methods based on simulated annealing and heuristics. In [17], a more efficient mixed ILP (MILP) formulation is proposed for unidirectional rings which results in significant reduction in computation time. The numerical results provided in [17] show that optimal or near-optimal solutions can usually be obtained in a few seconds or minutes for unidirectional rings with up to 16 nodes. The work of [17] is extended to bi-directional rings and dynamic traffic in [31]. The ILP formulation is also used in [18] to study the traffic grooming problem for mesh networks.

In order to illustrate the ILP approach, we will focus exclusively on uni-directional rings here; however, the interested reader can refer to the above references for more general formulations. Consider a uni-directional WDM ring with $N$ nodes. We assume that all available wavelengths have the same capacity and there may be multiple traffic circuits between a pair of end-nodes, but all traffic circuits have the same rate. The traffic granularity of the network is defined as the total number of low-rate traffic circuits that can be multiplexed onto a single wavelength. For example, if each circuit is OC-12 and the wavelength capacity is OC-48, then the traffic granularity is 4.

In designing a WDM ring, the key is to determine which ADMs are needed at each node. This mainly depends on how lower-rate traffic circuits are multiplexed onto high-rate wavelengths. An ADM for an individual wavelength is needed at a node only when the wavelength needs to be dropped at the node, i.e., when that wavelength is carrying one or more circuits that either originates or terminates at that node. If the wavelength only passes through the node, then no ADM for the wavelength is needed. Our objective is to find an optimal way to multiplex lower-rate traffic circuits so as to minimize the total number of ADMs required in the network. However, we can easily incorporate other considerations into our objective as well, such as the total number of wavelengths used in the network.

To present the ILP formulation, we need to introduce the following notation:

$N$: the number of nodes in the ring;

$L$: the number of wavelengths available;

$g$: the traffic granularity;

$m_{ij}$: the number of circuits from node $i$ to node $j$ $(i, j = 1, 2, \ldots, N)$;

$$x_{ijsl}: = \begin{cases} 1 & \text{if the } s\text{-th circuit between nodes } i \text{ and } j \text{ is multiplexed} \\ & \text{onto wavelength } l; \\ 0 & \text{otherwise;} \end{cases}$$

$$y_{il}: = \max_{s,j}(x_{ijsl}, x_{jisl})$$

$$= \begin{cases} 1 & \text{if any circuit with node } i \text{ being one of its end-nodes is} \\ & \text{multiplexed onto wavelength } l, \\ 0 & \text{otherwise;} \end{cases}$$

We note that if $y_{il} = 1$, then wavelength $l$ needs to be dropped at node $i$, which implies that an ADM for wavelength $l$ is required at node $i$. Since our objective is to minimize $\sum_{i=1}^{N} \sum_{l=1}^{L} y_{il}$, the total number of ADMs required in the ring, the traffic grooming problem can be formulated as the following integer linear programming (ILP) problem:

$$
\begin{aligned}
\min \quad & \sum_{i=1}^{N}\sum_{l=1}^{L} y_{il} & & \\
\text{s.t.} \quad & \sum_{i=1}^{N}\sum_{j=1}^{N}\sum_{s=1}^{m_{ij}} x_{ijsl} \le g & l = 1,2,\ldots,L \quad & (11.1)\\
& \sum_{l=1}^{L} x_{ijsl} = 1 & \forall\, i,j,s \quad & (11.2)\\
& y_{il} \ge x_{ijsl} & & \\
& y_{il} \ge x_{jisl} & \forall\, i,j,s,l \quad & (11.3)\\
& x_{ijsl}, y_{il} \text{ are all binary variables} & &
\end{aligned}
$$

The three constraints in the above ILP are:

(11.1): The total number of circuits multiplexed onto wavelength $l$ should not exceed $g$.

(11.2): Each circuit has to be assigned to one (and only one) wavelength.

(11.3): Given that the objective is to minimize $\sum_{i=1}^{N}\sum_{l=1}^{L} y_{il}$, it is equivalent to $y_{il} = \max_{s,j}(x_{ijsl}, x_{jisl})$.

In general, it is computationally infeasible to use the above ILP formulation to solve the traffic grooming problem for large rings. In [17], it is shown how the ILP formulation can be improved so that it can be solved more efficiently. For example, the binary integer constraint on $x_{ijsl}$ can be relaxed, resulting a mixed ILP which can be solve rather easily. The ILP formulation can be easily applied to bi-directional rings ([17]) and mesh networks ([18]). Other extensions of the ILP formulation include: a) non-uniform traffic ([17]), b) minimizing the number of wavelengths, or a weighted summation of the number of ADMs and the number of wavelengths ([17, 31]), and c) dynamic traffic ([31]). In [31], it was shown how the ILP formulation can be used in combination with heuristics to solve the traffic grooming problem.

### 11.2.2 Heuristic Algorithms

While the general topology design problem is known to be intractable, the traffic grooming problem is a special instance of the virtual topology design problem for which, in certain circumstances, a solution can be found. For example, [23] considers traffic grooming for a unidirectional ring and [27] considers the same problem for a bi-directional ring. Both [23] and [27] show that significant savings in the number of ADMs can be achieved through efficient traffic grooming algorithms. For example, shown in Figure 11.3 is the number of ADMs required when using the traffic grooming algorithm developed in [23] for the unidirectional ring with uniform traffic (single OC-3 between each pair of nodes groomed onto an OC-48 ring). This number is compared to the number of ADMs required when no grooming is used (i.e., all wavelengths are dropped at all nodes). It is also compared to a lower bound on the number of ADMs. As can be seen from the figure, the algorithms developed in [23] are not far from the lower bound, and achieve significant ADM savings.

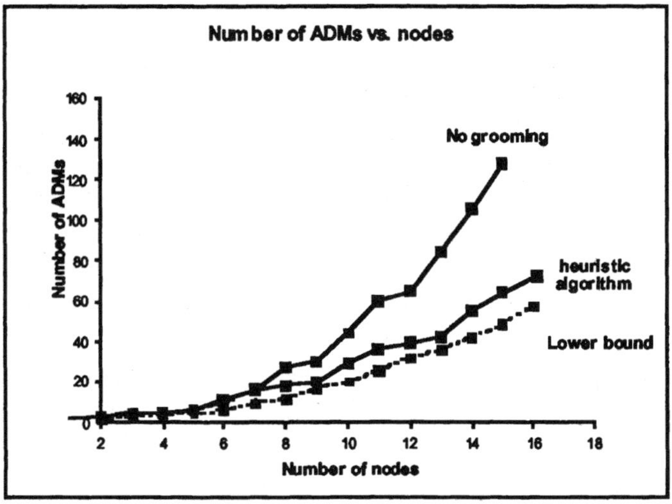


*Figure 11.3.* ADM savings in a unidirectional ring network.

The algorithms in [23, 27] consider three different traffic scenarios: 1) uniform traffic in a unidirectional and bi-directional ring, 2) distance dependent traffic where the amount of traffic between node pairs is inversely proportional to the distance separating them, and 3) hub traffic where all of the traffic is going to one node on the ring. All of those cases yielded elegant algorithms that are nearly optimal. The algorithms in [23, 27] are based on efficient "grouping" of circuits, where all circuits belonging to a group are assigned to the same wavelength and that wavelength is dropped at all of the nodes that belong to that group. Groups are chosen in such a way that the ratio of the number of circuits in a group to the number of nodes is maximized. This approach aims at making efficient use of ADMs. In fact, the algorithms are shown in [23] to be nearly optimal for uniform traffic and these "grouping" algorithms can also serve as the basis for solving the traffic grooming problem for general traffic. Of course, the general traffic grooming problem with arbitrary traffic is much more challenging. As stated earlier, the general problem can be formulated as an integer program. However, these integer programs are typically very computationally complex and can only be solved for very small problems that are often impractical.

Zhang and Qiao [30] make an attempt at solving the problem by separating the problem into two parts. In the first part, the heuristic packs the traffic demands (e.g., OC-3's) into "circles" where each circle has capacity equal to the tributary rate (OC-3) and contains non-overlapping demands. As many circles as needed are constructed to include all traffic demands. The second part of the heuristic groups circles into wavelengths (e.g., sixteen OC-3 circles in one OC-48 ring). Note that this algorithm is different than the two-step process mentioned in the previous section. There, the two steps are 1) grouping of tributaries into lightpaths, and 2) RWA of lightpath segments. Here, the two parts are 1) fitting tributaries onto a circle, and then 2) grouping of the circles. For this algorithm, the number of ADMs needed for a particular wavelength equals the number of "end nodes" involved. An end node is a node that terminates a connection in the circle. To minimize the number of ADMs, the heuristic attempts to match as many end nodes as possible when grouping the circles. This two part algorithm can achieve good performance for uniform traffic as long as the grooming factor is reasonably large (e.g., OC-3's onto OC48 wavelengths). Even for non-uniform traffic, this two part algorithm performs reasonably well if a good end-node matching algorithm is utilized. A similar two-step approach is also used in [31] for the traffic grooming problem with dynamic traffic, where the first step is solved based on the ILP formulation (instead of heuristic algorithms).

More recently, a number of researchers have developed heuristic algorithms for the traffic grooming problem with provable "worst case" performance bounds. These algorithms are known as approximation algorithms. For example, [10] and [5] consider the traffic grooming problem in a bidirectional ring with loopback protection and develop polynomial-time algorithms with a worst case performance of $8/5$ (i.e., the algorithms developed are within $8/5$ of the optimal).

## 11.3 Grooming Dynamic Traffic

Most earlier work on the grooming problem considered static traffic. Static traffic is common for many applications where a service provider designs and provisions network resources based on some estimate of the traffic. In many cases, however, the traffic changes over time. Such changes can be due to slow changes in traffic demands over a long period of time. More recently such changes can be attributed to the more rapid dynamics of Internet traffic. It is therefore important to design networks that are able to efficiently accommodate changes in traffic. There are three different models that have been used to characterize dynamic traffic:

**Stochastic Model** In this model, traffic requests (between a pair of nodes) arrive according to a stochastic point process and each request may last a random amount of time.

**Deterministic Model** In this model, traffic is represented by a set of different traffic requirements that the network needs to satisfy, but at different times. Each traffic requirement contains a set of demands. For example, the different traffic requirements can be a result of traffic fluctuation in different operation periods (morning, afternoon, and evening). Note that the static traffic case becomes a special case of this model in which there is only one traffic requirement for the network (and it never changes).

**Constrained Model** In this model, traffic demands between nodes are not specified. Rather, only a set of constraints on the traffic requirement are provided, such that the total amount of traffic at each node does not exceed a certain limit and/or the total capacity requirement on each fiber link does not exceed a certain limit.

To the best of our knowledge, the traffic grooming problem with dynamic stochastic traffic has not been studied in literature, though the stochastic traffic model has been used in the study of other design problems for optical networks, such as the problem of wavelength conversion and blocking (e.g., see [1, 2, 28] and references therein). The constrained traffic model is used in [3, 15, 14, 26], where the focus is on obtaining lower and upper bounds on network costs (such as the number of ADMs required). The model in [3] defines a class of traffic called $t$-allowable which allows each node to source up to t circuits. These $t$ circuits can be destined to any of the nodes in the network without restriction, and the destinations of the circuits can be dynamically changed. The approach taken is to design a network so that it can accommodate any $t$-allowable traffic matrix in a non-blocking way. The problem is formulated as a bipartite graph matching problem and algorithms are developed to minimize the number of wavelengths that must be processed at each node. These algorithms provide methods for achieving significant reductions in ADMs under a variety of traffic requirements. The deterministic traffic model is first considered in [3]. In [17], the traffic grooming problem with dynamic deterministic traffic is formulated as an integer linear programming problem for unidirectional rings. In [31], an approach based on a combination of ILP and heuristics is proposed to study the traffic grooming problem with dynamic deterministic traffic. First, the ILP formulation was used to solve a slightly different traffic grooming problem in which the objective is to minimize the total number of wavelengths. This problem is much easier to solve than the traffic grooming problem whose objective function is the total number of ADMs. Once a traffic grooming solution with minimum number of wavelengths is obtained, it can then be used to construct a solution with as few ADMs as possible based on a heuristic method. As we mentioned earlier, this two-step approach of minimizing the number of ADMs was first used in [30] for the traffic grooming problem with static traffic. However, only heuristic algorithms were used in [30] to minimize the total number of wavelengths.

## 11.4 Grooming with Cross-Connects

Another approach for supporting dynamic traffic is to use a cross-connect at one or more of the nodes in the network. The cross-connect is able to switch traffic from one wavelength onto any other to which it is connected. Not only can the addition of a cross-connect allow for some traffic dynamics, but it can also be used to reduce the number of ADMs required. In [23] it was shown that using a hub node with a cross-connect is optimal in terms of minimizing the number of ADMs required and in [12] it was shown the cost savings can be as much as 37.5%. The proof in [23] is obtained by showing that any traffic grooming that does not use a cross-connect can be transformed into one that uses a cross-connect without any additional ADMs. In [13] various network architectures with different amount of cross-connect capabilities are compared.

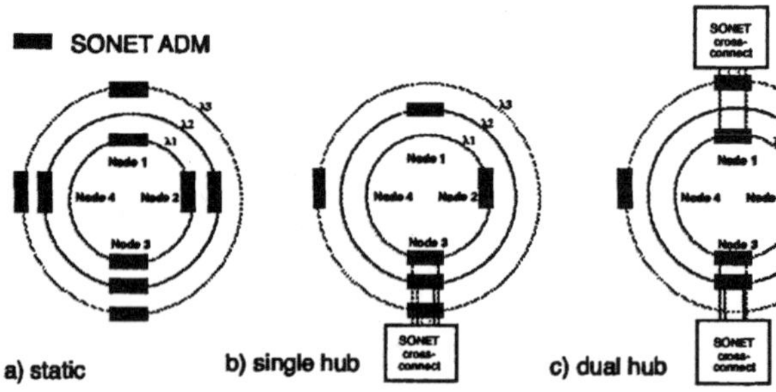


*Figure 11.4.* Grooming with cross-connect.

To illustrate the benefits of a cross-connect architecture, consider three possible ring architectures for the purpose of efficient grooming: a static ring without cross-connects, a single-hub ring, and a ring with multiple-hubs. With the static architecture no cross-connecting is employed, hence each circuit must be assigned to a single wavelength that must be processed (dropped) at both the source and the destination. The single hub architecture uses a large cross-connect at one hub node. The cross-connect is able to switch any low rate circuit from any incoming wavelength to any outgoing wavelength. With this architecture, each node sends all of its traffic to the hub node where the traffic is switched, groomed and sent back to the destination nodes. In the multiple hub architecture, $K$ hub nodes are used on the ring. Each hub node has a small cross-connect that can switch traffic among the wavelengths dropped at that node. Each node on the ring sends a fraction of its traffic to one of the hub nodes, where it is properly groomed and relayed to its destination. These three architectures are depicted in Figure 11.4. Shown in Figure 11.4a is the static grooming solution where one wavelength is used to support traffic between nodes 1, 2 and 3, another for traffic between 2, 3 and 4, and a third wavelength for traffic between 1, 3 and 4. The hub architecture shown in Figure 11.4b has each node send all of its traffic to the hub located at node 3, where the traffic is groomed and relayed back to its destination. Finally shown in Figure 11.4c is the multiple hub architecture where each node can send its traffic to one or more of the hubs.

To illustrate the potential benefit of the multiple hub architecture, consider a unidirectional ring with 9 nodes where each wavelength supports an OC-48 and traffic demand is uniform with two OC-12's between each pair. In this case each node generates 16 OC-12's or four wavelengths of traffic. With the single hub solution, each node can send all four wavelengths worth of traffic to be groomed at the hub at say node 1. In this case, each node would use 4 ADMs, and the hub would use $8 \times 4 = 32$ ADMs for a total of 64 ADMs. In a 2-hub architecture each node would send two wavelengths worth of traffic to each hub (at nodes 1 and 5) and an additional wavelength would be used for traffic between the two hubs, resulting in 58 ADMs. Finally a 4-hub architecture can be used where each node sends one wavelength to each of four hubs and some additional ADMs are used to handle the inter-hub traffic. Using the grooming algorithm given in [21] and [22], a 4-hub architecture can be found that requires only 26 wavelengths and 49 ADMs. Notice that in this case the number of hubs is equal to the number of wavelengths generated by a node. Also notice that in increasing the number of hubs from 1 to 4 the required number of wavelengths in the ring is reduced from 32 to 26. Thus the 4-hub architecture is more efficient in the use of wavelengths as well as ADMs.

It was shown in [20] and [22] that significant savings could be obtained by distributing the cross-connect function among multiple nodes. In [22] a lower bound on the number of ADMs is given as a function of the number of switching nodes (i.e., nodes with cross-connect capability), and algorithms that very nearly meet the lower bounds are provided. In fact, for uniform traffic, [21] shows that the number of electronic ports is reduced when the number of switching nodes (hubs) used is approximately equal to the number of wavelengths of traffic generated by each node. These savings are significant in two ways. First, the use of multiple cross-connects can reduce the number of ADMs needed. Second, using multiple smaller cross-connects rather than one large cross-connect at the hub reduces the cost of the cross-connects. The above papers all conclude that the use of cross-connects for grooming adds flexibility to the network over a static solution that does not use a cross-connect. This flexibility allows traffic to be provisioned dynamically thereby reducing the need to know the exact traffic requirements in advance. Another benefit of this flexibility is that the network will be more robust to node failures.

## 11.5 Grooming in a General Mesh Network

Most of the early work on grooming has focused on the ring topology. This is largely due to the fact that many networks employ SONET technology that is most often used in a ring topology. However, due to the growth in Internet traffic, an increasing number of networks are being arranged in a general mesh topology. This is because in many cases mesh networks have a compelling cost advantage over ring networks. Also, mesh networks are more resilient to various network failures and more flexible in accommodating changes in traffic demands (e.g., see [7, 16] and references therein). Therefore, there is a need to extend the grooming work to general mesh networks. In general, the traffic grooming problem for mesh networks has to be considered in combination with RWA problem, which we call the GRWA problem.

Typically, the cost of a nation-wide optical network is dominated by optical transponders and optical amplifiers. If one assumes that the fiber routes are fixed, then the amplifier cost is constant, in which case one should concentrate on minimizing the number of transponders in the network. Multiplexing and switching costs should also be considered. However, under realistic assumptions of either a low-cost interconnect between multiplexing equipment and transport equipment, or integrated (long-reach) transponders on the multiplexing equipment (as is typical of SONET ADMs), the relative cost of the grooming switch fabric is negligible, and minimizing transponders is still the correct objective. In addition, the advent of Ultra Long-Haul transmission often permits optical pass-through at junction nodes, hence, requiring transponders only at the end of lightpaths.

In early work on the RWA problem (e.g., see [25, Chapter 8] and references therein), the issue of grooming has largely been ignored, i.e., it has been assumed that each traffic demand takes up an entire wavelength. The traffic grooming problem for mesh networks is only recently considered in [19, 32, 18]. In [19], an attempt is made at solving the general grooming problem by formulating it as a 0/1 multi-commodity network flow problem with the goal of minimizing the number of links used. Clearly, minimizing link-usage is equivalent to minimizing the number of transponders because each link represents a lightpath and each lightpath requires the appropriate transponders for terminations and processing of the terminated traffic. Unfortunately, the 0/1 multi-commodity network flow problem is NP-complete, and very few algorithms have been developed for the problem. We should also point out that the issue of wavelength assignment was not considered in [19]. In [32], the objective considered is either to maximize the network throughput or to minimize the connection-blocking probability, which are operational network-design problems. In [18], the problem of GRWA with the objective of minimizing the number of transponders in the network is considered. The problem is first formulated as an ILP problem. Unfortunately, the resulting ILP problem is usually very hard to solve computationally, in particular for large networks. To overcome this difficulty, a decomposition method was proposed that divides the GRWA problem into two smaller problems: the traffic grooming and routing (GR) problem and the wavelength assignment (WA) problem. In the GR problem, one only needs to consider how to groom and route traffic demands onto lightpaths (with the same objective of minimizing the number of transponders) and the issue of how to assign specific wavelengths to lightpaths can be ignored. Similar to the GRWA problem, the GR problem is again formulated as an ILP problem. The size of the GR ILP problem is much smaller than its corresponding GRWA ILP problem. Furthermore, one can significantly improve the computational efficiency for the GR ILP problem by relaxing some of its integer constraints, which usually leads to near-optimal solutions for the GR problem. Once the GR problem is solved, one can then consider the WA problem with the goal of deriving a feasible wavelength assignment solution, that in many cases is quite easy to obtain.

## 11.6 Grooming in IP Networks

In future IP networks, SONET ADMs may no longer be needed to multiplex traffic onto wavelengths. Instead, future IP networks will involve routers that are connected via wavelengths using WDM cross-connects as shown in Figure 11.6a. Since the SONET multiplexers have been eliminated, the function of multiplexing traffic onto wavelengths has now been passed onto the IP routers. Unless optical bypass is intelligently employed, with the new architecture, all of the traffic on all fiber and on all wavelengths (which amounts to multiple Tera-bits) will now have to be processed at every IP router. Routers of this size and capacity far exceed any near-term prospects; and even when such routers could be built, they are likely to be very costly. This situation can be alleviated through the use of a WDM cross-connect to provide optical bypass as shown in Figure 11.6b. In order to achieve maximum efficiencies, one would need to bundle traffic onto wavelengths so that the number of wavelengths that have to be processed at each router is minimized. This objective results in both reducing the number of ports needed on the routers (one per wavelength add/dropped at the router) as well as reducing the total switching capacity of the router.

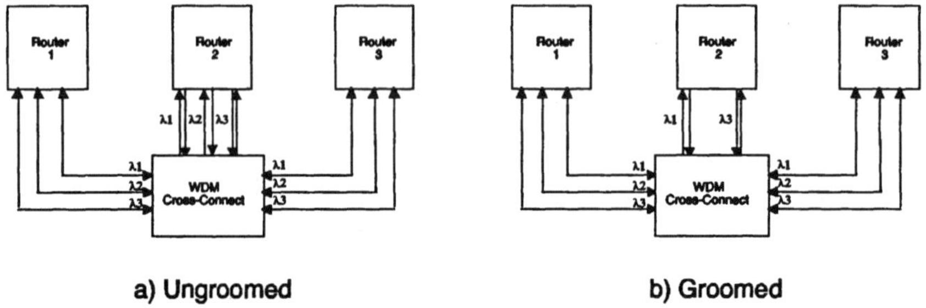


*Figure 11.5.* Grooming in an IP/WDM architecture.

This problem is similar to that of grooming of SONET streams described earlier. However, a number of important differences arise when considering the grooming of router traffic. First, unlike SONET networks, that are typically arranged in a ring topology, IP networks are arranged in a more general topology and hence the earlier grooming results cannot be applied directly. Second, SONET circuits are typically provisioned well in advance and remain for very long periods of time. As a result, in the case of a SONET network, the traffic grooming problem can be solved in advance, and network equipment can be laid-out accordingly. Most previous work on grooming for SONET rings considered particular traffic patterns (typically uniform traffic) for which a solution to the grooming problem was obtained. In the case of an IP network, not only is a uniform traffic pattern inappropriate, but also the traffic patterns are highly dynamic and hence a static solution would not be of much use.

## 11.7 The Impact of Tunable Transceivers

Here we consider the benefits of tunability in reducing electronic port counts in WDM/TDM networks (TDM stands for time division multiplexing). For a given traffic demand, we consider the design of networks that use the minimum number of tunable ports, where a tunable port refers to the combination of a tunable optical transceiver and an electronic port. Consider a network with $N$ nodes. On each wavelength in the network, up to $g$ low-rate circuits can be time division multiplexed, where $g$ is the traffic granularity. A static traffic requirement for the network is given by an $N \times N$ matrix $[m_{i,j}]_{N \times N}$, where $m_{i,j}$ is the number of circuits required from node $i$ to node $j$. Each node in the network is assumed to have a set of tunable ports, where each port includes a tunable optical transmitter and a tunable optical receiver. To illustrate the potential advantages of tunability, consider the following simple example of a unidirectional ring with $N = 4$, $g = 3$, and $m_{i,j} = 1$ for all $i, j$. In this case, the minimum number of wavelengths is 2, and there is a total of $N(N-1) = 12$ circuits that need to be assigned to the wavelengths. With $g = 3$, as many as 6 circuits can be assigned to each wavelength; this can be accomplished by assigning both circuits for each duplex connection to same time-slot. The traffic demand can then be supported by finding an assignment of each duplex connection to one of the $g$ time-slots in the TDM frame, on one of the wavelengths in the ring. Without the possibility of tunable transceivers, the assignment of circuits to wavelengths corresponds to the standard traffic grooming problem considered so far, for which the optimal grooming solution is given in Table 11.5.

*Table 11.5.* An optimal traffic assignment for fixed tuned transceivers.

| | $\lambda_1$ | $\lambda_2$ |
|---|---|---|
| Slot 1 | (1-2) | (2-3) |
| Slot 2 | (1-3) | (2-4) |
| Slot 3 | (1-4) | (3-4) |

*Table 11.6.* Optimal traffic assignment with tunable transceivers.

| | $\lambda_1$ | $\lambda_2$ |
|---|---|---|
| Slot 1 | (1-2) | (3-4) |
| Slot 2 | (1-3) | (2-4) |
| Slot 3 | (1-4) | (2-3) |

However, it was shown in [4] that using tunable transceivers can help reduce the number of transceivers significantly. For example, consider the traffic assignment given above. Notice that node 3 only transmits and receives one wavelength at any given time (i.e., wavelength 2 in slot 1, wavelength 1 in slot 2 and wavelength 2 in slot 3). Hence if node 3 were equipped with a tunable transceiver, it would only need one transceiver rather than 2 and a total of 6 transceivers would be required. In the above assignment of circuits to slots, nodes 2 and 4 must transmit on both wavelengths in the same slot and hence must each be equipped with two transceivers. Alternatively, a more clever assignment, shown in Table 11.6, requires each node to transmit only on one wavelength during each slot and hence each node need only be equipped with a single tunable transceiver.

In this example, we show that the number of transceivers can be reduced from 7 to 4 by proper slot assignment. In this case, the optimal assignment can be found by inspection; however in larger networks this may be a non-trivial combinatorial problem. In fact, it was shown in [4] that in general the optimal assignment problem with tunable transceivers is NP-complete. The approach in [4] transforms the traffic grooming with tunable transceivers problem into a graph edge-coloring problem. While the graph coloring problem is known to be NP-complete, in many cases an exact solution can be found. For example, in the uniform traffic case, it was shown in [4] that with the use of tunable transceivers, each node can use the minimum number of transceivers, i.e., no more transceivers than the amount of traffic that it generates. This result hold for general traffic as well, as long as the number of wavelengths is not limited. With limited wavelengths, [4] provides algorithms that are very nearly optimal and significantly reduce the number of transceivers as compared to the fixed tuned transceivers case.

## 11.8 Summary

In this article, we attempted to expose the reader to various aspects of the traffic grooming problem. For a more comprehensive survey of the grooming literature the reader is referred to [9]. We start with a discussion of the static traffic grooming problem. The static problem, at this point is rather well understood. In the most general case, the problem can be formulated as an ILP and solved using various heuristics. However, many aspects of traffic grooming remain largely unexplored. Those include the problem of grooming of stochastic traffic, as well as grooming traffic with tunable transceivers [4]. The latter problem begins to expose a fundamental aspect of optical networking, whereby through the use of optical time division multiplexing (TDM) techniques, electronic processing, both in the form of switching and line terminal processing, can be drastically reduced in the network.

## References

[1] D. Banerjee and B. Mukherjee, "A practical approach for routing and wavelength assignment in large wavelength-routed optical networks," *IEEE Journal on Selected Areas in Communications,* Vol. 14, No. 5, pp. 903-908, 1996.

[2] R. Barry and P. Humblet, "Models of blocking probability in all-optical networks with and without wavelength changes," *IEEE Journal on Selected Areas in Communications,* Vol. 14, No. 5, pp. 858-865, 1996.

[3] R. Berry and E. Modiano, "Reducing Electronic Multiplexing Costs in SONET/WDM Rings with Dynamically Changing Traffic," *IEEE Journal of Selected Areas in Communications,* October 2000.

[4] R. Berry and E. Modiano, " On the Benefit of Tunability in Reducing Electronic Port Counts in WDM/TDM Networks," *IEEE Infocom,* Hong Kong, March, 2004.

[5] G. Calinescu, O. Frieder, P. J. Wan, " Minimizing electronic line terminals for automatic ring protection in general WDM optical networks," *IEEE Journal on Selected Areas in*

*Communications,* pp. 183-189, January, 2002.

[6] T. Y. Chow and P. J. Lin, "The Ring Grooming Problem," submitted to SIAM J. Discrete Math., available on Los Alamos ArXiv, http://www.arxiv.org/abs/math.OC/0101091.

[7] L. A. Cox, J. Sanchez, and L. Lu, "Cost savings from optimized packing and grooming of optical circuits: mesh versus ring comparisons," *Optical Networks Magazines,* pp. 72-90, May/June, 2001.

[8] R. Dutta and G. Rouskas, "On optimal traffic grooming in WDM rings," in *Proc. of SIGMETRICS/PERFORMANCE 2001,* (ACM, Cambridge, MA, June 2001), pp. 164-174.

[9] R. Dutta and G. N. Rouskas, "Traffic grooming in WDM networks: past and future," *IEEE Network,* pp. 46-56, Nov.-Dec. 2002.

[10] T. Eilam, S. Moran, S. Zaks, "Lightpath arrangement in survivable rings to minimize the switching cost," *IEEE Journal on Selected Areas in Communications,* pp. 183-189, January, 2002.

[11] O. Gerstel, P. Lin, and G. Sasaki, "Wavelength assignment in a WDM ring to minimize the cost of embedded SONET rings," in *Proc. of Infocom,* San Francisco, CA, Apr. 1998, pp. 94-101.

[12] O. Gerstel, P. Lin, and G. Sasaki, "Combined WDM and SONET network design," in *Proc. Infocom,* New York, NY, Mar. 1999, pp. 734-743.

[13] O. Gerstel and R. Ramaswami, "Cost effective Grooming in WDM Rings," in *Proc. of Infocom,* San Francisco, CA, April, 1998.

[14] O. Gerstel, R. Ramaswami, and G. Sasaki, "Cost-effective traffic grooming in WDM rings," *IEEE/ACM Transactions on Networking,* Vol. 5, No. 5, pp. 618-630, 2000.

[15] O. Gerstel, G. Sasaki, S. Kutten, "Worst-case analysis of dynamic wavelength allocation in optical networks," *IEEE/ACM Transactions on Networking,* Vol. 7, No. 6, pp. 833-845, 1999.

[16] W. Grover, J. Doucette, M. Clouqueur, D. Leung, "New options and insights for survivable transport networks," *The Communications Magazine,* Vol. 40, No. 1, pp. 34-41, January 2002.

[17] J. Q. Hu, "Traffic grooming in WDM ring networks: A linear programming solution," *Journal of Optical Networks,* Vol. 1, pp. 397-408, 2002.

[18] J. Q. Hu and B. Leida, "Traffic Grooming, Routing, and Wavelength Assignment in Optical WDM Mesh Networks," manuscript, 2002. (available http://people.bu.edu/hqiang/papers/grw.pdf)

[19] V. R. Konda and T. Y. Chow, "Algorithm for traffic grooming in optical networks to minimize the number of transceivers," in *Proceedings of IEEE 2001 Workshop on High Performance Switching and Routing,* 2001.

[20] E. Modiano and R. Berry, "Using grooming cross-connects to reduce ADM costs in SONET/WDM ring networks," *Proc. of OFC 2001,* Anaheim, CA, March 2001.

[21] E. Modiano and R. Berry, "The Role of Switching in Reducing Network Port Counts", in *Proceedings of the 39th Annual Allerton Conference on Communication, Control, and Computing,* Allerton, Illinois, September, 2001.

[22] E. Modiano and R. Berry, "The Role of Switching in Reducing the Number of Electronic Ports in WDM Networks", in *IEEE Journal of Selected Areas in Communications,* to appear, 2004.

[23] E. Modiano and A. Chiu, "Traffic Grooming Algorithms for Minimizing Electronic Multiplexing Costs in Unidirectional SONET/WDM Ring Networks," CISS 98, Princeton, NJ, February, 1998. Extended version appeared in *IEEE Journal of Lightwave Technology,* January, 2000.

[24] E. Modiano and P. J. Lin, "Traffic grooming in WDM networks," *IEEE Communications Magazine,* pp. 124-129, July, 2001.

[25] R. Ramaswami and K. Sivarajan, *Optical Networks: A Practical Perspective,* Morgan Kaufmann Publishers, 1998.

[26] G. Sasaki, O. Gerstel, and R. Ramaswami, "A WDM ring network for incremental traffic," in *Proceedings of The Thirty-Sixth Annual Allerton Conference on Communication, Control, and Computing,* Monticello, IL, September 23-25, 1998.

[27] J. Simmons and A. Saleh, "Quantifying the benefit of wavelength add-drop in WDM rings with distance-independent and dependent traffic," *IEEE J. Lightwave Technology,* vol. 17, pp. 48-57, Jan. 1999.

[28] S. Subramaniam, A. Somani, M. Azizoglu, and R. Barry, "A performance model for wavelength conversion with non-Poisson traffic," in *Proc. Infocom,* (IEEE, Kobe, Japan, April 1997), Vol. 2, pp. 499-506.

[29] J. Wang, W. Cho, V. Vemuri, and B. Mukherjee, "Improved approaches for cost-effective traffic grooming in WDM ring networks: ILP formulations and single-hop and multihop connections," *Journal of Lightwave Technology,* Vol. 19, No. 11, pp. 1645-1653, 2001.

[30] X. Zhang and C. Qiao, "An effective and comprehensive approach to traffic grooming and wavelength assignment in SONET/WDM rings," in *SPIE Proc. Conf. All-Opt. Networking,* vol. 3531, Boston, MA, Sept. 1998.

[31] C. M. Zhao and J. Q. Hu, "Traffic Grooming for WDM Rings with Dynamic Traffic," manuscript, 2003. (available http://people.bu.edu/hqiang/papers/dynamic.pdf)

[32] K. Zhu and B. Mukherjee, "Traffic grooming in an optical WDM mesh network," *Proc. of IEEE International Conference on Communications,* pp. 721-725, Helsinki, Finland, June, 2001.

Chapter 12

# A PRACTICAL AND COST-EFFECTIVE APPROACH TO EFFICIENT TRAFFIC GROOMING IN WDM MESH NETWORKS

Harsha V. Madhyastha[1] and C. Siva Ram Murthy[2]

[1]*Department of Computer Science and Engineering, University of Washington, Seattle, WA 98105*

[2]*Department of Computer Science and Engineering, Indian Institute of Technology, Madras, Chennai 600036, INDIA*

Email: harsha@cs.washington.edu, murthy@iitm.ac.in

**Abstract** In this chapter, we present a new scheme for traffic grooming in WDM mesh networks. We propose a new node architecture which brings together all the three qualities desired: practical feasibility, cost-effectiveness and efficient grooming capability. None of the models considered so far in the literature have managed to satisfy all three criteria. We achieve these three ideals by considering a combination of groomers at multiple traffic granularities. We also present an algorithm for efficient traffic grooming with this new architecture. We justify the need for this new algorithm by imposing our node architecture on existing algorithms and comparing with them through a wide range of simulations.

**Keywords:** WDM optical mesh network, Grooming architecture, Traffic grooming, Integer linear Programming, Logical topology

## 12.1 Introduction

The advent of Wavelength Division Multiplexed (WDM) Optical Networks has made it possible for each physical link to carry traffic of the order of Tbps. Several wavelengths can be multiplexed on the same fiber with each wavelength capable of carrying traffic up to 10 Gbps. Yet, individual traffic demands are still of the order of Mbps. This requires efficient grouping of individual connections onto the same wavelength as dedicating a unique wavelength for each demand will lead to huge wastage of bandwidth. Intelligent grouping is also required because each wavelength has to be dropped at the source and destination of each of the connections assigned to it. Dropping a wavelength at any node involves conversion from optical to electronic domain, and the equipment for performing this is the main contributor towards the cost of the network. This grouping of connections and assigning wavelengths to these groups, so as to optimize on some objective such as throughput or network cost, is termed as "traffic grooming".

Traffic grooming in WDM optical networks has been the focus of research in much of the recent work. As the traffic grooming problem is known to be NP-hard for arbitrary traffic [7], most of the work has been limited to domains with constraints on traffic or physical topology. Almost all of the work has only looked at the traffic grooming problem in SONET/WDM rings. Initial research focused on uniform traffic and used circle construction techniques to minimize the number of wavelengths as well as the number of Add-Drop Multiplexers (ADMs) [11, 10]. Improving on this, [1] and [2] tried to minimize the overall network cost, considering parameters such as maximum number of physical hops, though they too concentrated on rings and laid emphasis mainly on uniform traffic. Non-uniform traffic, albeit with the constraint that the total traffic added or dropped at any node is lesser than some threshold, was handled for the first time in [5] and [6]. The first attempt to handle arbitrary traffic, though restricted to SONET rings, is seen in [9].

The recent surge in the industry to use mesh networks instead of SONET rings has breathed life into research on traffic grooming in mesh networks as well. The initial work in this direction, trying to minimize the number of transceivers by designing a suitable virtual topology, is found in [4]. Its main shortcoming is that it does not consider any limitations on the physical topology such as number of wavelengths. Grooming in survivable WDM mesh networks, assuming single-link failure, was considered in [8]; but it addressed the issue of dynamic grooming. The same problem was also addressed in [12], though without considering the survivability aspect. Dynamic grooming is the problem of routing and assigning wavelengths for a new demand, given the current state of the network, whereas in static grooming the traffic demands are known *a priori* and all of them have to be assigned routes and wavelengths to minimize required resources (wavelengths and grooming ports). Static grooming can also be viewed from the angle of maximizing the throughput given the constraints on resources. This problem has been addressed in the context of mesh networks in [13]. It outlines two different node architectures - MPLS/IP and SONET/WDM. It propounds that the former is more cost-effective and an ILP formulation for grooming with this architecture is presented. It also proposes two heuristics for traffic grooming with the MPLS/IP node architecture in a WDM mesh network. The main drawback in this work is that it assumes unlimited grooming capability (ability to switch traffic among streams) at each node as multiplexing is done in software in the MPLS/IP architecture. This is not a practical assumption as routing each packet by examining its header involves a large overhead, which makes the setup incapable of handling the large bandwidth of an optical WDM link. Hence, full-scale grooming, *i.e.*, as much multiplexing ability as required, at extremely fine granularities is not practically feasible. (Grooming at fine granularities involves switching streams which carry very low traffic while grooming at coarse granularities can only switch higher-rate traffic streams.) On the other hand, the SONET/WDM architecture also has its own limitations. Here, the switching cost of the groomer is proportional to the square of the number of ports on it (the number of ports on a groomer is the number of streams it is capable of grooming). So, though using the SONET/WDM architecture will lead to lesser grooming equipment cost as the number of traffic streams it can groom is limited, the overall cost will be high as grooming can only be done at coarse granularities, which will lead to the need for a greater number of wavelengths. In short, though MPLS/IP is efficient, it is infeasible and not cost-effective because of high processing overhead and SONET/WDM, though feasible, is neither efficient nor cost-effective because of grooming at coarse levels and the high switching cost involved.

In this chapter, we propose a new node architecture which does away with the shortcomings of the above two architectures and combines their advantages to achieve the right combination of feasibility, efficiency and cost-effectiveness. We do so by having groomers at multiple granularities at each node. The concept of using a multi-layer node architecture was also considered in [3]. But, it failed to identify the full potential of grooming at multiple levels. In the architecture used in [3], any add-drop traffic has to pass through the complete hierarchy from bottom to top. Due to this dependence between levels, switching cost benefits are obtained only at the intermediate nodes of lightpaths. Our node architecture ensures saving in switching cost at all nodes as the groomers at different layers are completely independent, which makes use of the true strength of grooming at multiple granularities.

The remaining part of the chapter is organized as follows. A detailed description of the node architecture we propose is given in Section 12.2. The exact specification of the traffic grooming problem we attempt to solve in this chapter is clearly stated in Section 12.3. To find the solution to this problem, we first give an ILP formulation in Section 12.4 and then propose a heuristic algorithm in Section 12.5. The working of our heuristic is illustrated in Section 12.6 with the help of an example and in Section 12.7, we present the results of the various simulations performed to study the performance of our heuristic. Finally in Section 12.8, we conclude and provide directions for future work.

## 12.2 Node Architecture

Our proposed novel node architecture involves the use of two groomers - one at a coarse level and the other at a finer level of granularity, which we call the higher level and lower level groomer, respectively. To make this setup practically feasible, unlike the MPLS/IP architecture, we work with the practical assumption that the number of ports on the lower level groomer is limited. Though limited, the capability to groom at finer levels helps in efficient grooming by reducing the number of required wavelengths compared to that possible with the higher level groomer alone. The additional cost of the lower level groomer is more than offset by the decrease in infrastructure cost due to fewer wavelengths. In addition to the coarse and fine granularity groomers, our node architecture also makes use of a mapper, which has negligible cost as it does no processing; it just multiplexes/demultiplexes the add/drop traffic assuming best possible packing of the lower level streams into the higher level streams. Its low cost is due to the fact that it does not perform any switching.

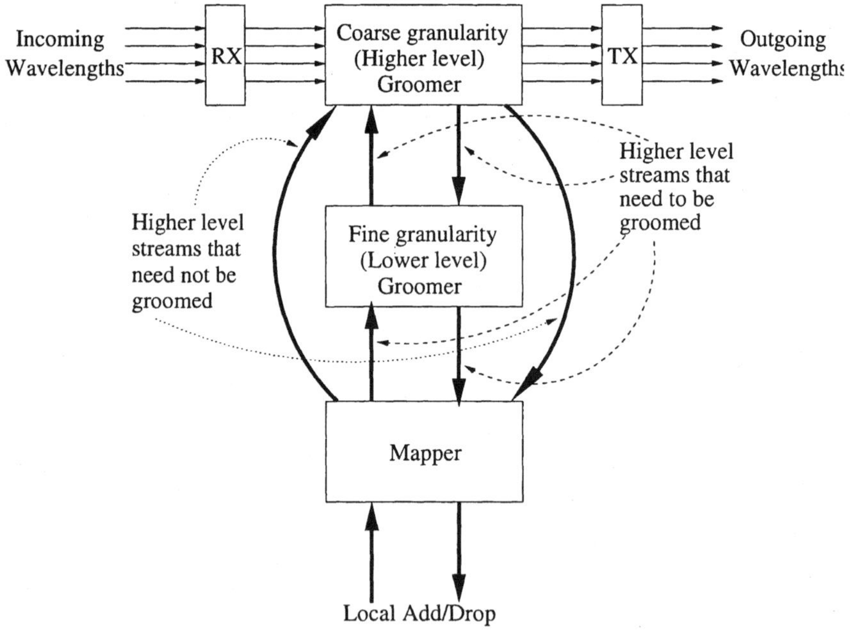


*Figure 12.1.* Mixed groomer node architecture.

The mixed groomer node architecture we present is shown in Fig. 12.1. This architecture is a very generic one and can be used on any hierarchy of traffic streams, for example, OC-48/OC-12/OC-3 or STM-16/STM-4/STM-1 or STM-1/VC-3/VC-12. From now on, for the sake of convenience, we will refer to a wavelength as OC-48, a higher level stream as OC-12 and a lower level stream as OC-3. So, in the node architecture shown, the OC-48s that need to be groomed are converted from optical to electronic form by the Receiver Array (RX) and fed as input to the higher level groomer. The function performed by the higher level groomer is to switch OC-12s among the different OC-48s it receives as input. The OC-12 groomer also receives OC-12s which do not need to be groomed (because they might be completely packed with OC-3s setup between the same source-destination pair), padded up to OC-48s, as input from the mapper. The mapper can also be implemented such that every OC-12, in which all the OC-3s on it are between the same source-destination pair, can be directed from the mapper to the OC-12 groomer. However, doing so when the OC-12 is not completely packed entails higher implementation complexity (as detection of padded up OC-3s is required). Hence, in our proposal, we only require the mapper to redirect OC-12s completely packed with OC-3s between the same source-destination pair to the OC-12 groomer.

If there is also a need to switch OC-3s among the OC-12s, then the OC-12 groomer feeds the corresponding OC-12 streams as input to the OC-3 groomer. Also, among the OC-12s generated by the mapper from the add/drop traffic, the ones which are not completely packed are routed to the OC-3 groomer. The streams between the OC-12 and the OC-3 groomer are essentially OC-48s, but only the OC-3 groomer can index the OC-3s within each OC-48 and switch them if required. The mapper receives the local add/drop traffic as input in the form of OC-3s padded up to OC-12s, and tries to pack them into OC-12s optimally. It does this by taking groups of 4 (the *groom factor* in this case) OC-12s and mapping the single OC-3s on them onto one OC-12. This has very low processing overhead as the OC-3s can be statically mapped to respective OC-12s. Since the mapper receives OC-3s padded up to OC-12s as input, if some OC-12 is assigned just one OC-3, then that OC-12 can be directly padded up to a OC-48 by the mapper and sent to the OC-12 groomer, bypassing the OC-3 groomer. The outgoing traffic from the node is converted from electronic to optical domain by the Transmitter Array (TX).

Essentially, the mixed groomer architecture can be divided into two logical units - the multiplexing/demultiplexing section (mapper) and the switching section (OC-3 and OC-12 groomers). The add/drop traffic that goes in and out of the groomer is in the form wherein each OC-3 is on a distinct OC-12. The mapper performs the task of multiplexing the OC-3s which are on the OC-12s which constitute the add traffic. This multiplexing is carried out to ensure best possible packing, *i.e.*, the OC-3s on every 4 OC-12s are multiplexed into 1 OC-12. The drop traffic is also similarly packed in the best possible manner. The mapper demultiplexes the OC-12s which constitute the drop traffic such that each of the OC-3s on these OC-12s is on a distinct OC-12. The task of switching traffic is completely handled by the OC-3 and OC-12 groomers. Since the mapper does not perform any switching, its cost is negligible in comparison with that of the groomers. The role of the OC-3 groomer is to switch OC-3s among OC-12s. Similarly, the function of the OC-12 groomer is to switch OC-12s among OC-48s. The number of switching ports taken up on the OC-3 groomer is the number of OC-12 streams it has to switch traffic amongst. Hence, from Fig. 12.1, it is clear that the number of OC-3 switching ports required is the sum of two quantities. The first being the number of OC-12s between it and the mapper. And, the second is the number of OC-12s between it and the OC-12 groomer. From the above explanation of how the mapper works, the number of OC-12 streams between the OC-3 groomer and the mapper is equal to $\lceil$ (Total add/drop traffic in terms of OC-3s)/(groom factor) $\rceil$. Similarly, the number of ports required on the OC-12 groomer is also the sum of two quantities. In this case, the first is the number of OC-48s fed as input to the groomer. The second is the number of OC-48s onto which it has to switch OC-12s, which are then fed as input to the OC-3 groomer. At the maximum, the value both these quantities take up is the number of OC-48s supported on the links incident at the node.

In our node architecture, the number of ports on the OC-3 groomer is constrained as this is a major contributor towards the cost of the setup. On the other hand, the number of ports on the OC-12 groomer can be assumed to be practically unlimited as grooming at a coarse level is comparatively inexpensive. Moreover, the number of ports required for full-scale grooming is lesser. To get an estimate of this, consider an OC-768 backbone, *i.e.*, each node in the network can handle bandwidth equivalent to OC-768. Since OC-768 is equal to 16 OC-48s, unlimited grooming capability at the OC-12 level would require 32 ports. This is because 16 ports would be required for the OC-48s on the link and another 16 for the add/drop traffic. On the other hand, since OC-768 is equivalent to 64 OC-12s, the number of ports required on the OC-3 groomer for input from the OC-12 groomer is 64. Also, 64 ports would be required for the OC-12s received from the mapper. This implies that a total of 128 ports are required on the OC-3 groomer. This quantity is 4 times as many as that on the OC-12 groomer. Since switching cost is proportional to the square of the number of ports, the switching cost at the OC-3 level is more than 16 times that at the OC-12 level[1]. On the whole, this clearly makes the cost of full-scale grooming at the OC-12 level negligible compared to that at the OC-3 level.

Let us now look at the advantages of the mixed groomer architecture over that of an OC-3 groomer or OC-12 groomer alone. If an OC-12 groomer alone is employed, it does not have the capability to switch OC-3s among OC-12s. So, the add/drop traffic in the form of OC-3s padded up to OC-12s cannot be multiplexed together. Each of these OC-12s will have to be assigned as they are to OC-48s on the link. Hence, as each OC-12 can only have one OC-3, the maximum traffic that can be supported is $\frac{1}{4}$, *i.e.*, 1/(groom factor) of the total bandwidth. On the other hand, using the OC-3 groomer alone suffers from two disadvantages. Firstly, due to the absence of the OC-12 groomer, if OC-12s among two OC-48s need to be swapped (switching at the OC-12 level), this has to be done by swapping each of the OC-3s on these OC-12s. This is costlier as switching needs to be done at a finer granularity. Also, since there is no OC-12 groomer to pick out the specific OC-12s, all the OC-12s on these OC-48s will have to be fed as input to the OC-3 groomer. More importantly, the OC-3 groomer directly receives OC-3s padded up to OC-12s as input. This implies that the number of ports consumed due to the add/drop traffic is equal to the number of OC-3s in the add/drop traffic. Note that in the mixed groomer, this number was $(\frac{1}{4})^{th}$ of the add/drop traffic because multiplexing/demultiplexing is performed by the mapper. So, the mixed groomer architecture derives its efficiency by the combination of the OC-3 and OC-12 groomers and also, maintains practical feasibility and cost-effectiveness by the constraint on the number of ports on the OC-3 groomer.

### 12.2.1 Example

The following example clearly brings to the fore the advantages of using a combination of groomers in place of having an OC-3 or OC-12 groomer alone. Consider the 6-node network shown in Fig. 12.2(a) with demands of 3, 1 and 4 OC-3s between the (source, destination) pairs *(1, 4)*, *(3, 5)* and *(2, 6)*, respectively. As outlined above switching cost at the OC-12 level is negligible to that at the OC-3 level which only depends on the number of OC-12s on the link. So, from here on, we consider OC-12 as a wavelength. The state of the network in each of the three cases explained below is as shown in Fig. 12.2.

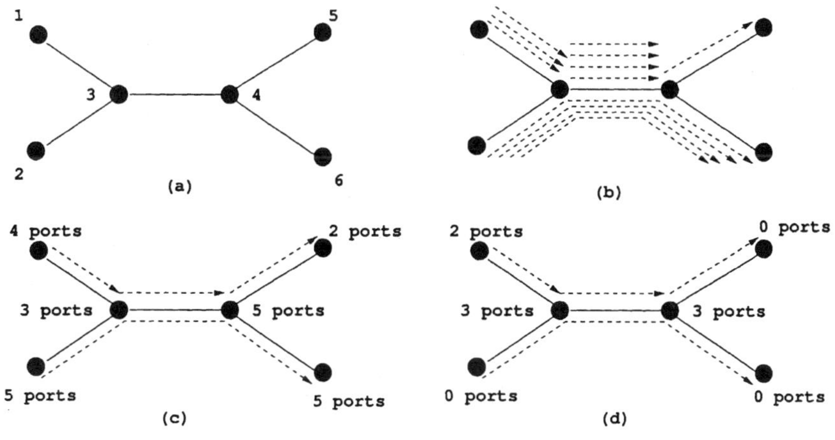


*Figure 12.2.* (a) Example 6-node physical topology and network state with (b) OC-12 groomer, (c) OC-3 groomer and (d) mixed groomer.

- OC-12 groomer alone: When only an OC-12 groomer is available at each node, there is no grooming capability at the OC-3 level at any node. So, multiple OC-3s cannot be groomed onto the same OC-12, which implies that each OC-3 has to be carried on a new OC-12. This in turn implies that the number of wavelengths required on a link is equal to the total number of OC-3s transmitted along the link. As shown in Fig. 12.2(b), though there is no switching cost at any of the nodes, the overall network cost is high due to the large number of wavelengths required to satisfy the traffic demand. In this example, at least 8 wavelengths are required on link *(3, 4)*.

- OC-3 groomer alone: If each node has an OC-3 groomer with full-scale grooming capability, then optimal grooming can be performed as shown in Fig. 12.2(c). But, the downside of this scheme is the high switching cost borne due to the large number of grooming ports required at each node, as shown in Fig. 12.2(c). As explained before, each add/drop OC-3 consumes a port on the OC-3 groomer and hence, the add/drop traffic itself consumes 3, 4, 1, 3, 1 and 4 ports at nodes 1, 2, 3, 4, 5 and 6, respectively. Also, at every node, among all the OC-12s on the links incident at that node, every OC-12 that needs to be groomed consumes an OC-3 grooming port at that node. An OC-12 needs to be groomed if some OC-3s on it need to be either dropped or switched to other OC-12s. All these properties together necessitate as many as 5 OC-3 grooming ports at nodes 2, 4 and 6. So, though the number of wavelengths required is reduced from 8 to 2 in comparison with the previous case, the grooming cost introduced keeps the network cost high.

- Mixed groomer architecture (OC-12 groomer + OC-3 groomer + Mapper): The network state achieved with the mixed groomer node architecture (shown in Fig. 12.2(d)) clearly highlights its merits because as in the case with the OC-3 groomer alone, the number of wavelengths required is 2 but with much lower switching cost. The maximum number of ports needed at any of the nodes is 3 and three of the nodes do not even require an OC-3 groomer. The 4 OC-3s from node 2 to node 6 can be routed on the same OC-12 without consuming any OC-3 grooming ports as an OC-12 which is completely packed with OC-3s between the same (source, destination) pair directly goes from the mapper to the OC-12 groomer. Also, no ports are required for the OC-12 from node 4 to node 5 as a single OC-3 is put onto it. Lesser number of ports are also taken up at nodes 1, 3 and 4 because the mapper multiplexes/demultiplexes the add/drop traffic and hence, the number of ports consumed on the OC-3 groomer by the add/drop traffic is only $(\frac{1}{4})^{th}$ the number of add/drop

OC-3s, which in this example translates into only one port at each of these nodes.

This example shows that our mixed groomer node architecture brings together the beneficial features of both a coarse granularity and a fine granularity groomer, *i.e.*, lower switching cost and lesser number of wavelengths required, respectively.

## 12.3 Problem Statement

The problem we address in this chapter is that of static grooming, with the objective of maximizing throughput given the various constraints on the available resources. We account for the following resource constraints:

1 The maximum number of distinct wavelengths on which traffic can be routed on each physical link - $\mathbf{W}_{max}$. In our problem setting, this is the maximum number of OC-48s that can be carried on any link. But, since the grooming capability on the OC-12 groomer is practically unlimited, this can be equivalently seen as the maximum number of OC-12s on each link. We assume the same $\mathbf{W}_{max}$ to hold over all physical links.

2 The number of ports on the OC-3 groomer at each node - $\mathbf{P}_{max}$. This places a limit on the number of OC-12 streams that can be groomed at each node, *i.e.*, the number of OC-12s which require OC-3s on them to be either dropped at that node or switched to other OC-12s.

The parameters given as input to the static grooming problem are:

1 The number of nodes **N** in the network. Each node is assumed to have a groomer with the mixed groomer node architecture. We assume that all physical nodes have groomers of the same size.

2 The physical topology of the network is given in the form of the adjacency matrix **L**, where $\mathbf{L}_{i,j}$ = $\mathbf{L}_{j,i}$ = 1 or 0 if a link exists or does not exist between nodes **i** and **j**, respectively. This is called the *"single fiber"* scenario.

3 As our mixed groomer architecture is a generic one, we also take the groom factor **G**, *i.e.*, the ratio between the bandwidths of the higher level stream and the lower level stream as input. In our case where we consider OC-12 and OC-3, the groom factor is 4. So, the total traffic (in terms of OC-3s) that can be loaded on any physical link is $\mathbf{W}_{max} \times G$.

4 The traffic matrix **T** gives the traffic demand $\mathbf{T}_{i,j}$ with node **i** as source and node **j** as destination. In the mixed groomer architecture, a OC-12 packed with **G** OC-3s between the same (source, destination) pair is

routed directly from the mapper to the OC-12 groomer (refer the previous section on Node Architecture) and so, does not take up any ports on the OC-3 groomer. As we are placing a constraint on the ports only on the OC-3 groomer, all entries in **T** (specified in units of OC-3s) are assumed to be lesser than **G**. If any entry $\mathbf{T}_{i,j}$ is greater than **G**, then $\lfloor \frac{\mathbf{T_{i,j}}}{\mathbf{G}} \rfloor$ OC-12s can be padded to OC-48s and directly put through the OC-12 groomer, leaving behind the entry $\mathbf{T}_{i,j}$ mod **G**, which is lesser than **G**. Thus, even if no restriction is placed on the traffic matrix entries, the problem can be reduced to an equivalent one wherein each demand is lesser than **G** OC-3s.

Given the above inputs and the limitations on the infrastructure, we aim to maximize the throughput, *i.e.*, maximize the percentage of successfully routed traffic. We give an Integer Linear Programming formulation of this grooming problem in the next section. If solved, the answer to the ILP formulation gives us the optimal solution to our problem but, solving any ILP problem entails exponential complexity. So, the ILP formulation can be used only to optimally solve the grooming problem for networks with very few nodes dealing with sparse traffic matrices. As the static grooming problem is known to be NP-hard for arbitrary traffic even for ring networks [7], it is clearly NP-hard for mesh networks as well due to the increase in complexity of the problem at hand. Hence, in order to obtain solutions for large networks, we propose a heuristic algorithm for solving it near-optimally. We demonstrate the need for this algorithm by comparing its performance with the only alternatives available - the two heuristics proposed in [13]. On executing all the three algorithms on a wide variety of traffic patterns, the results clearly show that grooming with the mixed groomer architecture necessitates the use of our heuristic as it realizes much higher throughputs compared to the other two. We also demonstrate the near-optimality of our heuristic by comparing its performance with that obtained by solving the ILP formulation on small networks.

## 12.4 ILP Formulation

Our objective in solving the above problem is to determine which are the demands that can be successfully routed to maximize throughput. This problem is usually broken up into the following two sub-problems:

- Determination of the logical topology - Which are the lightpaths to be set up? The set of lightpaths are seen as the links over which connections are routed.

- Routing of individual connections - Which are the demands to be satisfied and how is each demand routed over the logical topology?

We follow the above approach in our ILP formulation, but we look at the same problem from a different angle in our heuristic. In it we take up the Routing and Wavelength Assignment (RWA) approach. We view the problem as determining the connections to be routed, assigning routes to each one of them and then allocating wavelengths for them on each physical link along their assigned route. We adhere to the logical topology approach in our ILP formulation as it facilitates easier counting of ports - each lightpath set up consumes one port each on the groomers at its source and destination.

Here, we present an ILP formulation of the static grooming problem explained in the previous section. This formulation is much on the same lines as that given in [13]. The difference comes in due to the replacement of the MPLS/IP architecture with the more efficient mixed groomer architecture. As the constraint on the number of ports on the lower level groomer needs to be imposed, we need to count the number of ports assigned on the lower level groomer at each node. Hence, the logical topology section of the formulation is identical to that in [13], but the remaining sections of the formulation differ.

The following definitions are used:

**m, n**: The nodes at either end of a physical link. The link (**m, n**) is considered to be a directed edge from **m** to **n**. **m, n** $\in$ [1, **N**]

**i, j**: The source and destination of a lightpath, which might traverse multiple physical links. **i, j** $\in$ [1, **N**]

**s, d**: The source and destination of a routed connection, which might span multiple lightpaths. **s, d** $\in$ [1, **N**]

**k**: Any general node in the network. **k** $\in$ [1, **N**]

**t**: index of an individual OC-3 among the different OC-3s established between the same (source, destination) pair. If $\mathbf{T}_{3,5} = 6$, then 6 OC-3s with 3 as source and 5 as destination are enumerated from 1 to 6.

The variables used in our formulation and their physical interpretation are as follows:

- $\mathbf{V}_{i,j}^{w}$ = Number of lightpaths established on wavelength **w**, with node **i** as source and node **j** as destination.

- $\mathbf{P}_{m,n}^{i,j,w}$ = Number of lightpaths setup between node **i** and node **j** on wavelength **w** which are routed through the physical link (**m, n**).

- $\mathbf{R}_{i,j,k,w}^{s,d,t}$ = 1, if the $\mathbf{t}^{th}$ OC-3 from **s** to **d** is routed through the lightpath from **i** to **j** on wavelength **w** which has its first physical hop following **i** as **k**, else it is 0. Note that the tuple (**i, j, k, w**) refers to a unique lightpath since at most one physical link is allowed between two nodes (from the specification of **L** in the problem definition).

- $\mathbf{g}^w_{i,j,k}$ = 1, if the lightpath from **i** to **j** on wavelength **w**, which has its first physical hop following **i** as **k**, needs to be groomed at the OC-3 level at nodes **i** and **j**, else it is 0.
- $\mathbf{S}^t_{s,d}$ = 1, if the $\mathbf{t}^{th}$ OC-3 from **s** to **d** is established, else it is 0.

The objective function is defined as: Maximize $\sum_s \sum_d \sum_{t=1}^{T_{s,d}} \mathbf{S}^t_{s,d} \quad (\mathbf{s} \neq \mathbf{d})$

The constraints are:

- There should be no lightpaths from node **i** to node **j**, passing through a link incoming into **i**.

$$\sum_m \mathbf{P}^{i,j,w}_{m,i} = 0 \quad \forall \mathbf{i}, \mathbf{j}, \mathbf{w} \tag{12.1}$$

- There should be no lightpaths from node **i** to node **j**, passing through a link outgoing from **j**.

$$\sum_n \mathbf{P}^{i,j,w}_{j,n} = 0 \quad \forall \mathbf{i}, \mathbf{j}, \mathbf{w} \tag{12.2}$$

- The number of lightpaths on wavelength **w** from node **i** to node **j**, passing through a link outgoing from **i**, should be equal to the number of lightpaths on wavelength **w** established from **i** to **j** in the logical topology.

$$\sum_n \mathbf{P}^{i,j,w}_{i,n} = \mathbf{V}^w_{i,j} \quad \forall \mathbf{i}, \mathbf{j}, \mathbf{w} \tag{12.3}$$

- The number of lightpaths on wavelength **w** from node **i** to node **j**, passing through a link incoming into **j**, should be equal to the number of lightpaths on wavelength **w** established from **i** to **j** in the logical topology.

$$\sum_m \mathbf{P}^{i,j,w}_{m,j} = \mathbf{V}^w_{i,j} \quad \forall \mathbf{i}, \mathbf{j}, \mathbf{w} \tag{12.4}$$

- For any node **k**, other than **i** and **j**, the number of lightpaths from **i** to **j** on wavelength **w** routed through links incoming into it should be equal to the number of lightpaths from **i** to **j** on wavelength **w** routed through links outgoing from it.

$$\sum_m \mathbf{P}^{i,j,w}_{m,k} = \sum_n \mathbf{P}^{i,j,w}_{k,n} \quad (\mathbf{k} \neq \mathbf{i}, \mathbf{j}) \ \forall \mathbf{i}, \mathbf{j}, \mathbf{w}, \mathbf{k} \tag{12.5}$$

- A lightpath can pass through a link only if the link exists. Also, at most one lightpath on a particular wavelength can be routed through a physical link.

$$\sum_{i,j} \mathbf{P}^{i,j,w}_{m,n} \leq \mathbf{L}_{m,n} \quad \forall \mathbf{m}, \mathbf{n}, \mathbf{w} \tag{12.6}$$

In our problem setting $\mathbf{L}_{m,n}$ is restricted to values 0 or 1 and so, the values taken by $\mathbf{P}_{m,n}^{i,j,w}$ are also restricted to 0 or 1. But, the same equation holds if multiple physical links are allowed between two nodes.

- If the $\mathbf{t}^{th}$ OC-3 from **s** to **d** is setup, then it must be routed through some lightpath originating at **s**.

$$\sum_{j,k,w} \mathbf{R}_{s,j,k,w}^{s,d,t} = \mathbf{S}_{s,d}^{t} \ \ (\mathbf{k} \neq \mathbf{s}) \ \ \forall \mathbf{s}, \mathbf{d}, \mathbf{t} \tag{12.7}$$

- If the $\mathbf{t}^{th}$ OC-3 from **s** to **d** is setup, then it must be routed through some lightpath terminating at **d**.

$$\sum_{i,k,w} \mathbf{R}_{i,d,k,w}^{s,d,t} = \mathbf{S}_{s,d}^{t} \ \ (\mathbf{k} \neq \mathbf{i}) \ \ \forall \mathbf{s}, \mathbf{d}, \mathbf{t} \tag{12.8}$$

- Any traffic with **s** as the source cannot be routed on a lightpath which terminates at **s**.

$$\sum_{i,k,w} \mathbf{R}_{i,s,k,w}^{s,d,t} = 0 \ \ (\mathbf{k} \neq \mathbf{i}) \ \ \forall \mathbf{s}, \mathbf{d}, \mathbf{t} \tag{12.9}$$

- Any traffic with **d** as the destination cannot be routed on a lightpath which originates at **d**.

$$\sum_{j,k,w} \mathbf{R}_{d,j,k,w}^{s,d,t} = 0 \ \ (\mathbf{k} \neq \mathbf{d}) \ \ \forall \mathbf{s}, \mathbf{d}, \mathbf{t} \tag{12.10}$$

- On any node **k**, other than **s** and **d**, if the $\mathbf{t}^{th}$ OC-3 from **s** to **d** is routed on some lightpath terminating at **k** then it must also be routed on some lightpath originating at **k**.

$$\sum_{i,j,w} \mathbf{R}_{i,k,j,w}^{s,d,t} = \sum_{i,j,w} \mathbf{R}_{k,j,i,w}^{s,d,t} \ \ (\mathbf{k} \neq \mathbf{s}, \mathbf{d}) \ \ \forall \mathbf{s}, \mathbf{d}, \mathbf{t}, \mathbf{k} \tag{12.11}$$

- Traffic can be routed on a lightpath only if it exists and the total traffic routed on it cannot exceed the capacity of a wavelength.

$$\sum_{s,d,t} \mathbf{R}_{i,j,k,w}^{s,d,t} \leq \mathbf{G} \times \mathbf{P}_{i,k}^{i,j,w} \ \ (\mathbf{k} \neq \mathbf{i}) \ \ \forall \mathbf{i}, \mathbf{j}, \mathbf{k}, \mathbf{w} \tag{12.12}$$

- The lightpath from **i** to **j** on wavelength **w**, which has its first physical hop after **i** as **k**, needs to be passed through the OC-3 groomer at nodes **i** and **j** if more than one OC-3 has been routed on it.

$$\sum_{s,d,t} \mathbf{R}_{i,j,k,w}^{s,d,t} \geq 2 \times \mathbf{g}_{i,j,k}^{w} \ \ (\mathbf{k} \neq \mathbf{i}) \ \ \forall \mathbf{i}, \mathbf{j}, \mathbf{k}, \mathbf{w} \tag{12.13}$$

$$\sum_{s,d,t} \mathbf{R}_{i,j,k,w}^{s,d,t} \leq (\mathbf{G} - 1) \times \mathbf{g}_{i,j,k}^{w} + 1 \ \ (\mathbf{k} \neq \mathbf{i}) \ \ \forall \mathbf{i}, \mathbf{j}, \mathbf{k}, \mathbf{w} \tag{12.14}$$

- The total number of lightpaths passing through the OC-3 groomer at each node must be lesser than the number of ports on the groomer.

$$\sum_{j,k,w} \mathbf{g}_{i,j,k}^{w} + \mathbf{g}_{j,i,k}^{w} \leq \mathbf{P}_{max} \quad (\mathbf{k} \neq \mathbf{i}, \mathbf{j}) \ \forall \mathbf{i} \tag{12.15}$$

  Though we assume in our problem setting that the number of ports on the OC-3 groomer is same at all nodes, the above given constraint can be easily modified to handle the case where $\mathbf{P}_{max}$ varies across nodes by replacing $\mathbf{P}_{max}$ by $\mathbf{P}_i$ in the above constraint, where $\mathbf{P}_i$ is the number of ports on the OC-3 groomer at node **i**.

## 12.5 Heuristic

If the ILP formulation given in the previous section is solved, the optimal solution to any instance of the static grooming problem we are considering can be obtained. But, since the number of variables and constraints in the formulation increases exponentially with increase in the size of the problem, practical considerations force us to take up heuristic approaches to obtain near-optimal solutions. A couple of heuristics - *Maximizing Single-Hop Traffic* (MST) and *Maximizing Resource Utilization* (MRU) - were proposed in [13]. We put forward another heuristic which is tailored to suit the mixed groomer architecture. We justify the need for this new heuristic by comparisons with those proposed in [13] which clearly show the superiority of our approach.

As outlined before, the approach we are going to follow is to determine the connections to be established and assign routes and wavelengths to them rather than build the logical topology and route the connections on it. We perform this iteratively by maintaining a partition of the set of connections, **A** and **B**, such that only those in **A** have been assigned, and in every iteration, one connection in set **B** is assigned and moved into set **A**. Since the main resource constraint limiting us from obtaining a 100% throughput is the limit on the number of ports on the OC-3 groomer, we assign the connection whose establishment would lead to the least increment in the number of OC-3 grooming ports used over all the nodes in the network. To make this decision, we need to determine for each connection in set **B** the route and corresponding wavelength assignment that would lead to the least increase in used OC-3 grooming ports among all possible route and wavelength assignments. We keep performing this iteratively until no connection can be established due to the constraints on the number of wavelengths ($\mathbf{W}_{max}$) and on the number of ports ($\mathbf{P}_{max}$) available. To evaluate the increase in ports at each stage, we determine the new lightpaths that need to be established and the old lightpaths that need to be split in order to setup the required route and wavelength assignment. Then we use the property that one OC-3 port each is taken up on the groomer at its source and destination by each lightpath carrying more than one OC-3.

Determination of the *"least-port-increase"* route and wavelength assignment for each connection would entail performing a search over all possible routes from the source to destination of that connection and over each possible wavelength assignment for each route. This search space is clearly exponential in size and as there is no possibility for pruning, the complexity involved in performing this search is exponential. First we try and reduce the search space in terms of number of routes to be examined. While considering just the shortest physical hop path from source to destination would contradict the very purpose of the search, searching over all possible routes is exponential. So, as a trade-off between complexity and optimality, we pre-determine the **k**-shortest paths for every (source, destination) pair and search over these **k** routes. **k** is a parameter which can be decreased or increased depending on whether faster execution or proximity to optimal solution is desired. Even though we have cut down on the complexity significantly by considering the **k**-shortest paths, the search remains exponential as we need to evaluate the increment in used ports over all possible wavelength assignments for each of these **k** routes. Hence, we resort to the following approach. Though $\mathbf{W}_{max}$ wavelengths are available on each physical link in the network, we start off our grooming heuristic assuming the network to have only 1 wavelength. Once the process of iteratively assigning connections stops because no more connections can be established, we increment the number of wavelengths to 2 and continue assigning connections. This process of grooming, incrementing number of wavelengths and then again grooming is performed until all $\mathbf{W}_{max}$ wavelengths have been used. This approach reduces the complexity because as the number of wavelengths is increased, the number of possible wavelength assignments does not increase by much due to the fact that the traffic on many wavelengths could have already been fully allotted on several physical links to connections assigned until then. All these modifications bring the complexity down to practical levels without adversely affecting the efficacy of the algorithm.

Until now, our heuristic neither has any look-ahead nor any adaptive component other than the property that the *"least-port-increase"* route and wavelength assignment depends on the current state of the network. To incorporate look-ahead, we modify our policy for selecting the connection to be assigned. After evaluating the least increment in ports involved in setting up each connection, we select the minimum of these and pick out the connections corresponding to this minimum. For each such connection **C**, we determine the set of connections **S** that could be added without any additional ports being consumed if **C** were to be assigned. We now find, for each **C**, the total traffic carried by **C** and by all the connections in its corresponding set **S**. The maximum value of this traffic is found and one of the connections corresponding to this maximum is assigned. This look-ahead helps us drive the search in the direction of greater throughput. The connection selection policy can be further improved by assigning the connection whose *"least-port-increase"* route has the least number of physical hops among all the connections which correspond to the maximum *"look-ahead traffic"*. The motivation behind this step is to favor lesser use of physical resources. The adaptiveness of the heuristic is further enhanced by trying to reroute the assigned connections at each stage. Once a connection is assigned, we consider each connection **C** which was assigned before this stage. We remove the connection **C** and determine its *"least-port-increase"* route and wavelength assignment in this new state of the network. If one of the following two conditions is satisfied, connection **C** is assigned to the new route, else it is put back to its old route.

- Changing the assigned route for connection **C** would lead to a decrease in the overall number of used OC-3 grooming ports in the network.

- Changing the assigned route for connection **C** would keep the number of used OC-3 grooming ports same but the rerouting would modify the state of the network to permit some connections to be added without additional increase in ports, which should not have been possible without the rerouting.

Having gone through the complete logical development of our heuristic for solving the static grooming problem, our heuristic can now be concisely put down as follows:

1 Set **NumWavs** = 1. Determine the **k**-shortest paths between every pair of nodes and store them in **Paths**. Add all desired connections (as given in **T**) to the set **B** and initialize set **A** as a null set.

2 If set **B** is empty, 100% throughput has been achieved and therefore, stop.

3 For each connection **(s, d)** in the set **B**, evaluate the increase in the number of used OC-3 ports in the network corresponding to each of the routes in **Paths(s, d)** and each corresponding feasible wavelength assignment. Using this information, determine the route and wavelength assignment which leads to the least increase in ports and store this least increase in **IncrPorts(s, d)**. If no feasible route and wavelength assignment exists, set **IncrPorts(s, d)** to $\infty$.

4 Find the minimum value of **IncrPorts(s, d)** for all connections **(s, d)** in the set **B**. If this minimum is $\infty$, skip to step 8, else store all the **(s, d)** pairs corresponding to this minimum in the set **S**.

5 For each **(s, d)** pair in **S**, determine the subset of **B** - {**(s, d)**} which can be allotted without additional consumption of ports if the connection

**(s, d)** is assigned along its *"least-port-increase"* route and wavelength assignment. Sum up the traffic of connection **(s, d)** along with those in its corresponding subset and assign this value to **AddTraffic(s, d)**.

6 Find the maximum value of **AddTraffic(s, d)** for all connections **(s, d)** in the set **S**. Assign the connection **(s, d)** corresponding to this maximum along its *"least-port-increase"* route and wavelength assignment. If more than one connection takes this maximum value for **AddTraffic(s, d)**, assign any one whose route has the least number of physical hops. Move this assigned connection from set **B** to set **A**.

7 Consider each connection **(s, d)** in set **A** in increasing order of traffic. Remove the connection **(s, d)** and evaluate the decrease in the number of used OC-3 ports - **DecrPorts(s, d)**. Now, determine the *"least-port-increase"* route and wavelength assignment for **(s, d)**. If either the increase in number of ports associated with this new route is lesser than **DecrPorts(s, d)** or if the increase in number of ports is equal to **DecrPorts(s, d)** but assigning **(s, d)** to the new route would facilitate allocation of more traffic without consuming additional ports (which should not have been possible if the rerouting had not been done), then assign **(s, d)** to the new route. Else, put it back to the previously existing route and wavelength assignment. Go back to step 2.

8 If **NumWavs** $<$ $\mathbf{W}_{max}$, then increment **NumWavs** and go back to step 3, else stop.

## 12.6 Illustrative Example

To offer a better understanding of our heuristic and to show a glimpse of how our heuristic outperforms those proposed in [13], we consider an example. The 9-node network considered has a physical topology as shown in Fig. 12.3. In this example, we take the groom factor **G** = 18, the number of wavelengths $\mathbf{W}_{max}$ = 2 and the number of ports $\mathbf{P}_{max}$ = 2. The traffic matrix **T** is as shown in Table 12.1.

When the *Maximizing Single-Hop Traffic* (MST) heuristic from [13] is executed on the above example (without the $\mathbf{P}_{max}$ constraint as it does not consider the mixed groomer architecture), the logical topology shown in Fig. 12.4(a) is setup which can be established only if $\mathbf{P}_{max} \geq 5$. On routing the individual connections on this logical topology, a throughput of 98% is obtained even though both the wavelengths have been utilized on most of the links. Similarly, the *Maximizing Resource Utilization* (MRU) heuristic from [13] was executed on this example and the logical topology setup was that shown in Fig. 12.4(b), which requires $\mathbf{P}_{max} \geq 3$. Though *MRU* consumes lesser grooming ports than *MST*, the throughput obtained also decreases to 67%.

*Table 12.1.* Example traffic matrix

| Node | 1 | 2 | 3 | 4 | 5 | 6 | 7 | 8 | 9 |
|---|---|---|---|---|---|---|---|---|---|
| 1 | | 5 | 4 | | | | | | |
| 2 | | | | 2 | | | | | |
| 3 | | | | | 14 | | | | |
| 4 | | | | | | 2 | | | 1 |
| 5 | | | | | | | | 4 | |
| 6 | | | | | | | 7 | 2 | |
| 7 | | 4 | | | | | | 2 | |
| 8 | 4 | 5 | | | | | | | |
| 9 | | | | | | | | | |

On the other hand, our heuristic manages to achieve 100% throughput making use of only one wavelength and with the constraint of having only 2 ports on the OC-3 groomer at each node. The various states of the network as the execution of our heuristic progresses are shown in Fig. 12.5.

**(a)** Initially, the *"least-port-increase"* route for each connection is the shortest path from source to destination. As no lightpath presently exists, the **IncrPorts** value for each connection is 2 except for **(4, 9)**, whose **IncrPorts** value is 0 as a lightpath with just one OC-3 does not pass through the OC-3 groomer (refer section on Node Architecture). Hence, **(4, 9)** gets assigned.

**(b)** All the unassigned connections now need at least 2 ports and so, the one with maximum traffic among them - **(3, 5)** - gets assigned.

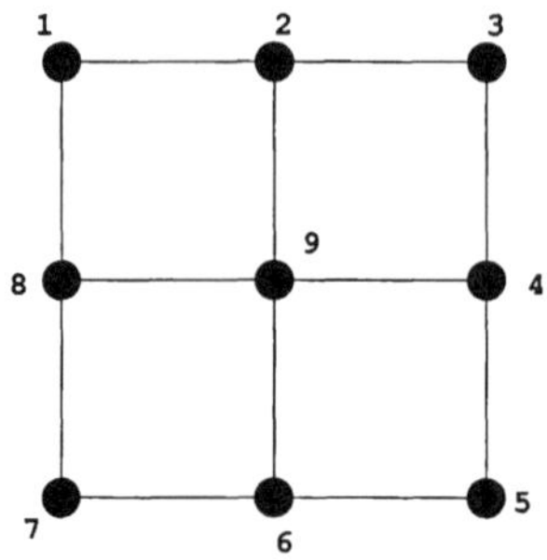


*Figure 12.3.* 9-node physical topology.

**(c)** Even now, the least value of **IncrPorts** is 2 and as many lightpaths have not yet been setup, no rerouting is beneficial. So, the connection with the next highest traffic - **(6, 7)** - gets established.

**(d)** Continuing with the trend, the connection carrying the highest traffic among all the unassigned connections - **(8, 2)** - gets assigned. One point to note at this stage is that the **AddTraffic** value for both **(7, 8)** and **(6, 8)** is 4 because if any one of them is setup, the other connection can be routed without consuming additional ports. Yet, **(8, 2)** got assigned as **T(8, 2)** $= 5 > 4$.

**(e)** Next, the connection**(7, 8)** gets assigned as its **IncrPorts** value is the current minimum 2 and its **AddTraffic** value is 8 (because if **(7, 8)** is setup, then the connections **(6, 8)** and **(7, 2)** can be routed without taking up more ports). After **(7, 8)** is setup, both **(6, 8)** and **(7, 2)** get assigned as the **IncrPorts** value for both is 0.

**(f)** Even now, the least value of **IncrPorts** is 2 and among the connections corresponding to this minimum, **(1, 2)** has the highest traffic and so, is selected for assignment.

**(g)** Having introduced **(1, 2)**, we can see that it is beneficial to reroute connection **(8, 2)**, changing its route from **8** $\rightarrow$ **9** $\rightarrow$ **2** to **8** $\rightarrow$ **1** $\rightarrow$ **2**. The motivation behind this is that the rerouting does not increase the number

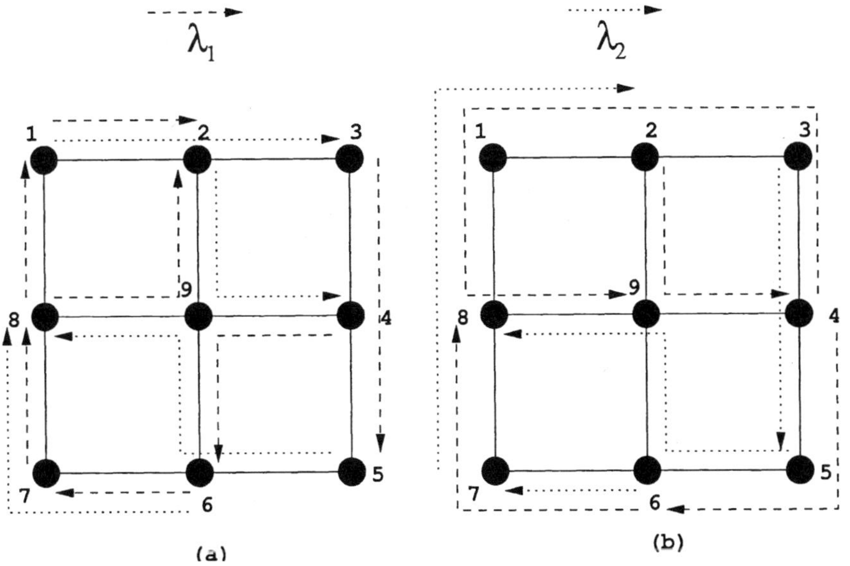


*Figure 12.4.* Logical topology setup by: (a) MST heuristic and (b) MRU heuristic.

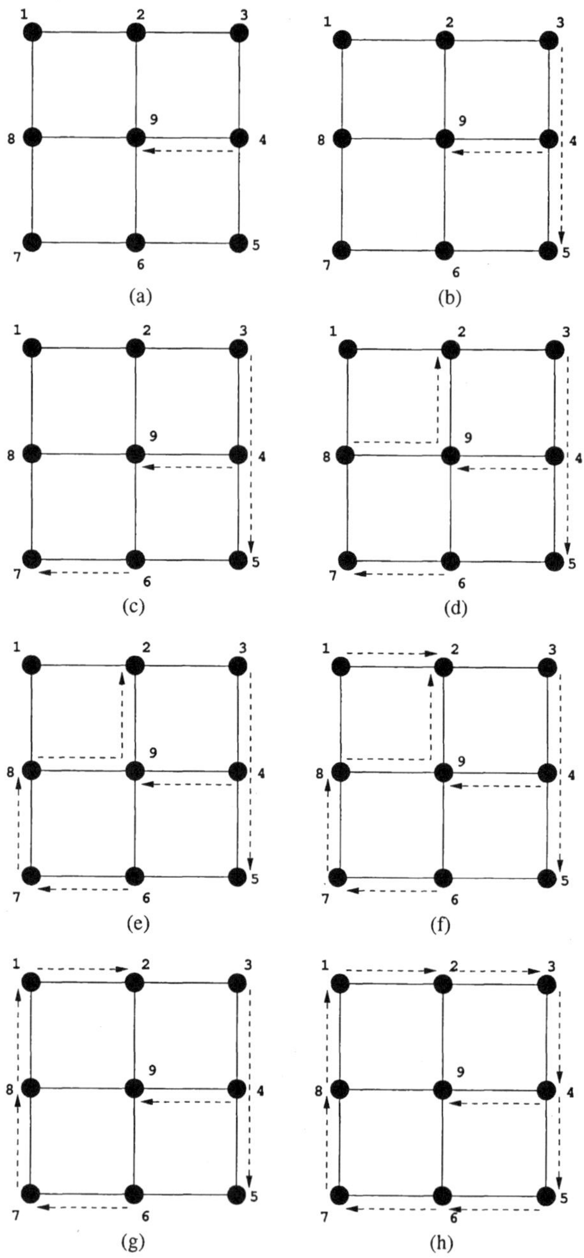


*Figure 12.5.* Development of logical topology in our heuristic.

of OC-3 grooming ports used on the whole but permits the connection

**(8, 1)** to be routed without further usage of ports. So, the route for **(8, 2)** is changed and now, since **IncrPorts** value for **(8, 1)** is 0, it is assigned.

**(h)** Progressing in the same manner as described until now, the final logical topology setup is as shown in Fig. 12.5(h).

This example not only clearly outlines the working of our heuristic but also shows that it is better than both *MST* and *MRU* as a higher throughput was achieved utilizing lesser number of wavelengths and under a tighter constraint on the number of ports.

## 12.7 Simulations and Results

In this section, we present the results of various simulations that we have conducted. These simulations can be broadly classified into three groups, based on their objectives:

1 Comparison with the optimal solution obtained by solving the ILP formulation

2 Demonstrating the efficiency of our heuristic

3 Comparison of our heuristic with *MST* and *MRU* heuristics

In our simulations, we used the 6-node network and the 15-node network, whose physical topologies are as given in [13]. These two networks were used for comparison of our heuristic's solution with that yielded by the ILP formulation, and the *MST* and *MRU* heuristics, respectively. The traffic matrices used for these simulations were obtained by generating each demand as a uniformly distributed random number in the range 0 to 5. As our problem formulation requires each traffic demand to be lesser than the groom factor **G**, the value of **G** is taken to be greater than 5 in all our simulations.

### 12.7.1 Comparison with ILP

Solving the ILP formulation we presented in Section IV gives the optimal solution for a particular instance of the static grooming problem. Hence, to demonstrate the near-optimality of our heuristic we compared the solution it provided with that obtained on solving the corresponding ILP formulation. We also determined the results given by the *MST* and *MRU* heuristics to show that our heuristic's solution is much nearer to the optimum. Since solving the ILP formulation entails very high complexity, these simulations could be carried out only on the 6-node network given in [13].

In these simulations, the number of wavelengths was varied from 1 to 4, with the groom factor varying from 6 to 8 for each wavelength. For each combination of number of wavelengths and groom factor, the *MST* heuristic

*Table 12.2.* Comparison of throughput with ILP and MST

| $W_{max}$ | G | ILP | Our heuristic | MST |
|---|---|---|---|---|
| 1 | 6 | 42% | 35% | 26% |
| 2 | 6 | 64% | 50% | 47% |
| 3 | 6 | 76% | 71% | 70% |
| 4 | 6 | 88% | 86% | 80% |
| 1 | 7 | 47% | 40% | 26% |
| 2 | 7 | 65% | 64% | 48% |
| 3 | 7 | 91% | 84% | 71% |
| 4 | 7 | 95% | 93% | 81% |
| 1 | 8 | 51% | 45% | 26% |
| 2 | 8 | 68% | 65% | 50% |
| 3 | 8 | 91% | 89% | 73% |
| 4 | 8 | 100% | 96% | 82% |

*Table 12.3.* Comparison of throughput with ILP and MRU

| $W_{max}$ | G | ILP | Our heuristic | MRU |
|---|---|---|---|---|
| 1 | 6 | 42% | 35% | 23% |
| 2 | 6 | 67% | 56% | 50% |
| 3 | 6 | 76% | 71% | 63% |
| 4 | 6 | 91% | 89% | 81% |
| 1 | 7 | 47% | 40% | 24% |
| 2 | 7 | 70% | 68% | 53% |
| 3 | 7 | 91% | 84% | 68% |
| 4 | 7 | 96% | 93% | 87% |
| 1 | 8 | 51% | 45% | 28% |
| 2 | 8 | 78% | 75% | 54% |
| 3 | 8 | 91% | 89% | 74% |
| 4 | 8 | 100% | 100% | 92% |

was executed and the maximum of groomer ports used among all nodes was determined, as this is the size of the groomer required. Using this groomer size, the corresponding throughput yielded by our heuristic and by solving the ILP formulation were obtained. These simulations were again repeated with the *MRU* heuristic. The results of these are shown in Table 12.2 and Table 12.3, from which it can be clearly observed that the solution yielded by our heuristic is very close to the optimal solution in most of the cases. It is also seen that the throughput given by the *MST* and *MRU* heuristics is lesser than that given by our heuristic in all the cases considered. The better performance of our heuristic is due to our approach of routing connections over the *"least-port-increase"* route and the adaptability incorporated through rerouting of connections.

Over and above this, we also used our ILP formulation to demonstrate the benefits of using our mixed groomer node architecture in place of a coarse granularity or fine granularity groomer alone. For the case of a coarse granularity groomer, the only change required in the formulation is to set the value of $\mathbf{P}_{max}$ to 0, whereas for a fine granularity groomer, we need to set $\mathbf{P}_{max}$ to $\infty$ and add variables to count the actual number of ports used. The same methodology as that used in the comparison of our heuristic with the optimal solution was used here too. Not only was the throughput obtained in the 3 cases (coarse groomer, fine groomer and mixed groomer) measured but the number of ports consumed in the fine groomer case was also determined. The value of $\mathbf{P}_{max}$ was taken to be 5 for the mixed groomer. The results of these comparisons are shown in Table 12.4. We observe that in all the considered instances, the throughput yielded by utilizing the mixed groomer node architecture is much higher than that given by the coarse groomer. Though higher throughput is achieved with the fine granularity groomer, in most of the cases this increase is insignificant in comparison with the associated increase in network cost (increase in number of groomer ports required). Though this could not be verified due to the high complexity involved in solving ILP formulations, we believe that utilizing the fine groomer in larger networks with more dense traffic will entail even higher increase in network cost without much advantage in the throughput compared to the mixed groomer case. These results re-emphasize the need for our mixed groomer node architecture.

### 12.7.2 Efficiency of Our Heuristic

Any good grooming algorithm must satisfy the basic condition that the throughput must increase with increase in available resources. In our problem setting, the two basic resources are:

1 Total bandwidth available in the network - measured in terms of number of wavelengths $\mathbf{W}_{max}$ and groom factor $\mathbf{G}$.

2 Grooming capacity of the groomer - measured in terms of number of ports $\mathbf{P}_{max}$ on the OC-3 groomer in the mixed groomer architecture.

To show that our heuristic yields higher throughput with greater available bandwidth, we fixed the number of wavelengths at 10 and the number of OC-3 grooming ports at 15, and then increased the groom factor from 6 to 15. Sim-

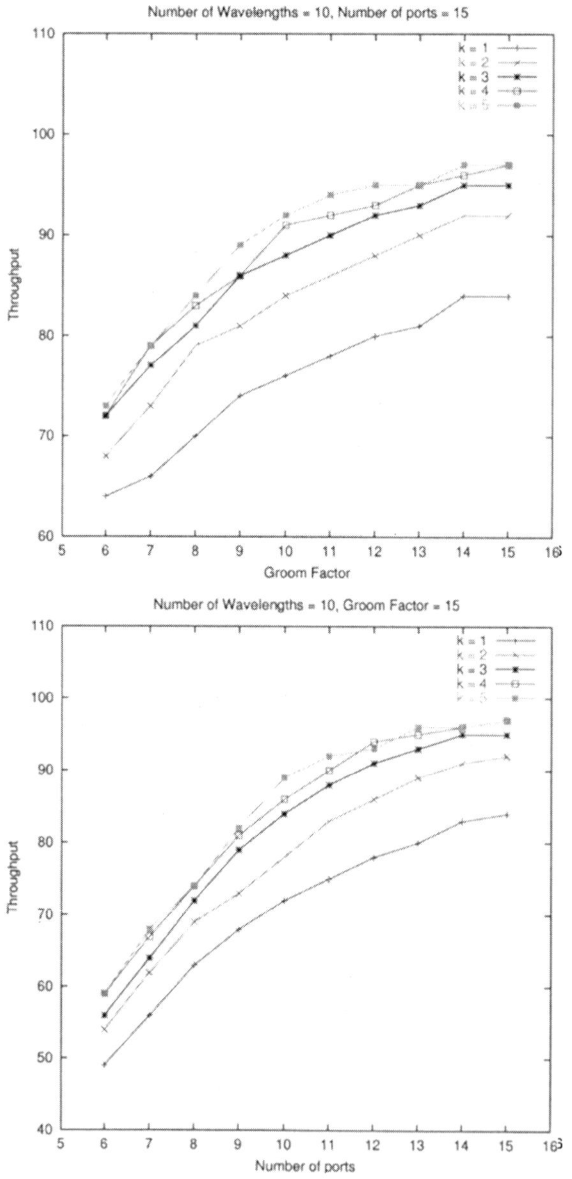


*Figure 12.6.* Increasing throughput with: (a) increasing groom factor and (b) increasing number of ports.

*Table 12.4.* Comparison of throughput of mixed groomer with coarse groomer and fine groomer

| $W_{max}$ | G | Coarse groomer | Fine groomer | | Mixed groomer |
|---|---|---|---|---|---|
| | | | Throughput | No. of ports | |
| 1 | 6 | 10% | 41% | 3 | 41% |
| 2 | 6 | 11% | 63% | 6 | 62% |
| 3 | 6 | 12% | 78% | 7 | 76% |
| 4 | 6 | 15% | 97% | 11 | 86% |
| 1 | 7 | 10% | 47% | 3 | 47% |
| 2 | 7 | 14% | 74% | 6 | 71% |
| 3 | 7 | 15% | 85% | 7 | 77% |
| 4 | 7 | 20% | 97% | 10 | 90% |
| 1 | 8 | 10% | 51% | 3 | 51% |
| 2 | 8 | 15% | 79% | 6 | 74% |
| 3 | 8 | 16% | 95% | 7 | 82% |
| 4 | 8 | 22% | 98% | 8 | 93% |

ilarly, to demonstrate increase in throughput with greater grooming capability, we fixed the number of wavelengths at 10 and the groom factor at 15, and then increased the limit on the number of OC-3 ports from 6 to 15. Both these simulations were repeated for values of **k** (where **k**-shortest paths were used for determining *"least-port-increase"* route) varying from 1 to 5. The graphs in Fig. 12.6(a) and Fig. 12.6(b) not only show increase in throughput as desired but also demonstrate that the performance of the heuristic saturates even with the small values of **k** considered. So, even though the heuristic considers only the **k**-shortest paths, the throughput is almost as good as that obtained by a comprehensive search over all routes. The property of throughput increasing with increase in number of wavelengths is shown by the results obtained in the next section.

### 12.7.3 Comparison with MST and MRU

Having proposed the mixed groomer node architecture, we also presented an algorithm for grooming with this setup as we cannot expect the *Maximizing Single-Hop Traffic* (MST) and *Maximizing Resource Utilization* (MRU) heuristics to perform well in this new scenario. To justify the need for our heuristic, we performed various simulations comparing its performance with that of *MST* and *MRU*, and the results clearly highlight the superiority of our heuristic.

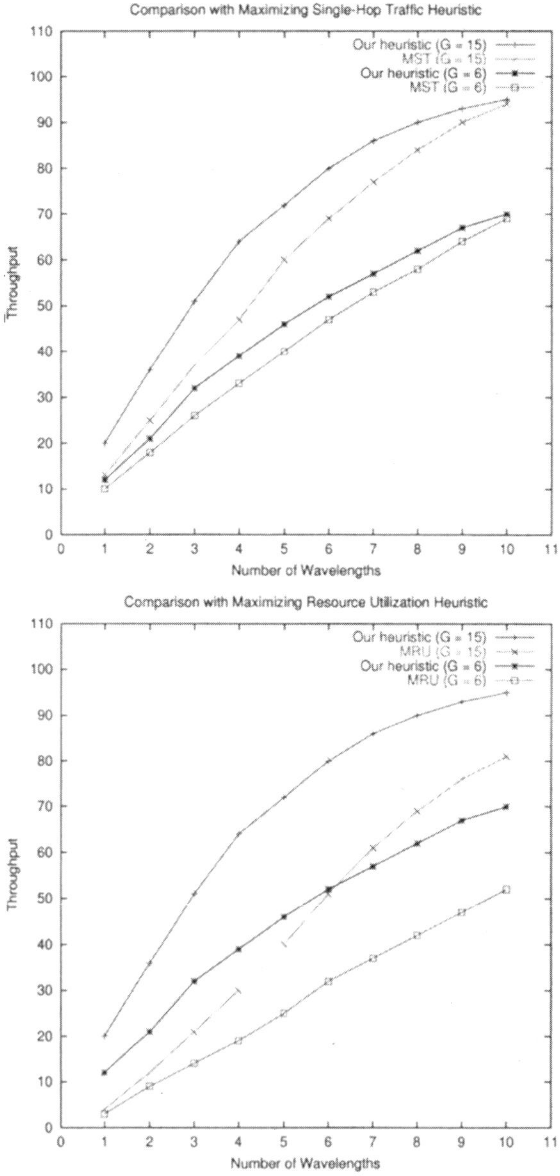


*Figure 12.7.* Comparison of throughput with: (a) MST and (b) MRU.

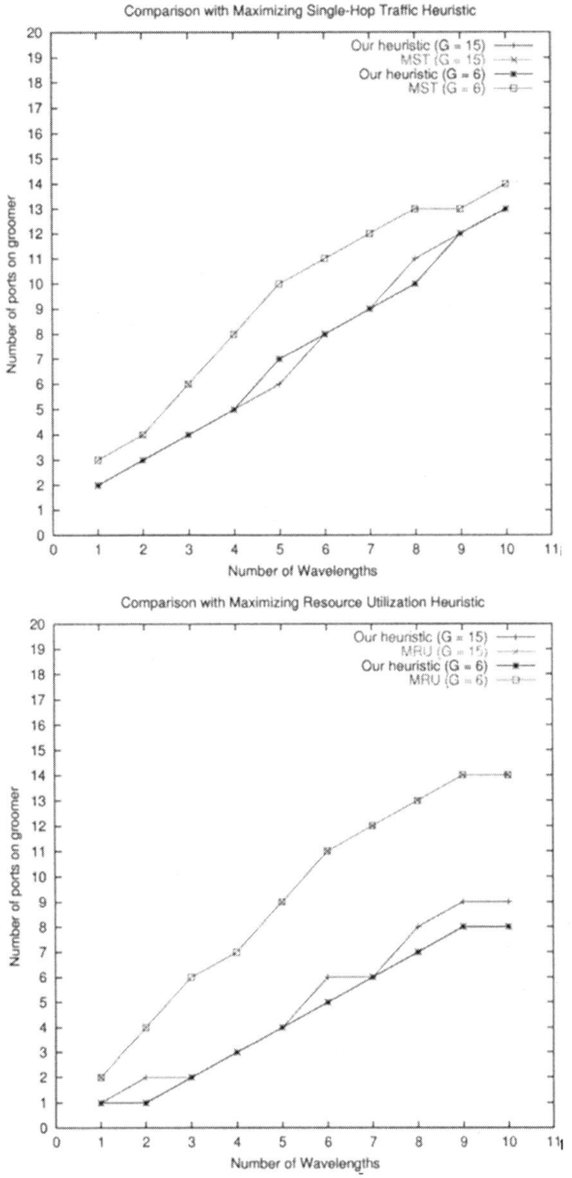


*Figure 12.8.* Comparison of number of ports with: (a) MST and (b) MRU.

Since the objective of our heuristic is to generate higher throughput given the constraint on grooming capability, the obvious way of displaying better performance is by showing higher throughput under the same grooming constraints. For this purpose, during each run of the simulations, we executed the *MST* heuristic, and obtained the throughput it yields and the number of ports taken up on the groomer at each node. Here too, we evaluated the maximum ports taken up among all nodes as that would be the groomer size required at each node. Using the same groomer size, we executed our heuristic and obtained the corresponding throughput. We conducted these simulations with wavelengths varying from 1 to 10. For each wavelength, the value of the throughput was determined with the value of the groom factor as 6 and 15, in order to demonstrate better performance under both sparse and dense traffic scenarios. The results of this experiment are shown in Fig. 12.7(a). Similar comparisons were carried out with the *MRU* heuristic and the corresponding results are shown in Fig. 12.7(b). These results reflect the fact that with the same amount of resources available, our heuristic performs much more efficient grooming than the *MST* and *MRU* heuristics. This has been shown in scenarios of both dense and sparse traffic.

An alternative way of looking at our problem of obtaining better throughput under grooming constraints is to minimize the grooming capability required to obtain a specific throughput. In light of this new view, we compared our heuristic with *MST* and *MRU* heuristics in terms of number of ports required on the OC-3 groomer to obtain the same throughput. As before, we executed the *MST* heuristic and determined the throughput obtained and the groomer size required to obtain it. We then executed our heuristic repeatedly to find the minimum groomer size required to obtain a throughput greater than that obtained by the *MST* heuristic. Here too, we performed the simulations with wavelengths varying from 1 to 10 and the number of ports corresponding to each wavelength was determined with the groom factor taking the values 6 and 15. The results of these simulations are shown in Fig. 12.8(a). The *MST* heuristic was seen to consume the same number of ports for a given number of wavelengths, irrespective of the groom factor. Hence, the plots with $\mathbf{G} = 6$ and $\mathbf{G} = 15$ for the *MST* heuristic are seen to coincide in Fig. 12.8(a). Results of similar comparisons with the *MRU* heuristic are shown in Fig. 12.8(b) and here too, the plots with $\mathbf{G} = 6$ and $\mathbf{G} = 15$ for the *MRU* heuristic are seen to coincide. The results of this section again indicate the higher efficiency of our heuristic in cases of both dense and sparse traffic as it is able to generate the same throughput as that given by the *MST* and *MRU* heuristics even with far less resources at hand. All the results obtained in this section substantiate the fact that the policy of assigning connections to their *"least-port-increase"* route and rerouting them to save on resources helps our heuristic to obtain excellent results.

## 12.8 Conclusion

In this chapter, we proposed a novel node architecture for traffic grooming in WDM optical networks. We listed out the advantages of the mixed groomer architecture in comparison with the MPLS/IP and SONET/WDM node architectures, and outlined its features in the light of practical feasibility, cost-effectiveness and efficient grooming capability. We presented an ILP formulation and also proposed a heuristic for the static grooming problem with the objective of maximizing throughput. We performed a wide range of simulations to demonstrate the efficiency of our heuristic and to display better performance in comparison with the *MST* and *MRU* heuristics. The results obtained in these simulations clearly substantiate our claims.

In the future, we intend to address the issue of dynamic grooming with our mixed groomer node architecture. Also, the concept of survivability can be brought into the focus of the grooming problem, irrespective of the static or dynamic setting.

## Acknowledgments

This work was supported by the Department of Science and Technology, New Delhi, India. The mixed groomer node architecture was first conceived when the first author undertook a summer internship at Tejas Networks, Bangalore during the period May to July 2002. The authors thank Dr. Sarath Kumar and Dr. Kumar N. Sivarajan at Tejas Networks for their invaluable contributions through insightful discussions.

## Notes

1. Though the number of ports on the OC-3 level is *exactly* 4 times that at the OC-12 level, we say switching cost is "more than" 16 times because the intrinsic cost of switching traffic streams increases as we go to finer levels of granularity.

## References

[1] Gerstel, O., Lin, P., and Sasaki, G. (1999). Combined WDM and SONET network design. In *Proc. IEEE INFOCOM'99*, pages 734–743.

[2] Gerstel, O. and Ramaswami, R. (2000). Cost effective traffic grooming in WDM rings. *IEEE/ACM Transactions on Networking*, 8(5):618–630.

[3] Hiuban, G., Pérennes, S., and Syska, M. (2002). Traffic grooming in WDM networks with multi-layer-switches. In *Proc. IEEE ICC'02*.

[4] Konda, V. R. and Chow, T. (2001). Algorithm for traffic grooming in optical networks. In *Proc. IEEE HPSR'01*, pages 218–221.

[5] Modiano, E. and Berry, R. (1999). Minimizing electronic multiplexing costs for dynamic traffic in unidirectional SONET ring networks. In *Proc. IEEE ICC'99*, pages 1724–1730.

[6] Modiano, E. and Berry, R. (2000). Reducing electronic multiplexing costs in SONET/WDM rings with dynamically changing traffic. *IEEE Journal on Selected Areas in Communications*, 18(10):1961–1971.

[7] Modiano, E. and Chiu, A. (2000). Traffic grooming algorithms for reducing electronic multiplexing costs in WDM ring networks. *IEEE Journal on Lightwave Technology*, 18(1):2–12.

[8] Thiagarajan, S. and Somani, A. K. (2001). Traffic grooming for survivable WDM mesh networks. In *Proc. SPIE OptiComm'01*, pages 54–65.

[9] Wan, P. J., Calinescu, G., Liu, L., and Frieder, O. (2000). Grooming of arbitrary traffic in SONET/WDM rings. *IEEE Journal on Selected Areas in Communications*, 18(10):1995–2003.

[10] Zhang, X. and Qiao, C. (1999). On scheduling all-to-all personalized connections and cost-effective designs in WDM rings. *IEEE/ACM Transactions on Networking*, 7(3):435–445.

[11] Zhang, X. and Qiao, C. (2000). An effective and comprehensive approach to traffic grooming and wavelength assignment in SONET/WDM rings. *IEEE/ACM Transactions on Networking*, 8(5):608–617.

[12] Zhu, H., Zang, H., Zhu, K., and Mukherjee, B. (2002). Dynamic traffic grooming in WDM mesh networks using a novel graph model. In *Proc. IEEE Globecom'02*.

[13] Zhu, K. and Mukherjee, B. (2002). Traffic grooming in an optical WDM mesh network. *IEEE Journal on Selected Areas in Communications*, 20(1):122–133.

V

# PROTECTION AND RESTORATION

# Chapter 13

# A SURVEY OF SURVIVABILITY TECHNIQUES FOR OPTICAL WDM NETWORKS

Mahesh Sivakumar, Rama K. Shenai and Krishna M. Sivalingam

*Department of CSEE, University of Maryland at Baltimore County (UMBC), Baltimore, MD 21250*

Email: {masiv1,shenai1,krishna}@csee.umbc.edu

**Abstract** In an optical wavelength division multiplexed (WDM) network, link or node failures may result in huge amounts of lost data, due to the enormous fiber throughput. This requires that optical WDM network be designed to be resilient to failures. Thus, survivability can be defined as the ability to respond gracefully to such failures. This chapter presents a comprehensive survey of various mechanisms proposed to achieve survivability. The survey considers different topologies, different failure models, implementation issues, signaling issues and quality-of-protection issues.

**Keywords:** Survivable Optical WDM Networks, Protection and Restoration, Single and dual-link failures, Node failures, Channel failures, Implementation, GMPLS-based Signaling, Quality of Protection.

## 13.1 Introduction

Optical fiber based networks, characterized by a bandwidth of over 50 terabits per second (Tb/s), have emerged as the transmission medium of choice for high speed communication due to their capacity, reliability, cost and scalability. Recent advances in optical technology and in particular wavelength division multiplexing (WDM), a multiplexing technique that partitions the bandwidth provided by an optic fiber into individual multi-gigabit channels, have been identified as enabling technologies that will allows us to fully and effectively utilize the fiber capacity [1, 2]. Current optical technology demonstrations have shown a feasibility of 160 channels, each operating at 10 gigabits per second (Gb/s), and future networks are expected to operate at 40 Gb/s per channel or higher.

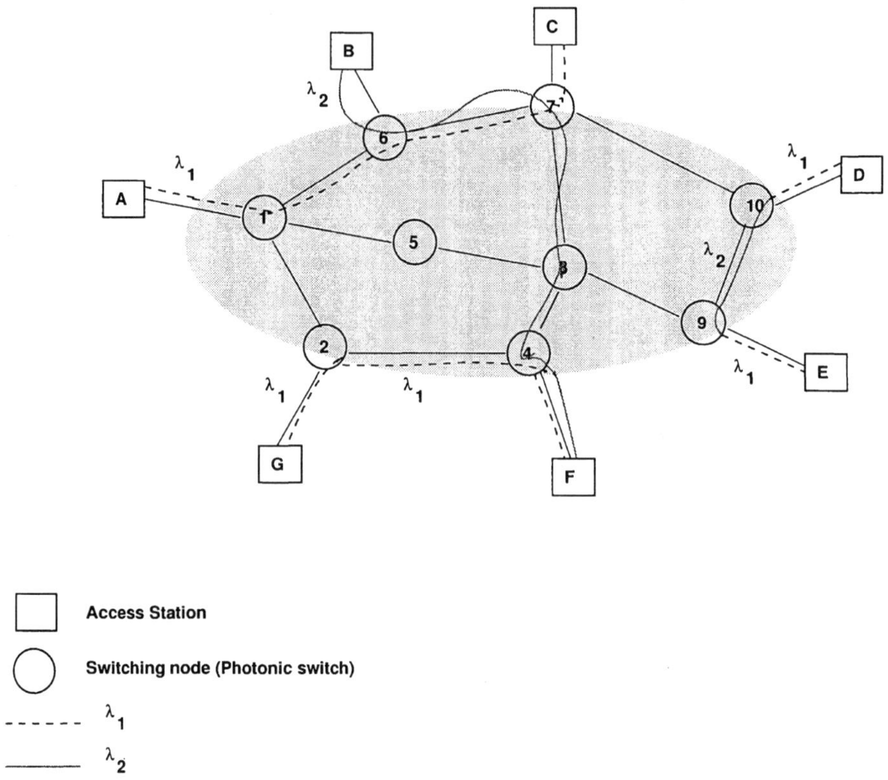


*Figure 13.1.* A wavelength-routed optical WDM network.

The advent of wavelength-routing enabled the design of wide area networks and the concept of all-optical lightpaths from source to destination [3]. Figure 13.1 shows a typical optical WDM wavelength routed network. The network consists of optical routing nodes called wavelength selective or wavelength interchanging cross-connects [4]) interconnected by fiber links (uni-directional or bi-directional). The connection between the source and the destination node is realized by a *lightpath*, an end-to-end optical path. The procedure of setting up a lightpath between any source-destination pair involves choosing an appropriate route and assigning the required wavelength(s) on the route selected. This problem is referred to as the *Routing-and-Wavelength Assignment* (RWA) problem. The network lightpaths can be established in two ways. They can be established in a *static* manner, where the set of lightpaths are determined before network operation begins. Alternatively, the lightpaths can be established in a *dynamic* manner, where the lightpaths are established on a demand basis, where lightpaths enter and leave the network based on some arrival process.

There are several issues that have to be carefully examined in order to fully exploit the potential of such networks. One important issue relates to the network's ability to provide continuous service in the presence of failures. This problem is more pronounced in an optical network architecture since the failure of a network component or a fiber link would lead to a failure of all the lightpaths traversing the link. Since each lightpath is expected is expected to operate at a rate of several Gb/s, such a failure could lead to a severe disruption in the network traffic. These failures essentially pose a requirement for WDM network systems to be fault-tolerant or survivable. Optical network *survivability* can be defined as the ability of the (WDM) network to respond gracefully to unexpected failures.

There are several techniques that have been proposed in literature to realize a survivable optical network [5, 6, 7, 8, 9]. This chapter presents the various survivability mechanisms designed for different topologies and different failure scenarios. We analyze the various failure scenarios, the appropriate survivability schemes and their related implementation issues in addition to a discussion on the recent work in this area.

The rest of the chapter is organized as follows. Section 13.2 presents a taxonomy of the failure models and provides a classification of the different survivability techniques employed for optical ring and mesh networks. This is followed by a detailed discussion of the various survivability techniques that are aimed at specific failure scenarios in Sections 13.3–13.5. Section 13.6 presents models that provide variable protection guarantees. A discussion of the implementation techniques and related issues are presented in Section 13.8. Section 13.7 examines the current state of issues that surround extending IP-based MPLS protocols to optical networks. Section 13.9 concludes the chapter.

## 13.2 Basic Survivability Techniques

This section discusses the different failure models and basic survivability techniques.

**Failure models.** Traffic disruption in an optical network can be caused by the failure of several components. The commonly studied failures in the literature include the following[10, 11].

- Link failures: This is the most commonly addressed failure scenario and is attributed to the high frequency of fiber cuts. This also includes failure of link components such as line amplifiers and regenerators. For example, a recent report by the Federal Communications Commission (FCC) state that 136 cable cuts were reported in 1997 alone [10].

- Node failures: can either occur due to an operator error or power outages leading to component failures in a node. Although less common, they are traditionally more difficult to handle compared to link failures.

- Channel failures: can be caused by the failure of transceiver equipment operating on specific WDM channel(s) on a link.

- Other failures: Optical layers may also be required to handle failures at other layers, such as optical-client interface and the client equipment.

The two common topologies considered are ring and mesh topologies. Rings are the typical choice for metropolitan area networks while mesh topologies are used for wide area networks. However, wide area networks are often implemented using a ring-of-rings topology. The ring topology is often chosen due to its better understood protection properties, but is typically less resource efficient compared to the mesh topology.

**Survivability techniques.** A taxonomy of the several survivability techniques proposed in literature is shown in Figures 13.2 and 13.3. They are broadly classified as *Protection* (Pro-active) and *Restoration* (Reactive) techniques depending on whether the following survivability-related functions are performed before or after the occurrence of a failure:

- Calculation of the backup path (or restoration route) for affected connection(s).

- Determination of the wavelength(s) that need to be assigned to the backup paths.

- Reconfiguration of the nodes involved in setting up the backup paths.

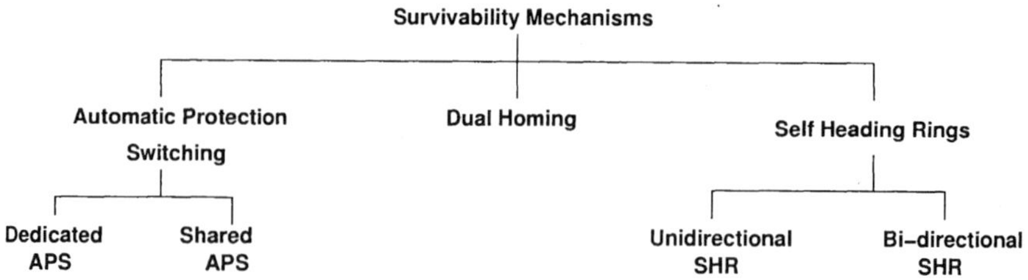


*Figure 13.2.* Classification of survivability mechanisms for optical WDM ring networks.

### 13.2.1 Protection

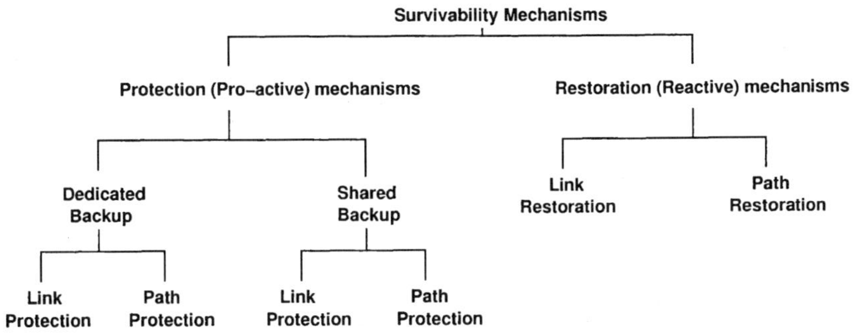


*Figure 13.3.* Classification of survivability mechanisms for optical WDM mesh networks.

Protection techniques are pro-active and require the reservation of backup resources either at connection setup time or at network design time. This approach has the advantage of fast and guaranteed recovery in the event of a failure. However, this leads to inefficient utilization of resources since the backup resources are kept idle in the absence of a failure. Protection techniques have been studied for both ring and mesh network topologies.

**Protection in Ring Networks.** Much of the current ring infrastructure are based on SONET/SDH rings, which are also called *self-healing* since they incorporate protection mechanisms to survive different types of link and node failures. The mechanisms used to provide protection in ring networks include (i) Automatic protection switching (APS) techniques, (ii) Self-healing ring (SHR) mechanisms and (iii) Dual Homing techniques. These mechanisms are also incorporated by SONET/SDH rings in order to provide survivability. Fig. 13.2 summarizes the different survivability techniques employed in ring networks.

**Automatic Protection Switching (APS).** APS is a signaling mechanism that is used to achieve protection against link failures. APS can either be dedicated or shared based on the assignment of protection resources. As the name indicates, *dedicated APS* involves the provisioning of a dedicated backup for every link or path in the network. Dedicated APS can be realized by the following techniques:

- *1+1 protection architecture:* Here, traffic is simultaneously transmitted on both the primary and protection paths or with a slight delay on the backup path. The destination, in the event of a failure, switches from the primary to the backup path. The delay on the backup path is to ensure that no data is lost during the time it switches from the primary to the backup.

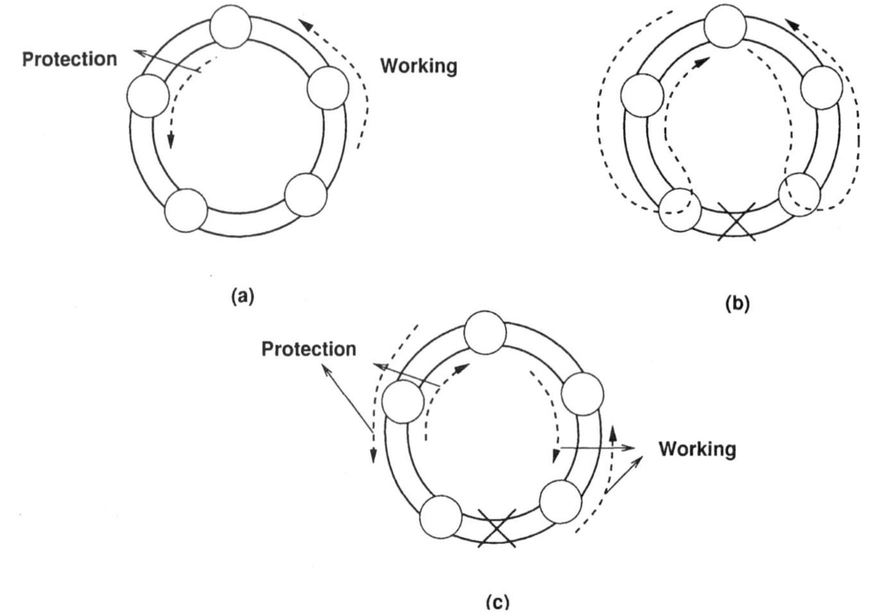


*Figure 13.4.* Example of Uni-directional self healing rings (USHR). (a) USHR (b) USHR/L - Line protection switched USHR (c) USHR/P - Path protection switched USHR.

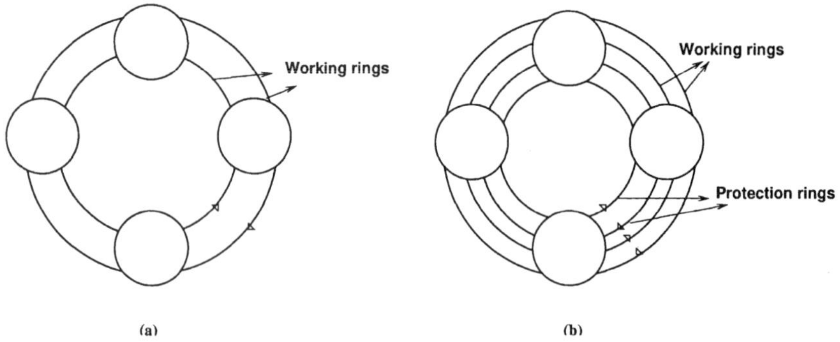


*Figure 13.5.* Example of Bi-directional self healing rings (BSHR). (a) BSHR/2 – Two fiber BHSR. (b) BSHR/4 – Four fiber Bidirectional Self Healing ring.

- *1:1 protection architecture:* In this case, data is only transmitted on the primary lightpath, and the backup lightpath is used in the event of a failure. The advantage of this technique is that the backup path could be used to route low-priority pre-emptive traffic in the absence of failures. The disadvantage is that when failure occurs, there is a possible loss of data due to the delay in switching to the backup path.

With the *Shared APS* approach, backup resources are not dedicated but shared among multiple primary paths. Shared APS can be realized by using the following techniques:

- *1:N protection architecture:* Here $N$ working links share a single protection link thereby providing protection to one of the $N$ working links at a given time. This architecture efficiently utilizes the backup resources. Unlike dedicated APS, in order to handle any future failures, the traffic on the protection path should be switched back onto the working link, once it is repaired.

- *M:N protection architecture:* This is the generalized version of the 1:N scheme, where M protection links are shared among N working links. When more than M of the N working links fail simultaneously, the traffic routed to the protection links can be decided according to pre-assigned priorities.

**Self-Healing Rings (SHR).** A more effective technique that is used to protect against both link and node failures in ring topologies is the *self-healing ring* (SHR) technique shown in Figures 13.4 and 13.5. Their effectiveness is attributed to the developments in add/drop multiplexing (ADM) technology and the simplicity of the control mechanisms used. SHRs are available in unidirectional (USHR) and bidirectional (BSHR) flavors depending on the direction of traffic flow under normal operation.

- *Unidirectional SHR:* Figure 13.4(a) illustrates a unidirectional SHR. Normal traffic flows in one direction only. In the event of a failure of a link or node in the working direction, traffic is switched on to the protection path which carries it in the opposite direction. USHR can either provide: (i) *line protection (UHSR/L) or loopback*, where, the two nodes adjacent to the failure handle the responsibility of switching the traffic from the working to the protection fiber, or (ii) *path protection (UHSR/P)*, which is similar to the 1+1 APS discussed earlier. These two variations are shown in Figure 13.4(b) and 13.4(c) respectively.

- *Bidirectional SHR:* Here, normal traffic here flows in both directions. Figures 13.5(a) and 13.5(b) illustrate the two architectures used in

BHSR. In *BSHR/2*, each ring allocates half its capacity for protection. In the event of a failure, the two nodes adjacent to the failure switch the traffic to the protection bandwidth on both the rings. Alternatively, *BSHR/4* uses twice the number of rings used in BHSR/2. Here, two fibers are dedicated for protection and working. When a failure occurs, traffic is switched from the working fibers to the protection fibers.

**Dual Homing.** Node failures in ring networks become critical when the node happens to be a major hub or a central office and could be a sink or source for a large volume of traffic. Dual homing techniques address the problem of handling node failures. These techniques address the single point-of-failure issue by introducing a backup node (i.e. by making use of two hub nodes in place of a single one). Thus if one node fails, the other node can take over the operation, since connections are placed between each non-hub node and the protected (and its backup) node.

**Protection in Mesh Networks.** Protection schemes aimed at making the mesh network survivable do so either at the path level (termed *path-level protection*) where they offer services to each lightpath individually or at the line or signal level (*link-level Protection*) where they protect lightpaths on a per-link basis.

**Path Protection.** In path protection, for each connection request, a primary lightpath and a backup lightpath between the source and destination nodes are used. Essentially path protection is performed individually for each lightpath. In the event of a link or a node failure in the lightpath, the traffic will be switched on to the appropriate backup path. The protection path should be link or node disjoint respectively. Also, the protection paths used for different connections using the same working paths can be different [11]. Path protection can either be *dedicated* or *shared* as explained below:

- *Dedicated path protection:* In this protection scheme, each request is given a primary lightpath and a backup lightpath, which are dedicated to it, at connection time. This is to say that the resources on the backup path are not shared with any other backup or primary paths, and is used only for the chosen connection request. Dedicated path protection can be realized similar to *1+1 APS* or *1:1 APS*.

- *Shared path protection:* This protection scheme is similar to the dedicated case in that each connection request is provided a primary path and a disjoint backup path and wavelength(s). However the backup resources (wavelength(s)) can be shared by one or more paths as long as they do not require them simultaneously. Depending on the type of paths

(Primary or backup paths) that share the backup resources, we can have two schemes (i) *Backup or shared multiplexing*: Here the backup resources are shared between one or more backup paths that do not need them simultaneously. (ii) *Primary backup multiplexing*: In this case, the backup resources can be shared between a primary path and one or more backup paths that do not need them simultaneously [12]. While this technique reduces the connection blocking probability, it could result in a potential reduction in restoration guarantee. Moreover this technique is suitable for a dynamic traffic scenario where lightpaths are setup and torn down frequently.

**Link Protection.** Link protection protects each network link independently. Depending upon the type of traffic demand, backup paths and wavelengths are reserved around each link either at the connection-setup time (dynamic traffic demand) or at the network design time (static traffic demand). In the event of independent single link failures, all the lightpaths traversing the failed link will be re-routed around that link via the corresponding predetermined backup path. Unlike path protection where the end nodes of the connection need to be signaled to handle the failure, the backup path signaling is handled the end nodes of the link. This will lead to a lower restoration time for link protection when compared to that of path protection. However, restoration routes in link protection are generally longer and fewer in number making them less flexible. As in the case of path protection, link protection can be either dedicated or shared.

## 13.2.2 Restoration

Restoration is a reactive mechanism that handles a failure after it occurs. The routing and wavelength assignment (RWA) algorithm calculates a backup path (route and wavelength) used to reroute the traffic after the occurrence of a failure. Restoration aims to effectively utilize backup resources by using them only in the event of a failure. However, it does not provide any survivability guarantee since it may not be always possible to determine backup paths for all affected lightpaths on a link. In addition, the restoration time could be much higher compared to that of protection due to the additional computation and signaling involved. Restoration can be done at the path level (where they restore individual lightpaths) or at the line level (where they restore on a per-link basis).

Restoration or reactive techniques are attractive due to the efficiency of the backup routes they provide and the spare capacity savings they attain. One of the main benefits of optical layer protection is the speed with which they restore affected failures. Hence, reactive algorithms can be effective only if their restoration times are comparable to or better than those of higher layers. In general, algorithms that tackle fault tolerance after the occurrence of a failure, target the following objectives:

- *Generality:* The need for the techniques to be independent of the underlying topology. As an example, an arbitrary two-connected mesh topology could be considered for this purpose.
- *Scalability:* The need for restoration algorithms to adapt (provide similar services) to rapid pace of change and aggregation.
- *Reliability:* The need for a distributed nature in operating the scheme due to the absence of single entity managed infrastructures.
- *Speed:* As mentioned earlier, the need to compute effective restoration routes in less time (For example. at least as fast as 50ms as required by SONET self-healing rings(SHR)).

In the following sections, we describe the path-level and link-level restoration mechanisms.

**Path restoration.** In this restoration scheme, the source and the destination nodes of the each connection request (lightpath) affected by a failure run a distributed RWA algorithm to dynamically determine the backup path and wavelength(s) on an end-to-end basis. If the algorithm does find a backup path free, the traffic is then routed on that path on appropriate wavelength(s) after signaling its cross-connects. If not, the connection is said to be blocked.

A distributed path restoration scheme with the objective of minimizing restoration time and spare capacity used apart from optimizing the network restorability (fraction of paths restored) was proposed in [13]. The algorithm is based on the *interference principle.* Interference is caused by backup multiplexing wherein the use of a backup link (shared between backup paths) renders the other infeasible. The proposed algorithm tries to establish the backup path that has the least interference. Such a path is found by having the source and destination node of the failed lightpath emit a statelet (broadcast signal) on all its outgoing spans. At each intermediate node that receives the statelet, interference numbers are calculated for each of its span as the difference between the number of statelets intending to use it and the number of statelets it can support. The interference number of each span a statelet traverses is added to it. At some point, the statelets from the source and destination meet at a tandem node which denotes the end of the search. Once a match is obtained, reverse complimentary statelets are sent back to the sources. Of all the matches, the backup path with the least interference number is chosen.

**Link restoration.** Here, the end nodes of the failed link run the dynamic RWA algorithm to find a backup path around the failed link for each lightpath traversing the link. Again, the absence of such a backup path would block the connection request.

**Restoration failure scenarios.** We next analyze the failure characteristics of restoration algorithms and provide insight into the requirements for efficient restoration against failures. Such failures are characterized into three groups, namely (i) *Fundamental*, (ii) *Basic algorithmic* and (iii) *Practical algorithmic* in [14], while considering dual-link failure scenarios.

**Fundamental Failures.** Such failures are independent of the restoration algorithm deployed and occur due to network conditions at the time of recovery. *Disconnection failures* are the most fundamental ones and occur due to dual-link failures that partition the network (i.e., disconnect some nodes from the others). The primary sources of disconnection failures are nodes of degree two. Such failures depend on the design of the network and cannot be accounted for in the restoration or protection algorithms. Another fundamental failure relates to the unavailability of restoration routes due to capacity limitations (i.e., unavailability of link/node disjoint paths) and is termed as a *capacity failure*. The most common case of such a failure occurs due to network nodes of degree three and cuts of three edges. In these cases, no dual-link failure can be tolerated since both the failed links would require the third link for restoration.

**Basic Algorithmic Failures.** Such failures are characteristic of the recovery algorithm used. *Directional Failures* occur due to the partitioning of fibers into distinct primary and backup networks. The logical partitioning of the backup fibers from the primary fibers reduces the number of links carrying traffic in a particular direction across a cut and could worsen the effects of fundamental failures. A restoration algorithm could also fail due to the resulting hop-length of a recovery route being too long. Longer paths suffer from issues of signal regeneration and jitter and could even translate into the failure of the restoration algorithm. Such failures are called *path length failures.* Another basic failure caused by the restoration algorithm design relates to the class of failures termed as *path hit failures.* Dual-link failures where the failure of a second link occurs on the backup path of the first link failure belong to this class. Both path length and path hit failures can suffer from directional failures (directional path length and directional path hit failure respectively).

**Practical Algorithmic Failures.** These failures are similar to basic algorithmic failures in that they occur due to algorithm limitations but are specific to features of particular algorithms. The primary set of failures that belong to this class are the *Blocked path failures* which occur when the restoration routes of different primaries share links between them. Thus, failure of those two primaries cannot be recovered simultaneously (note that this scenario is not applicable to single link failures). Other possible failures may be due to those occurring with pre-planned routes (*pre-planning failures*) and those occurring as a result of topological constraints placed on restoration routes (*topological constraint failures*).

Experimental results show that path hit failures are the most dominant of all failures and blocked path failures are dominated by the effects of pre-planning and topological constraints. The vulnerability of links was used as the performance measure. A link A is said to be vulnerable to link B, if , after the recovery of link B, A's failure prevents complete recovery of one or both the links. Moreover, mesh restoration is found to be more beneficial than embedding ring protection in meshes and so is the need for using link restoration to optimize restoration time.

### 13.2.3 Metrics of Evaluation

In the previous sections, we provided a basic classification of the survivability mechanisms presented for optical WDM networks. This section describes the typical performance metrics used in algorithm evaluation. These include:

- *Protection switching time or Restoration time:* The protection switching time (or restoration time) is defined as the down-time the connection experiences due to a failure, and is indicative of the potential data and revenue losses [5].

- *Capacity utilization:* Capacity utilization is defined as the measure of additional backup resources that have been reserved by the specific survivability (Link/path protection/restoration) scheme [5].

- *Restoration efficiency:* Restoration efficiency or restoration performance can be defined as the ratio of the total number of restored connections to the the total number of failed connections.

- *Restoration guarantee:* The extent to which a protection/restoration mechanism can restore a failed connection is termed as the restoration guarantee. Dedicated protection mechanisms provide 100% restoration guarantee and a 100% restoration efficiency.

## 13.3 Single Link Failures in Mesh Networks

The simplest and the most likely failure that occurs in an optical wavelength division multiplexed (WDM) network is that of a link, which the optical layer clearly needs to handle. Common causes for link failures are fiber cuts and failure of optical line equipment [10]. *Single-link failure* scenarios involve the independent failure of a single link and is based on the assumption that at any instant of time, only one link failure exists in the network. This section surveys different approaches to survive single-link failures in the context of mesh network topologies.

As indicated previously, both protection and restoration mechanisms have been proposed to survive single-link failures. Protection techniques for single-link failures essentially involve reserving backup resources on a per-link or on a per-connection basis (link and path protection). Restoration schemes are dynamic in which they make use of intelligent cross-connects and controllers in the network. However restoration algorithms have a higher restoration time and operational complexity when compared to most protection algorithms [3, 11].

## 13.3.1 Embedded ring-based schemes

Protection techniques for single-link failures have been studied in detail. Initial approaches to protecting single-link failures involved extending ring protection techniques into the context of mesh networks. An example of such a technique involved decomposing the available mesh network topology into multiple self healing rings [15]. This mechanism in conjunction with WDM loopback recovery is used to handle single link and node failures [7]. The first mesh decomposition technique termed node cover involved decomposing the mesh network in to set of rings that were chosen in such a way that each node belonged to one or more rings and each link belonged to at most one ring. Such node covers do not protect all the links in the network since they do not cover all of them. To overcome this, the technique of ring cover was introduced. Ring covers imposed the constraint that each link must belong to at least one ring. Such a necessary condition assures that all the links are protected but does not constrain spare resource redundancy: each link could belong to any number of links [16]. The optimized version for this technique would hence relate to finding a minimal ring cover for a given topology.

Cycle cover is a four-fiber ring-cover technique, based on the assumption that each network link comprises a pair of counter-propagating working and a pair of counter-propagating protection fibers[17]. The technique involved identifying a family of directed cycles that cover all the protection fibers where each protection fiber is used only once. Although polynomial time simple algorithms exist for finding cycle covers for planar topologies, the problem of finding minimal cycle covers for non-planar topologies is an NP-hard problem and is not scalable (needs re-computation if a new node is added). An example of mesh decomposition using cycle covers is illustrated in Figure 13.6. While

Figure 13.6(a) shows the protection cycles formed by the cycle cover decomposition, Figure 13.6(b) shows how a connection affected by the failure of a link is rerouted.

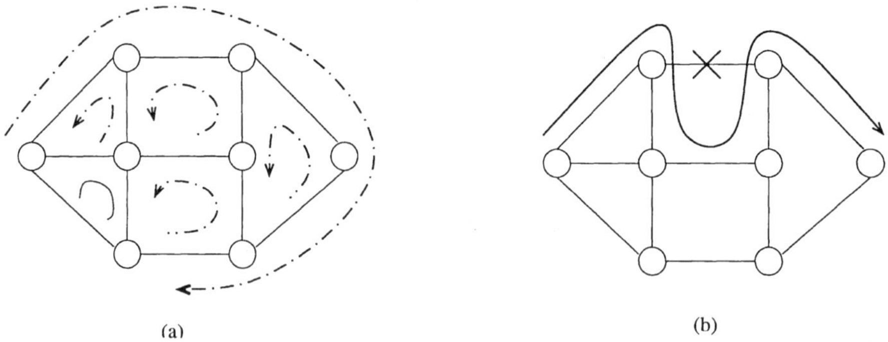


*Figure 13.6.* (a) Mesh topology decomposition by the cycle cover technique (b) Recovery of a lightpath after a link failure [16].

An effective alternative to ring-covers is the double cycle-cover technique for which effective polynomial time algorithms exist. The technique involves identifying cycles in the network such that each link appears in exactly two rings. Figure 13.7 depicts two possible mesh-topology decompositions that can be obtained by the double-cycle-cover technique. Theoretically, the technique can be used for both four-fiber and two-fiber implementations.

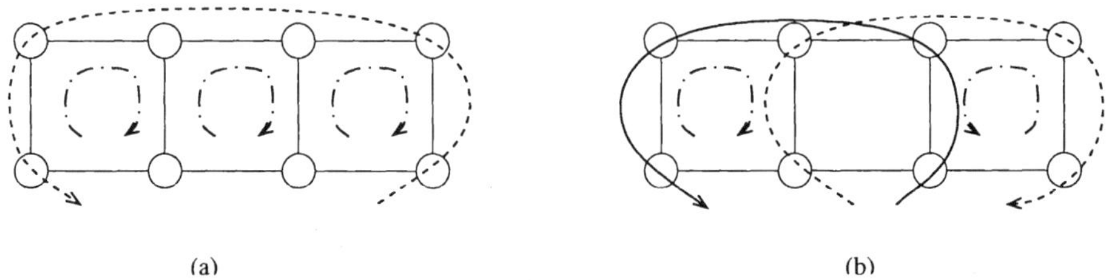


*Figure 13.7.* Two possible (planar) mesh-topology decompositions obtained by applying the double-cycle-cover technique([16]).

Another approach to single-link failure survivability involved the use of protection cycles. It is based on the property of a ring to protect not only its own links, but also any possible chordal link(a link connecting two non-adjacent ring nodes) [16]. P-cycle protection design is illustrated in Figure 13.8. The use of virtual protection cycles (*p-cycles*), established in the spare capacity of mesh networks, in order to recover from link and node failures was considered in [18]. Automatic protection switching was realized using protection cycles in [19]. The work proposes mechanisms to identify the protection cycles and imposes bounds on the number of simultaneous failures that may occur.

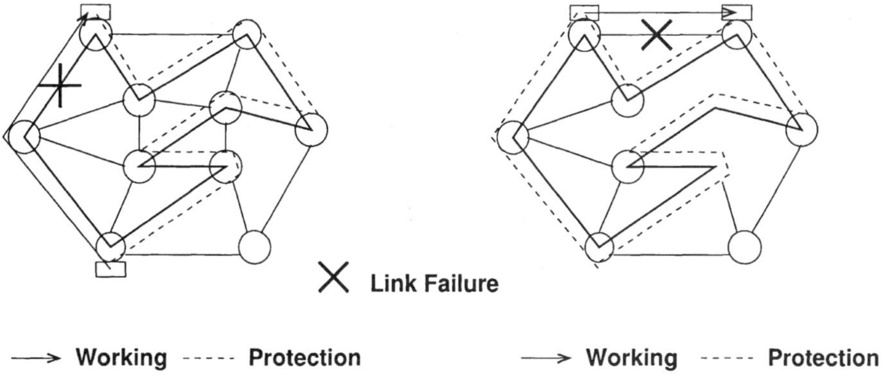


*Figure 13.8.* P-cycle protection design. A ring is able to recover either one of its links(a) or one of its chords(b)([16]).

## 13.3.2 Capacity Optimization

Protection mechanisms for single link failures model have been considered in conjunction with static traffic models, where Integer Linear Programming (ILP) formulations are used to optimize the capacity utilization of backup resources used in the protection techniques. Generally this capacity optimization can be approached in two ways: (i) Given a certain capacity, maximize the protected carried demand (ii) Given a certain demand, and a given 100% restoration requirement, minimize the total capacity used.

A detailed discussion on protection and restoration techniques specific to single-link failure models is presented in [5, 6], where the authors formulate a set of ILPs to minimize the capacity utilized by the protection resources for a given demand while satisfying the demand for a 100% restoration (guaranteed restoration). The ILPs are formulated for (i) Dedicated-Path protection (ii) Shared-Path protection and (iii) Shared-Link protection. Numerical results in [5] suggest that shared-path protection performs significantly better than dedicated-path and shared-link in terms of capacity utilization. A comprehensive comparison study of the protection/restoration techniques in terms of the protection switching times (the concept of protection switching times were discussed in an earlier section) is done in [6]. The numerical results indicate that the protection switching times are lowest for shared path protection and are highest for shared link protection.

Recent work on single link protection mechanisms involve optimizing the resource allocated for a backup path. Primary backup multiplexing introduced in [20] involves a single primary and multiple backup paths sharing the same channel. Here, the protection bandwidth on a backup path may be allocated to new primary paths, thereby reducing failure restoration guarantee of existing connections. This technique allows a primary lightpath and one or more backup lightpaths to share a channel. The goal is to satisfy an increased number of connections in exchange for a reduced failure restoration guarantee. The paper presents conditions for a connection's recoverability in this case in the event of a failure. Algorithms are proposed to maximize the recovery of a connection using primary backup multiplexing. A trade-off analysis between the reduction in guarantee and the blocking performance gain is also performed. The results show that a 90% performance gain at less than 10% guarantee reduction under low-load conditions is possible for the studied cases.

### 13.3.3 Other mechanisms

Another mechanism proposed in [21] considers the use of wavelength converters as a protection resource. Here wavelength converters are shared among multiple backup paths. The mechanism essentially looks to reduce the number of connections blocked due to the unavailability of a wavelength converter at a node. In a further bid to improve network utilization a backup path relocation scheme where backup paths are migrated to improve the provisioning of primary paths is considered in [21].

The benefits and tradeoffs in combining protection and restoration into a hybrid protection-restoration technique is considered in [22]. Here, the proposed survivability mechanism pro-actively reserves bandwidth similar to preplanned protection only for a subset of the network links in times of need as specified by a predetermined policy. The main objectives of this technique is to obtain a balance between a network's provisioning capacity and restoration efficiency specially under moderate and high loads.

### 13.3.4 Channel failures

Channel failures in optical WDM networks are defined to occur when a subset of wavelengths on a fiber become unavailable. The primary reason for the failure of a single wavelength channel is due to the failure of specific lasers. The usual mechanism to recover from such a situation is to switch to a backup laser emitting the same wavelength [23]. However, this may result in the maintenance of a large number of lasers. One solution to providing recovery from failed channels while reducing keeping the laser inventory is through the use of tunable lasers.

In [24], a local restoration scheme is used to handle the failure of a particular channel in a GMPLS-enabled optical mesh network. If the optical channel connecting two ports on neighboring OXCs fails, the local restoration mechanism uses a different port/channel combination on the OXCs to overcome the failure. In case the local restoration scheme fails, end-to-end restoration is used. The use or tunable lasers to provide restoration capability is also studied in [25], where a small number of tunable lasers are used to cater to both link and channel failures.

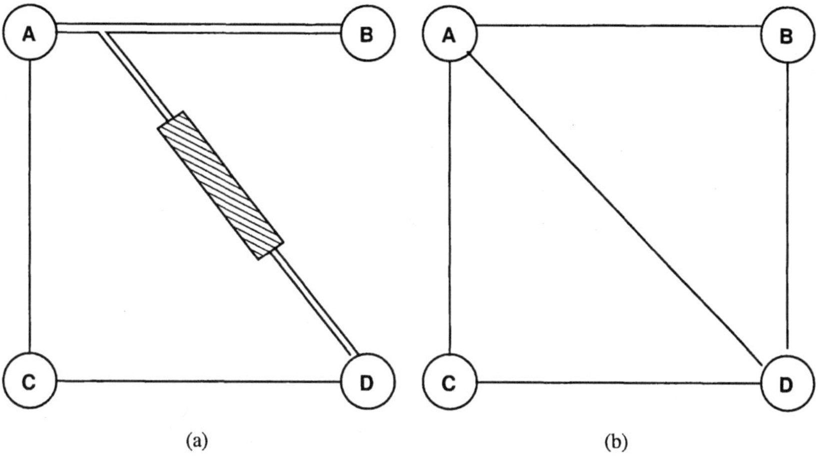


*Figure 13.9.* Comparison between (a) the physical topology and (b) its corresponding graph [26] illustrating the possibility of a simultaneous double link failure.

## 13.4 Dual-Link failures

Another failure model of interest considered in this section is the *dual-link failure* model, a failure model in which any two links in the network may fail in any unspecified or arbitrary order [26]. Most of the prior work on link failure models have considered the influence of independent single-link failures – a failure model that assumes the failure of a single network link at any given instance of time. However as indicated in [26], a couple of reasons have elevated the importance of the double link failure model. With an increasing number of fiber cuts it is highly possible for two or more fibers to be down at a given point of time. This possibility is specially enhanced due to the increased link repair times, which could last from hours to days. Another case cited in [26] exemplifies the possibility of a double-link failure. This scenario is shown in Figure 13.9, where the network graph (Figure 13.9(b)) captures only the connectivity of the network but not its physical layout. In such a case, link disruption as illustrated in Figure 13.9 (a) could cause a possible unexpected double link failure of links $BD$ and $AD$.

Dual-link failure survivability is considered in [26]. Here, three link-level protection approaches were presented with the predefine assumption that the network graph was 3-connected (A necessary requirement for dual link survivability, encompassed by Menger's theorem, which states that a graph is $k$-connected if and only if there exists $k$ edge-disjoint paths between every pair of nodes. The three approaches with specific cases are illustrated as four cases in Figure 13.10.

The three mechanisms are described below:

**Backup paths with link identification - Methods I and II.** Here, two edge-disjoint paths, the first backup path $b_1$(e) and the second backup paths $b_2$(e) are pre-computed for each edge $e$. On the failure of link $e$, the first (or primary) backup path $b_1$(e) is used as the backup to route traffic affected on link $e$. The scenario considered is the failure of a second link $f$, which occurs after the failure of link $e$. Four possible cases are discussed in [26, 27], which are as enumerated as follows:

*Case (i):* (Figure 13.10(a)) In this case, the primary backup path of link $f$, $b_1$(f) is not affected by the failure of link $e$. Since $b_1$(f) is disjoint with $b_1$(e), failure recovery of links $e$ and $f$ are independent. Hence, the primary backup paths, $b_1$(e) and $b_1$(f), will re-route the affected traffic across links $e$ and $f$ respectively.

*Case (ii):* (Figure 13.10(b)) Here, the primary backup path of link $f$, $b_1$(f) is affected due to the failure of link $e$. Since link $f$ does not lie on $b_1$(e), backup paths $b_1$(e) and $b_2$(f) will re-route the affected traffic on links $e$ and $f$ respectively.

*Case (iii):* (Figure 13.10(c)) In this case, the primary backup paths of links $e$ and $f$ (i.e $b_1$(e) and $b_1$(f)) are affected due to the failure of links $f$ and $e$ respectively. Here, the link recovery method **I** will employ the secondary backup path $b_2$(f) and $b_2$(e) to re-route the affected traffic on links $e$ and $f$. The link recovery method **II** on the other hand employs $b_2$(f) to re-route the traffic on $f$ as well as the traffic on the backup path $b_1$(e). Hence the net traffic re-routed in this method equals ($b_1$(e)-f) ∪ $b_2$(f)

*Case (iv):* (Figure 13.10(d)) This case is similar to cases (ii) and (iii) described previously. Here the primary backup path of link $e$, $b_1$(e) is affected by the failure of link $f$. Working in a similar fashion as in case (iii), method **I** employs $b_2$(e) and $b_1$(f) to re-route traffic on links $e$ and $f$, while method **II** employs $b_1$(f) to re-route the affected traffic on links $e$ and $f$. The net traffic re-routed in this method equals ($b_1$(e)-f) ∪ $b_1$(f)).

Capacity optimization of methods **I** and **II** have been proposed in [27] with the optimization formulated as an Integer Linear Programming (ILP) problem. A 10-15% capacity utilization savings is suggested by using shared protection over dedicated protection.

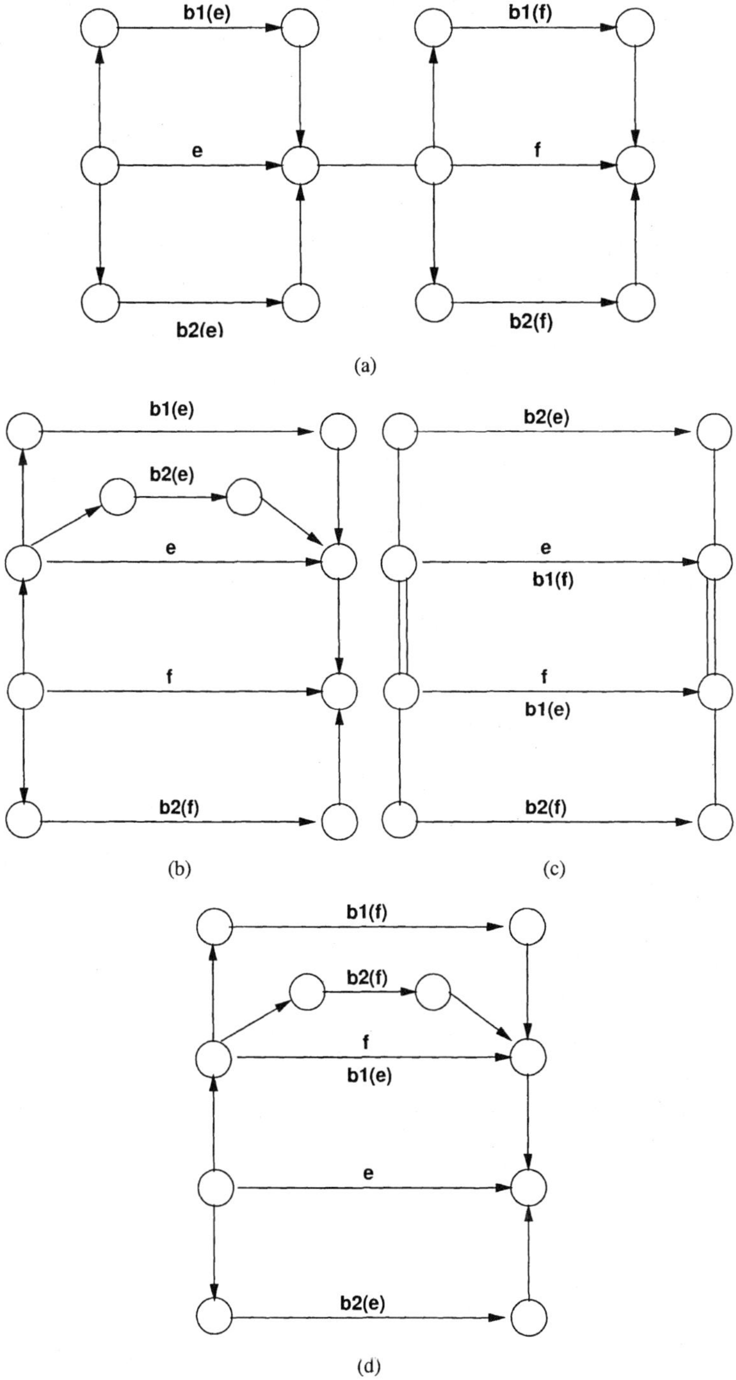


*Figure 13.10.* Four cases illustrating the re-routing of traffic on the affected links $e$ and $f$ [27].

**Backup paths without link identification - Method III.** In this method, a single backup path $b(e)$ is pre-computed for every link $e$. This is based on the assumption that for every link $f \in b(e)$, a backup path $b(f)$ that does not contain link $e$ can be found. Hence the failure of link $e$, followed by $f$, will cause $b(f)$ routes traffic from both links $e$ and $f$. An advantage of this method is that the link failure information is not signaled to all the nodes. A heuristic method to compute the restoration backup paths is also presented in [26]. Numerical results in [26] suggest a 100% recovery from double-link failures with a modest increase in backup capacity.

**Capacity Analysis.** An analysis of the capacity requirement for a span-restorable mesh networks to provide protection against dual link failures incident on a common node is presented in [9]. A *span* is defined as the set of transmission channels that directly connect adjacent cross-connects. Such a scenario belongs to a class of dual link failures that occur when logically distinct spans between cross-connects happen to share the same physical duct. However, they do so only for a few hundred meters. An example would be a bridge-crossing close to a common cross-connecting node. In such cases, damage to the duct at the point it is shared between the two spans could present itself as a dual link failure on distinct spans. These fiber optic spans that share such common ducts are referred to as *shared-risk link groups (SRLGs)* and the class of SRLGs that are incident on a common node are called *co-incident SRLGs* (which is the most common scenario and the model studied as mentioned earlier). To study the effect of SRLGs, the authors consider span protection (i.e. *link protection*) to provide fault tolerance motivated by the fact that link protection enables setting up backup resources once and for all during connection setup and does not require dissemination of information to the source or destination nodes. Shared Backup Path Protection (SBPP), on the other hand, requires a large number of network state data maintained and disseminated to every node on a lightpath basis.

The authors extend the *arc-path* ILP formulation for the Spare Capacity Allocation (SCA) problem to include SRLGs and calculate the capacity required to protect a span-restorable network for a known set of SRLGs. Further, they conduct experiments to analyze the effect of co-incident SRLGs on capacity requirement and identify the set of SRLGs, which when removed from the network, would provide considerable savings in cost. Results show that the presence of even a single co-incident SRLG could increase the spare capacity considerably. Moreover, SRLGs at the core required more spare capacity than the ones located around the edge due to the fact that more traffic is routed through the core. An important observation relates to the fact that co-incident SRLGs are much more expensive than arbitrary dual-link failures and non-coincident SRLGs. The problem of enhancing dual-failure restorability in path-protected mesh-restorable optical wavelength division multiplexed (WDM) networks was considered in [28]. The authors propose a hybrid mechanism that uses both protection and restoration to provides maximum (close to 100%) dual-failure restorability with minimum additional spare capacity. The basic premise of the algorithm is to identify scenarios in the dual-link failure model that necessitate additional spare capacity and provide protection for those scenarios only. Findings indicate that the proposed scheme achieves close to complete (100%) dual-failure restorability with only maximum of 3% wavelength-links needing two backups even at high loads. Furthermore, at moderate to high loads, the proposed scheme attains close to 16% improvement over the base model that provides complete single-failure restorability.

## 13.5 Node Failures

The previous sections discussed the various aspects and parameters involved in protecting a WDM optical network against channel and link failures. While node failures are less common when compared to link failures, they do tend to occur and deserve attention. However, they have not been addressed much in literature due to the complexity of the recovery mechanisms involved. This section discusses the importance of node failures and their implication on the survivability of the underlying network.

### 13.5.1 Failure Characteristics

Node failures are generally caused by failures in part or entirety of node equipment. Component failure at the nodes could be caused by power outages or operator error apart from natural calamities. Node failures are different from link failures in the following respects:

- Unlike link failures where the failure of a link would only require the lightpaths traversing the link to be rerouted, node failures would result in a failure of the lightpaths traversing all the links that are incident the node. A node failure would hence require the rerouting of all such lightpaths.

- Lightpaths that originate or terminate at that node cannot be protected without having spare nodes.

- The failure of a node would not only cut off the forwarding (routing) services but also the hardware present at the node. For example, wavelength converters are placed at nodes to reduce the blocking probability of connections. Hence, the failure of a node could require the other nodes to compensate for the loss of hardware to maintain the performance before failure.

- Protection mechanisms handling link failures usually provide backup paths that are link disjoint with the primary path. However, node failures would require backup paths that are both node and link disjoint. Since the number of node disjoint paths are fewer as compared to the number of link disjoint paths and more difficult to calculate, recovery mechanisms for node failures are complex in nature.
- Node failures can be modeled as individual link failures in cases where only part of a node fails ( i.e. some links still work). However, the nature and extent of the failure is difficult to comprehend and most protection mechanisms do not provide protection against multiple failures. As a result, such treatment is usually avoided.
- Node failures have been addressed using ring based algorithms but suffer from complex synchronization requirements to guarantee network connectivity in the wake of one or more failures.

### 13.5.2 Proposed Solutions

The work on the *Spare Capacity Allocation (SCA)* problem has been extended to include protection against single node failures in [29]. It also proposes a graph based algorithm to provide a solution to the SCA model. Given the virtual network topology (working paths in the network), the SCA minimizes the spare capacity or cost involved in provisioning spare capacity and finding backup routes to provide fault tolerance. As mentioned earlier, ILP models for survivable network design are NP-complete and hence necessitate heuristics that quickly find feasible solutions for fault scenarios. The paper extends *successive survivable routing (SSR)* [29], a previously proposed algorithm, to solve the SCA problem.

The problem considers *failure-independent* path restoration that requires all affected flows to be routed on their backup paths in the event of a failure (i.e. 100% restoration). The SCA model is based on a matrix based heuristic that is fast and near optimal. The objective function of the model minimizes the total cost of spare capacity on networks by determining the minimum spare capacity required (*spare provision matrix*) on links in the event of a failure. The spare provision matrix is determined directly from the path-link incident matrices for the working and backup paths or from their aggregate per-flow based information.

The model requires the calculation of a node-disjoint backup path for every source-destination pairs in the demand. Each iteration of the algorithm removes links from the node and tries to find a node-disjoint path in the resulting network. The algorithm works by detecting *trap links* (a trap link is a link which is on the working path and whose reversed direction link is in the path found in the new network) during each of its iterations until all of them are removed resulting in a node-disjoint path. To detect trap links, all the intermediate nodes of the working path are split to allow backup paths to at least traverse one trap link.

The Successive Survivable Routing (SSR) algorithm solves the SCA model by decomposition. Multi-commodity flows are first decomposed into multiple single flows. Each flow uses shortest path routing to route link-disjoint backup routes. Each flow *successively* updates its backup path with the one that has minimal additional reservation.

The comparison of two protection algorithms, namely the ring based *Double Cycle Cover (DCC)* and *generalized loopback* aimed at protecting double failure scenarios (two node failures or a link-node failure) is presented in [30]. DCC works by providing every link in the network by exactly two rings in opposite directions. As in rings, protection is provided by APS mechanisms that switch the traffic from the working ring to the protection ring in the event of a failure. However, the approach is not scalable as are previous ring-based approaches. Generalized loopback is similar to the DCC scheme in its use of an APS-like mechanism for protection. However, a pair of conjugate digraphs are calculated for routing primaries and reserving protection paths. The *primary digraph* is used to route primaries while the *secondary or backup digraph* is used to route the protection paths. Both the primary and secondary digraphs are calculated at call-setup or network design time. In the event of a failure, nodes adjacent to the failure switch the traffic on to the secondary digraph links. A *minimal backup digraph* that uses the minimal number of links to guarantee protection against single failures, is used for the backup path. The advantage of this approach is that the links not used in the minimal backup digraph (*noncritical links*) can be used to route unprotected low priority traffic.

The paper uses two measures of evaluation to study the two protection schemes. *Robust connectivity* represents the global end-to-end protection capability of the network and is calculated as the average percentage of nodes to which connections from a node can be routed in a way that is completely robust to any single link failure. It allows protection against single failures of any source destination pair. The penalty associated with providing robustness to failures is characterized by *Robust path-length expansion*, a parameter that calculates the penalty in terms of the increase in the number of hops for end-to-end paths. The average over all node pairs of the ratio of shortest robust path to the shortest unprotected path is the path-length expansion ratio.

Performance analysis of the schemes show that the generalized loopback mechanism is more robust and capacity efficient when it comes to node failures in all-optical networks. Moreover, it also provides for provision for utilizing the unused backup capacity to carry unprotected traffic without affecting the reliability of the network.

## 13.6 Quality of Protection

The previously proposed techniques considered two classical grades of survivability – guaranteed protection (For example, 100% protection against single-link failures) or best-effort restoration. This section presents a survey of different approaches that propose a generalized Quality of Protection (QoP) framework supporting varying degrees of survivability [31, 32, 33, 20]. These surveyed approaches aim to bridge the gap between the two well known protection grades.

The lack of a uniform paradigm for providing protection guarantees motivated the authors in [33] to formally introduce the concept of *Quality-of-Protection*. Multiple protection grades based on the amount of bandwidth utilized in protecting the bandwidth is considered. These common protection grades considered include:

(a) *Guaranteed protection:* the connection will be protected with a very high likelihood (99.999% is typical) of recovery.

(b) *Best effort protection:* where the connection will be protected using the available protection bandwidth.

(c) *Unprotected traffic:* no effort is made to protect the connection, in case of a failure.

(d) *Preemptable traffic:* traffic that normally uses protection bandwidth for classes (a) and (b) and is preempted when the protection bandwidth is used to protect a failure.

When failure occurs, the probability of a connection to recover from a failure is determined by its Quality-of-Protection (QoP). The proposed approach is based on providing an absolute guarantee versus the differentiated services model, where it is straightforward to cite that the guarantee of a lower priority class is dependent on the class of a higher priority. Such a framework proposed in [33] allows allows one to specify the probability with which a connection will be restored, providing the customer with a complete range of protection guarantees at different but fixed pricing.

Two approaches are considered below which can be incorporated into a QoP framework providing different grades of restorability. The first approach considers the benefits in network provisioning by reducing the restorability for certain lightpaths, while the second approach considers an added protection grade of dual failure restorability.

### 13.6.1 Reducing restoration guarantees

The work in [20] focuses on the concept of a *D-connection*, (dependable connection), which are connections where the protection bandwidth can be used by the working paths, D-connections are allowed to have a reduced failure restoration guarantee, thus allowing the connections to go unsupported in certain cases. The routing-and-wavelength algorithm (RWA) used to set up a D-connection is based on the concept of primary backup multiplexing technique. This technique allows a primary lightpath and one or more backup lightpaths to share a channel. (this idea is to satisfy an increased number of connections for a reduced failure restoration guarantee). Conditions are posted for a connection's recoverability in this case in the event of a failure. Algorithms are proposed to maximize the recovery of a D-connection using primary backup multiplexing. A trade-off analysis between the reduction in guarantee and the blocking performance gain is also performed. The authors suggest a 90% performance gain at less than 10% guarantee reduction under low-load conditions (the results indicate that a small reduction in protection guarantee can entail a good increase in blocking performance gain).

### 13.6.2 Improving restoration guarantees

In [31], the authors extend the model proposed in [33], by considering an additional protection grade of complete dual failure restorability. Providing dual failure restorability to all available lightpaths inherently leads to a high penalties in capacity utilization. In [31], the authors consider formulation by which a significant subset set of lightpaths (termed premium) are provided with complete dual failure restorability at little penalties to network provisioning. The proposed mechanisms essentially illustrates an economically viable and flexible way to support an added service class such as assured dual failure protection, with the available infrastructure. T

To summarize, these architectures aim to bridge the gap between the two well known protection grades, fully guaranteed and no-guaranteed connections, while studying the trade-off between a reduction in protection guarantee versus the blocking performance gain.

## 13.7 MPLS for Optical WDM networks: Challenges and Remedies

This section concentrates on issues relating to the adaptation of Multi-Protocol Label Switching (MPLS) [34] protocols to the optical layer. In particular, it discusses the possibilities of IP/MPLS over optical layer and MPLS features required to establish optical connections with restoration functionality.

The issue of reconfigurability in WDM optical networks is dependent on the efficient use of optical resources in the network. The need for exchanging information from different vendors becomes essential in such cases. This fact advocates the need for having a control plane that would build an effective platform for vendor interoperability. Multi-Protocol Label(lambda) Switching (MPLS) [34] is one such technique that could be deployed to manage optical connections. It consists of a set of distributed control protocols which are currently used for Internet Protocol (IP) networks. Besides being used for Virtual Private Networks (VPN), Quality of Service (QoS) and other different services, MPLS is also used for IP layer restoration which gives us the motivation to adapt them to optical networks as well. This necessitates additional modifications to the IP/MPLS layer in order to be used over the optical layer.

### 13.7.1 MPLS based IP-Networks

MPLS is essentially used to establish virtual connections termed *Label Switched Paths* (LSP). Packets in each LSP are given a unique label and the intermediate nodes (*Label Switched Routers (LSR)*) use label forwarding tables to route the incoming packet on appropriate output ports (with a new label). The process of distributing labels to establish an LSP is done through MPLS signaling protocols that do so by calculating the paths either at the source *Resource ReSerVation protocol (RSVP)* or on a per-hop basis *Label Distribution Protocol (LDP).* These protocols also contain information on bandwidth requirements of the LSP. Once the paths are calculated, they are then established using IP-related protocols such as the *Open Shortest Path First (OSPF).* MPLS also provides IP restoration either at the source of a flow or over a sub-path around the failure which are analogous to link and path protection in the optical layer.

### 13.7.2 IP/MPLS over WDM optical networks

The labels and LSRs in the MPLS domain may be considered analogous to the wavelength channels and Optical Crossconnects (OXC) in the optical domain. This analogy aids in the extension of MPLS protocols to the optical layer too. In such a configuration, end-to-end connections can be established as follows: (i) The IP based processor at each OXC would disseminate information to all its neighboring nodes with the help of interior gateway protocols (e.g OSPF extended to adapt to the optical layer); (ii) The RWA algorithm at the optical layer would then be invoked and would use this information to find optimal routes between s-d pairs; and (iii) The MPLS signaling protocol would configure the OXCs en route.

However, inherent differences between IP and the optical layer would require careful design when extending MPLS protocols. There are hence several challenges that need to be addressed to efficiently adapt MPLS to optical networks. Some of them are summarized below.

- While IP allows zero bandwidth paths when no packets are switched along the path using the bandwidth, the cross-connection of channels in optical transport systems (OTS) prevents such a scenario.

- Unlike IP paths, optical lightpaths are bidirectional. Absence of coherence in assigning channels to bidirectional links could result in the same channel being assigned in both directions. This phenomenon is termed as *glare* and is absent in unidirectional paths due to a single node having total control of the channels.

- Optical layer restoration has more stringent requirements as compared to that of IP.

- The channel selection process in the optical layer is more complicated if wavelength continuity is required.

### 13.7.3 Case Studies

In this section, we present specific case studies relating to the adaptation of MPLS protocols to the optical layer. In particular this section discusses current approaches that are considered in the IP/MPLS over optical layer to establish connection with restoration requirements.

There has been considerable amount of work that concentrates on IP over WDM. The enhancements in MPLS required to handle optical connections with restoration are discussed in [34]. Towards this end, it investigates the optimal protection approach in obtaining fast and efficient protection while also presenting the enhancements necessary in the MPLS layer. The paper presents an efficient solution that is neither fully pro-active nor fully re-active. The proposed approach calculates the working path and protection path at the same time but neither reserves channels nor configures the cross-connect on the protection path.

While using MPLS to provision optical lightpaths, an inherent issue that needs to be handled is the mapping of logical topology to connections. Failure of a single link in the optical domain could result in the logical topology at the higher layer being disconnected. Hence it is necessary to have a mapping of the links in the optical layer with the logical topology at the MPLS layer. The authors in [34] propose the idea of maintaining a *failure event set* $F$ that would contain the failure scenarios. Corresponding to each failure scenario $f$, we have the planning failure event set $F_p$, which represent the probabilistic events that have to be handled to meet the network restoration objective. The problem with this approach is the need to model every possible failure event in $F$.

Thc lightpath sctup algorithm in such a configuration would need to iterate through the failure event set and find a disjoint path for each working path in the set that is likely to fail for a particular failure scenario. This would mean that the algorithm should have access to the connections, the working and restoration paths for the connections, and the failure event set. This could be achieved either using a centralized approach to carry out the function mentioned above or use a simplified approach that would only calculate the restoration paths but at a frequent rate (for optimization). In [35], the authors proposed the idea of using a hybrid approach where the number of channels required for restoration is also provided. The algorithm is able to accurately predict if the restoration objective could be met.

In [36], the problem of building an integrated scheme where IP routers hold the responsibility of dynamically provisioning low-rate traffic streams is considered. Such *sub-lambda* connection requests are then routed over the WDM layer whose provisioning algorithms are extended for this purpose. This approach stems from the IP networks' need to provide service to users at a rate much lower than the full wavelength capacity at which optical lightpaths operate. The optical bypass would eliminate the need for SONET and aid in reducing the number of IP ports as well as the total switching capacity of the routers and the number of wavelengths required to achieve full connectivity. The paper considers dynamic traffic in a mesh topology where the IP and optical layers work independently in setting up their corresponding connections. They however coordinate between themselves through standard routing protocols. The IP layer begins by trying to establish a working and protection path (both paths could be single hop or multi-hop) for each connection request. Moreover the algorithm also has provision for backup multiplexing wherever possible and both the working and backup path can be routed on a single wavelength. If the IP layer is unable to find working or protection paths, the optical layer RWA mechanism is invoked to setup the required lightpath. The proposed scheme reduces the blocking probability as compared to current approaches. Moreover, as long as the wavelength availability is not a bottleneck, faster IP gives better performance.

An implementation of a distributed restoration method called Robust Optical Layer End-to-end X-connection (ROLEX) within the GMPLS control plane by appropriately extending RSVP can be found in [37]. The objective was to provide fast restoration capabilities to GMPLS. ROLEX supports the two and one-ended versions. Connections requests are handled as follows.

- RSVP sends its PATH message to establish the service (working) path.

- On receiving the PATH message, the intermediate nodes select available channels and forward the message. After forwarding the message, the nodes request cross-connects from the XC controller.

- The destination node finally sends back the acknowledgment through the RESV message.

- The same process is done for establishing the protection path too. However, the difference here is that the cross-connects are not configured (channels are not selected) and only the path is selected. This allows optimization of the backup path with provision for backup multiplexing.
- While the two-ended version does the process from both ends, the single-ended counterpart works from the source. The two-ended version ends when the PATH and RESV message meet.

The proposed restoration architecture was found to provide restoration times comparable to that of SONET. Moreover, it is independent of the cross-connect platform used.

## 13.8 Implementation issues

This section discusses the issues involved in the implementation of survivability (protection/restoration) techniques in the optical layer. Unlike most mechanisms that concentrate on the importance of optical layer protection and the various protection schemes, the authors of [10] attempt to investigate ways of implementing them in conjunction with the other layers. This work presents a discussion on the factors contributing to the complexity of optical protection schemes and then deals with the efficient interaction between the client and optical layer with respect to protection. It also presents the subtleties of some interesting optical protection implementation alternatives to SONET-like implementations.

### 13.8.1 Design Considerations

In dealing with the implementation choices for optical layer protection, [10] considers the following *failure scenarios*:

- Failure of client ports or client-optical interfaces that require client protection with or without some optical layer protection.
- Failure of optical layer hardware and client-optical interface hardware (transponders).
- Link failures (includes line amplifiers).
- Node failures (less likely).

The work also considers the following *service classes* based on [33].

### 13.8.2 Protection schemes: Complexity Issues

The complexity involved in implementing a protection scheme relates to the hardware and software in optical layer equipment. This in turn affects the architecture of the nodes. The factors that contribute to this complexity include:

- **Protection Switching Time:**, as mentioned previously which refers to the time taken to switch to the protection route on a failure and *reliability* of the scheme. The restoration time of the scheme is dictated by the applications they support. While voice traffic would be sufficient with restoration times of 60ms would be sufficient for voice traffic (SONET protection), data traffic on the other hand require stringent restoration times (optical layer protection). The choice of switching equipment largely depends on the aforementioned constraints.

- **Efficient use of protection bandwidth:** This could be achieved by multiplexing low-priority traffic on the otherwise idle protection bandwidth. This could however complicate the protection scheme by requiring reconfiguration of the protection path in the event of a failure.

- **Efficient approach to protection:** This considers the use of *Mesh Protection* as an alternative to *ring protection* (which suffers from being less bandwidth efficient). The complexity in interconnecting rings and the absence of centralized management further supports the need for mesh protection. However, the following issues need to be efficiently handled: (i) speed of signaling mechanisms that aid in configuring the protection route; (ii) size of protection tables and its efficient management. (iii) protection efficiency. (iv) sophisticated network planning and (v) need for a standard scheme.

- **Protecting all failure modes**, which includes *node failures*, *link failures* and *multiple concurrent failures*. The need for interconnecting rings at two common nodes and the need to find node-disjoint paths makes protection against node failures more complicated. Multiple concurrent failures are much more complicated due to the size of protection tables and the need to coordinate multiple protection paths. This can be accounted for by either using protection (centralized or distributed) or restoration schemes that handle the additional requirements. Another possible solution is to decompose the network in such a way that it results in a single failure per domain.

- **Including the effect of *wavelength converters***, which decrease blocking probability and increase wavelength utilization. Since converters are expensive, efficient limited converter based designs have been proposed. Node failures remove some converters from the network and hence would require other nodes to compensate for the lost conversion capability.

### 13.8.3 Client versus Optical Layer Protection: Interoperability issues

Efficient interaction between the client and optical layers could be in the form of a combined protection scheme or an efficient interface between the client and optical layer schemes leading to effective overall solutions. The following possibilities have been explored in past work on this issue.

**Mapping client requests to optical lightpaths.** The optical layer provides the necessary end-to-end communication abstraction between two clients communicating over a lightpath. Based on the type of abstraction, the following are possible options.

- The client provides protected links to an unprotected optical layer. This implies that the client protection is independent of the optical layer that merely provides the lightpath service for all the client links. This requires the optical layer to provide lightpaths in such a way as to not allow more than the number of failures the client layer is designed to protect.
- The working client uses a protected or unprotected optical lightpath and the client's protection link uses a low-priority lightpath which is preempted in case of protected lightpath failure. This scheme protects more failure modes than the previous case.
- Packet-level client protection is provided where the client is aware of the optical abstraction and sends high priority packets on protected paths while using low grade lightpaths for low-priority packets.

**Interactive multi-layer protection.** Protection can be provided by both the client and the optical layer provided they coordinate between them. This can be achieved by imposing a timer which would activate the client layer protection until the optical layer is done with its protection. The coordination can be done in one of the following ways:

- Imposing a control plane (MPLS) for both the layers thereby integrating both the layers.
- Using optical layer protection for simple link failures and allowing the higher layers to protect against more complex failures.
- Decoupling the two layers by allowing a 1+1 (or 1:N) client protection and providing appropriate (dedicated or shared) protection at the optical layer. There is a trade-off involved with reduced cost (using 1:N client) as opposed to increased availability (using 1+1 clients).

Another alternative is to allow protected and unprotected client traffic to be sent on different WDM systems or different bands on the same system thereby physically separating the two types. The nature of the different systems would be chosen based on the type of traffic traversing them.

### 13.8.4 Architecture based implementations: Examples

This section describes implementations the for various schemes discussed above.

**Pre-Split and Post-Split path protection.** This situation considers path protection for the optical lightpath with unprotected client requests which can be achieved either by splitting the signal before the transponder (pre-split) or by splitting the signal after the transponder (post-split). This reducing the number of transponders required, at the cost of reduced transponder availability in case of failures.

**Flexing bus architecture for line/path protection.** This solution provides a closed loop ring for each link/path with dedicated backups (using splitters and amplifiers/on-off switches). Upon failure of a link/path, traffic is switched on to the appropriate backup path by turning on the amplifier which is kept off during normal transmission. This implementation avoids single point failures and allows for adding/dropping lightpaths at every node. However, the cost of such an implementation will be high, as may be expected.

**Shared ring protection.** This solution provides ring based protection for lightpaths with backup multiplexing using an optical switch that switches between the working and protection path. Such a solution reduces the number of OADMs required but incurs the penalty of possibly long backup paths. Protection can either be provided on two separate fibers or by sharing bandwidth on the same fiber.

**Multi layer protection.** This scheme can protect from multiple failures (not concurrent) using both client and optical layer. While the client layer provides protection for a first failure, the optical layer could gear itself to protect any additional failure that could occur before client restoration is complete, thus hiding the failure from the client. This could again be subjected to careful optical layer design to avoid long alternate routes.

**Transponder Protection.** The transponders could be protected by having: (i) a *fixed spare* transponder which can be used at the transmitter and at the receiver in case of a failure, (ii) a *tunable spare* able to tune to any wavelength to replace the failed transponder signal thus avoiding the need for coordination between the transmitter and receiver as is required in the previous case, or (iii) a *provisional spare* that avoids the separate spare and uses an optical switch before the transponder to choose the spare transponder for each client link.

## 13.9 Conclusions

Optical wavelength division multiplexing (WDM) technology has enabled optical networks to realize transfer rates of the order several Gb/s of data on each channel. At the same time, network partitioning due to failures could afflict data losses of the order of a few tens to hundreds of Gb/s. Optical network survivability has thus emerged as one of the important problems to be handled in order to effectively implement a WDM optical network and provide valuable optical services to clients.

In this chapter, we analyzed the various aspects of survivability in WDM optical networks. We first studied the various failure models in such networks. We then presented a taxonomy of the classical survivability mechanisms, namely protection and restoration. Techniques to handle single link failures, double link failures, node and channel failures were then summarized. The concept of Quality of Protection (QoP) and related work on this concept were presented. This was followed by a discussion of signaling protocols including the use of MPLS mechanisms. Finally, we also analyzed the issues related to implementing such schemes in the network and ways of realizing them.

## Acknowledgments

The authors acknowledge the support of this research through grants from the National Science Foundation (Grant Number ANI-0322959), Cisco Systems and Intel Corporation. The authors also gratefully acknowledge the discussions with Manav Mishra, Christian Maciocco and Muthu Venkatachalam.

## References

[1] P. Green, "Progress in optical networking," *IEEE Communications Magazine*, vol. 39, pp. 54–61, Jan. 2001.

[2] K. Sivalingam and S. Subramaniam, eds., *Optical WDM Networks: Principles and Practice*. Boston, MA: Kluwer Academic Publishers, 2000.

[3] R. Ramaswami and K. N. Sivarajan, *Optical Networks: A Practical Perspective*. Morgan Kaufmann, 2 ed., 2001.

[4] B. Mukherjee, *Optical Communication Networks*. Addison Wesley, 1997.

[5] S. Ramamurthy and B. Mukherjee, "Survivable WDM mesh networks Part 1 - Protection," in *Proc. IEEE INFOCOM*, vol. 2, (New York, NY), pp. 744–751, Mar. 1999.

[6] S. Ramamurthy and B. Mukherjee, "Survivable WDM mesh networks, Part II - Restoration," in *Proc. International Conference on Communications (ICC)*, (Vancouver, Canada), pp. 2023–2030, June 1999.

[7] M. Medard, S. G. Finn, and R. A. Barry, "WDM loop-back recovery in mesh networks," in *Proc. IEEE INFOCOM*, (New York, NY), 1999.

[8] W. Grover, *Mesh-based Survivable Networks: Options and Strategies for Optical, MPLS, SONET and ATM Networking*. Prentice Hall PTR, 2003.

[9] J. Doucette and W. D. Grover, "Capacity Design Studies of Span-Restorable Mesh Networks with Shared-Risk Link Group (SRLG) Effects," in *Proc. SPIE OPTICOMM*, (Boston, MA), July 2002.

[10] O. Gerstel and R. Ramaswami, "Optical Layer Survivability-An Implementation Perspective," *IEEE Journal on Selected Areas in Communications*, vol. 18, pp. 1885–1923, Oct. 2000.

[11] D. Zhou and S. Subramaniam, "Survivability in optical networks," *IEEE Network Magazine*, vol. 14, pp. 16–23, Nov. 2000.

[12] G. Mohan, C. S. R. Murthy, and A. Somani, "Efficient algorithms for routing dependable connections in WDM optical networks," *IEEE/ACM Transactions on Networking*, vol. 9, pp. 553–566, Oct. 2001.

[13] R. R. Iraschko, W. D. Grover, and M. H. MacGregor, "A distributed real time path restoration protocol with performance close to centralized multi-commodity maxflow," in *Proc. International Workshop on Design of Reliable Communication Networks (DRCN)*, (Brugge, Belgium), May 1998.

[14] S. S. Lumetta and M. Medard, "Towards a deeper understanding of link restoration algorithms for mesh networks," in *Proc. IEEE INFOCOM*, (Anchorage, AK), Apr. 2001.

[15] J. B. Slevinsky, W. D. Grover, and M. H. MacGregor, "An algorithm for survivable network design employing multiple self-healing rings," in *Proc. IEEE GLOBECOM*, (Houston, TX), pp. 1568–1572, Nov. 1993.

[16] G. Maier, S. D. Patre, A. Pattavina, and M. Martinelli, "Optical network survivability: protection techniques in the WDM layer," *Photonic Network Communications*, vol. 4, pp. 251–269, July 2002.

[17] G. Ellinas, A. Hailermarian, and T. Stern, "Protection cycles in mesh WDM networks," *IEEE Journal on Selected Areas in Communications*, vol. 18, pp. 1924–1937, Oct. 2000.

[18] D. Stamatelakis and W. D. Grover, "IP Layer Restoration and Network Planning Based on Virtual Protection Cycles," *IEEE Journal on Selected Areas in Communications*, vol. 18, pp. 1938–1949, Oct. 2000.

[19] G. Ellinas, A. G. Hailemariam and T. E. Stern, "Protection Cycles in Mesh WDM Networks," *IEEE Journal on Selected Areas in Communications*, vol. 18, pp. 1924–1937, Oct. 2000.

[20] G. Mohan and A. K. Somani, "Routing dependable connections with specified failure restoration guarantees," in *Proc. IEEE INFOCOM*, (Tel-Aviv, Israel), Apr. 2000.

[21] S. Gowda and K. M. Sivalingam, "Protection Mechanisms for Optical WDM Networks based on Wavelength Converter Multiplexing and Backup Path Relocation Techniques," in *Proc. IEEE INFOCOM*, (San Francisco, CA), Mar. 2003.

[22] R. Shenai, C. Macciocco, M. Mishra, and K. Sivalingam, "Threshold based selective link restoration for optical WDM mesh networks," in *Proc. International Workshop on Design of Reliable Communication Networks (DRCN)*, (Banff, AB), Oct. 2003.

[23] "New Focus TLS 420C: Using Tunable Lasers to Improve Network Efficiency and Meet Customer Requirements." White Paper, http://www.newfocus.com, 2001.

[24] S. Sengupta and R. Ramamurthy, "From Network Design to Dynamic Provisioning and Restoration in Optical XC Mesh Networks: An Architecture and Algorithm Overview," *IEEE Network*, pp. 46–54, Jul/Aug 2001.

[25] H. Krishnamurthy, K. M. Sivalingam, and M. Mishra, "Restoration mechanisms based on tunable lasers for handling channel and link failures in optical WDM networks," in *Proc. SPIE OPTICOMM*, (Boston, MA), July 2002.

[26] H. Choi, S. Subramaniam, and H.-A. Choi, "On Double-Link Failure Recovery in WDM Optical Networks," in *Proc. IEEE INFOCOM*, (New York, NY), June 2002.

[27] W. He, M. Sridharan and A. K. Somani, "Capacity Optimization for Surviving Double-Link Failures in Mesh-Restorable Optical Networks," in *Proc. SPIE OPTICOMM*, (Boston, MA), July 2002.

[28] M. Sivakumar, C. Maciocco, M. Mishra, and K. Sivalingam, "A Hybrid Protection-Restoration Mechanism for Enhancing Dual-Failure Restorability in Optical Mesh-Restorable Networks," in *Proc. SPIE OPTICOMM*, (Dallas, TX), pp. 37–48, Oct. 2003.

[29] Y. Liu and D. Tipper, "Successive Survivable Routing for Node Failures," in *Proc. IEEE GLOBECOM*, (San Antonio, TX), pp. 25–29, Nov. 2001.

[30] S. Kim and S. S. Lumetta, "Addressing node failure in all-optical networks," *Journal of Optical Networking*, vol. 1, pp. 154–163, Apr. 2002.

[31] M. Clouqueur and W. Grover, "Mesh-restorable networks with enhanced dual-failure restorability properties," in *Proc. SPIE OPTICOMM*, (Boston, MA), July 2002.

[32] C. Vijayasaradhi and C. S. R. Murthy, "Routing differentiated reliable connections in single and multi-fiber WDM optical networks," in *Proc. SPIE OPTICOMM*, (Denver, CO), pp. 24–35, Aug. 2001.

[33] O. Gerstel and G. Sasaki, "Quality of protection (QoP): a quantitative unifying paradigm to protection service grades," in *Proc. SPIE OPTICOMM*, (Denver, CO), pp. 12–23, Aug. 2001.

[34] R. Doverspike and J. Yates, "Challenges in MPLS in optical network restoration," *IEEE Communications Magazine*, vol. 39, pp. 89–97, Feb. 2000.

[35] S. Chaudhuri, G. Hjalmtysson, and J. Yates, "Control of lightpaths in an optical network." IETF draft and OIF submission 2000.4, 2000.

[36] C. Assi, Y. Ye, Abdallah, S. Dixit, and M. Ali, "On the merits of IP/MPLS protection/restoration in IP over WDM networks," in *Proc. IEEE GLOBECOM*, (San Antonio, TX), pp. 65–69, 2001.

[37] G. Li, J. Yates, R. Doverspike, and D. Wang, "Experiments in Fast Restoration using GM-PLS in Optical/ Electronic Mesh Networks," in *Proc. of Optical Fiber Communications (OFC)*, (Anaheim, CA), Mar. 2001.

Chapter 14

# TRADEOFFS AND COMPARISON OF RESTORATION STRATEGIES IN OPTICAL WDM NETWORKS

Arun K. Somani
*Department of Electrical and Computer Engineering, Iowa State University, Ames, Iowa 50011*
Email: arun@iastate.edu

**Abstract** The problem of providing *dependable connections* in WDM networks is an important problem due to high traffic speed and high vulnerability of such networks. We call a connection with fault tolerant requirements as a dependable connection (D-connection). We consider *single-link* and *multi-link failure models* in our study. We consider a full redundancy-based schemes where the backup path for every accepted are identified at the and recommend to use a pro-active approach wherein a D-connection is identified with the establishment of the primary lightpath and a backup lightpath at the time of honoring the request (called restoration).

We discuss algorithms to select routes and wavelengths to establish D-connections with improved blocking performance. The algorithms use *backup multiplexing* and alternate *L+1 routing* techniques to efficiently utilize the wavelength channels. We also analyze the performance of these algorithms and quantify the tradeoffs in using different algorithms.

**Keywords:** WDM networks, fault tolerance, D-connection, restoration, single-link failure.

## 14.1 Introduction

Wavelength-division multiplexing (WDM) technology allows building of very large capacity, of the order of terabits per second wide area networks. Such networks provide low error rates and low delay and offer a viable solution to meet the bandwidth demand arising from several emerging applications. Wavelength division multiplexing divides the available bandwidth of a fiber into many non-overlapping channels, each channel carried on a different wavelength on the same fiber. All the channels can be used simultaneously.

Existing networks utilize WDM technology as a physical media for point-to-point transmission. Each channel in these networks provides the equivalent of a physical wire and the end nodes perform an electrical to optical and optical to electrical signal conversion. A source to a destination connection is created using multiple hops over the point-to-point connections. This increases delay and requires buffering at the intermediate nodes, not desired for high-speed connections.

A viable alternative to overcome the shortcomings of point-to-point WDM networks is to apply WDM technology to the path layer where a message is directly transmitted from the source to destination by using a *lightpath* without requiring any electro-optical conversion and buffering at the intermediate nodes. The intermediate nodes simply provide a route through them for the path. A lightpath is uniquely identified by a physical path and a wavelength used for that path on all point-to-point fibers. This is also known as *wavelength routing*. Thus a virtual topology may be created using several such lightpaths as needed to meet the traffic demands.

The architecture of a WDM network consists of *wavelength cross-connects* (WXCs) that are interconnected by fiber links. A WXC routes a message arriving on an incoming fiber on a wavelength to an outgoing fiber on the same wavelength. Use of the same wavelength is required in the absence of wavelength conversion at the intermediate node. An alternative is to use a *wavelength interchanging cross-connect* (WIXC) that employs wavelength converters. The latter results in higher performance and lower blocking probability in establishing connections. Another architectural concept is to use multiple fiber links on each link to minimize the need for converters [1].

Since most WDM networks impose wavelength continuity constraint, several good heuristic *routing and wavelength assignment* (RWA) algorithms have been developed in literature [2, 3, 4, 5]. The routing methods such as shortest path routing, fixed path routing, fixed paths alternate routing, least-loaded routing, and fixed path least congested routing methods have been analytically and experimentally evaluated [6, 5, 7]. The RWA algorithms consider static and dynamic scenarios and have the objective to assign lightpaths while minimizing the resources required [8]. Static scenarios are relatively easy to handle as the requirements for routing are well known in advance. In a dynamic traffic environment, the connection requests arrive to a network one by one in a random order. The dynamic traffic demand scenario is applicable to the circuit layer or to the path layer. Sometimes the dynamic traffic may require the reconfiguration of the network in response to changing traffic patterns [9]. Moreover, in IP – over – WDM networks, lightpaths may be established between two IP routers, changing the topology of the IP network. The IP layer and optical path layer may interact to route the traffic efficiently.

### 14.1.1 Lightpath Failure, Protection, and Restoration

WDM networks are prone to component failures. Catastrophic link failures in optical networks are, in fact, quite common. A variety of factors can lead to link failure including optical fiber, transmitter, receiver, amplifier, router and converter faults. A fiber cut, possibly the result of an errant excavation, has been estimated to occur, on average, once every four days by TEN, a pan-European carrier network [10]. It has been shown in [11] that detection, location and isolation of all of these fault scenarios is both very important and very possible. A link fault can be detected as easily as the receiver nodes detecting a loss of light on the link, and invoking a network management algorithm to first notify and then recover from the fault without causing a network failure.

A fiber-cut causes a link failure. When a link fails, all its constituent fibers will fail. A node failure may be caused due to the failure of the associated WXC. A fiber may also fail due to the failure of its end components. Since WDM networks carry high volumes of traffic, failures may have severe consequences. Therefore, it is imperative that these networks have fault tolerance capability. Failure monitoring, detection, location, and recovery mechanisms have to be part of design and operation [13]. For example, the nodes adjacent to the failed link can detect the failure by monitoring the power levels of signals on the link [13]. In [11], some mechanisms to detect and isolate faults such as fiber cuts, router and amplifier failures are discussed.

Fault-tolerance refers to the ability of the network to reconfigure and reestablish communication upon failure. A mechanism that includes set up of a backup path as part of connection establishment and be ready to replace a primary path when a failure is detected on a primary path node or link is called protection. Alternatively, the connection establishment mechanism may only identify a path that must be used when a primary path experiences a node or link failure reserve sufficient resources. The actual process is managed by a combination of hardware and software methods. The process of reestablishing communication through a lightpath between the end nodes of a failed lightpath is known as *lightpath restoration* [14]. A lightpath that carries traffic during the normal operation is known as the *primary lightpath.* When a primary lightpath fails, the traffic is rerouted over a new lightpath known as the *backup lightpath.* The objective of any algorithm is to minimize the *spare resources* (wavelengths, fibers) required and to minimize the connection blocking probability. We call a connection with fault tolerant requirements as a dependable connection (D-connection).

There have already been several approaches to link fault tolerance laid forth in literature. Three such strategies, presented in [15] and [16], require the usage of network resources to provide backup lightpath routing so that when a fault occurs there is an alternate path for the connection to use. Service in such an approach is only interrupted briefly to allow restoration to occur. The major drawback of this approach is the allocation of valuable system resources on typically unused backup lightpaths.

We consider *single-link* and *multi-link failure models* in our study. A *single-link failure model* assumes that at any instant of time, at most one link has failed. A *multi-link failure model* may be considered in two different ways, two simultaneous failures or recovery from two failures that occur sequentially. We consider the following two techniques: (1) a full redundancy-based scheme (called **protection**) where both the primary and backup paths are established and provisioned at the time of setting up a connection and (2) to use a pro-active approach (called restoration) wherein a D-connection is identified at the time of honoring the request and the primary lightpath is established but the identified backup lightpath is only provisioned when a failure actually occurs.

Due to the use of backup lightpaths, resource (wavelength channel) consumption becomes higher. We discuss algorithms to select routes and wavelengths to establish D-connections with improved blocking performance. The algorithms use *backup multiplexing* and alternate *subgraph-based L+1 routing* techniques to efficiently utilize the wavelength channels.

The idea behind *backup multiplexing* is to allow two backup lightpaths to share a wavelength channel, if their primary lightpaths do not fail simultaneously. Upon a link failure, all the failed paths find their backup lightpaths readily available. Thus, these algorithms ensure 100% *restoration guarantee.* We define the restoration guarantee as the guarantee with which a failed lightpath finds its backup readily available. An alternative to backup multiplexing with less than 100% guarantee is to use *primary-backup multiplexing* [16] that allows a primary lightpath and one or more backup lightpaths to share the same channel.

The idea behind the *subgraph-based L+1 routing* scheme is to plan network routing such that for any link failure, there exists an alternate path for every accepted request. When a link fails, all paths that get affected by the link failure are reassigned to their new path.

We discuss algorithms to select routes and wavelengths to establish D-connections with improved blocking performance. The algorithms use *backup multiplexing* and alternate *L+1 routing* techniques to efficiently utilize the wavelength channels. We also analyze the performance of these algorithms and quantify the tradeoffs in using different algorithm.

### 14.1.2 Characterizing Restoration Methods

The restoration schemes differ in their assumption about the functionality of cross-connects, traffic demand, performance metric, and network control. Moreover, a restoration scheme may assume either centralized or distributed control. In large networks, distributed control is preferred over the centralized control. A distributed control protocol requires several control messages to be exchanged between nodes.

The restoration methods are broadly classified into reactive and pro-active methods. The reactive method is a simple way of recovering from failures. When an existing lightpath fails, a search is initiated for finding a new lightpath which does not use the failed components. This has low overhead as no resources are pre-reserved or kept free. However, this method may not guarantee successful recovery. In case of distributed implementation, contention among simultaneous recovery attempts for various failed connections may occur. In a pro-active method, backup lightpaths are planned and resources are reserved at the time of establishing the primary connection. This method obviously guarantees 100% restoration. The restoration time of a pro-active method is much smaller than reactive method as the resources for restoration are already identified and reserved. In that sense, it is also a protection scheme.

A pro-active or reactive restoration method is either *link based* or *path based* [8, 12]. A link based method employs *local detouring* while the path based method employs *end-to-end detouring*. The two detouring mechanisms are shown in Figure 14.1. For a link based method, all routes passing through that link are transferred to a local rerouting path that replaces that link. This method is attractive for its local nature, however limits the choices for alternatives [8]. In case of wavelength selective networks, the backup path must necessarily use the same wavelengths for existing requests as that of their corresponding primary paths as the working segments are retained.

In a path based restoration method, each request that is using the failed link requires selection of a new lightpath. Such a path is selected between the two end nodes of the failed primary lightpath. The backup path may use any available wavelength as no operational segment of independent path is necessarily the part of the new path.

### 14.1.3 Backup Multiplexing - Based Restoration

A pro-active restoration method may use a dedicated backup lightpath for every primary lightpath. However it may be expensive in terms of resource utilization as shown in Figure 14.2. The figure shows two primary lightpaths $p_1$ and $p_2$ and their respective backup lightpaths $b_1$ and $b_2$ on a wavelength.

For better resource utilization, under the assumption of a single link failure, backup path resources may be used in such a way they are sufficient to restore all the failed paths. This is known as backup multiplexing technique. For example, if two primary lightpaths do not fail simultaneously, their backup lightpaths can share a wavelength channel as illustrated in Figurc 14.2. The same two lightpaths, $p_1$ and $p_2$, are link-disjoint. Hence they do not fail si-

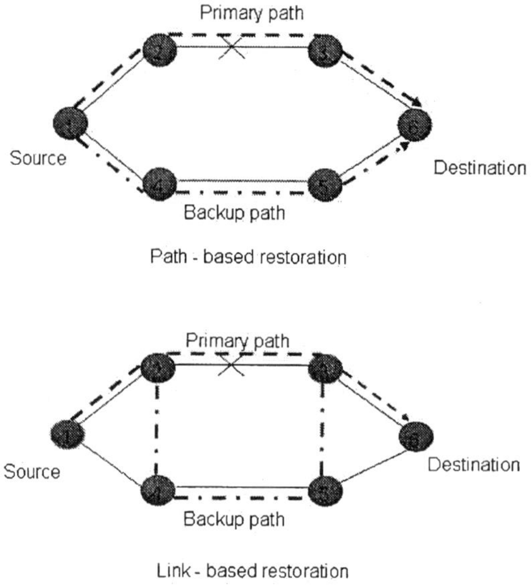


*Figure 14.1.* Path and link based restoration.

multaneously upon a single-link failure. Therefore, the corresponding backup paths $b_1$ and $b_2$ can share the wavelength. A pro-active method can employ *primary-backup multiplexing* to improve upon the resource utilization.

A recovery process is a single link-failure independent if two paths chosen to serve that request are link disjoint. Since we are dealing with only one link failure, in the primary-backup scheme or backup multiplexing scheme either the primary path is affected by the failure or it is not. In case it is affected, the alternate path is guaranteed to be not affected and hence can be used for recovery. This method of course uses more resources, but guarantees that the alternate path is available in case primary path is affected. An alternative to this method is to plan for recovery that is failure dependent and that is what the subgraph-based L+1 routing strategy uses, where for each failure, there is a different alternate path. This strategy can be used to tolerate node and multiple link failures as well.

The problem of restorable network design for a static traffic demand has been dealt with in [8, 14, 17, 18, 12, 19, 20, 21, 22, 23, 24]. Several other works [25, 16] also use dynamic scenarios.

A reactive method does not reserve any backup lightpath for a primary lightpath before failures actually occur. A pro-active method reserves resources without or with backup multiplexing. Therefore, they perform poorly when compared to a reactive method. On the other hand, the restoration time in a

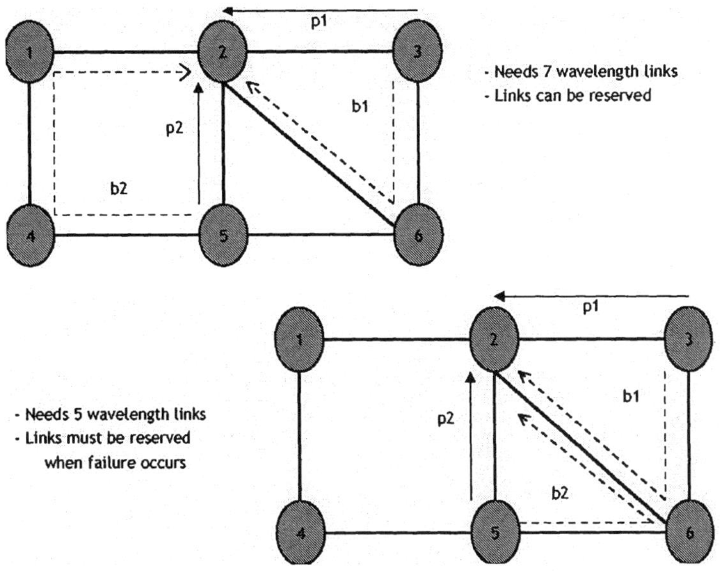


*Figure 14.2.* Dedicated vs backup path multiplexing reservation.

pro-active method is much shorter than a reactive method. A happy medium here could be to also allow primary-backup multiplexing where some resources that are needed for backup may be used by some low priority primary traffic. This does not guarantee 100% restoration, but may improve the overall performance of the network. A D-connection loses its recoverability only when the following three events occur simultaneously. 1) a link fails during the period of its existence 2) the failed link is used by its primary lightpath, and 3) a channel on its backup lightpath is used by some other primary lightpath. However, such a situation is less probable.

Such primary-backup multiplexing technique is depicted in Figure 14.3. It shows three primary lightpaths $p_1, p_2$, and $p_3$ and their respective backup lightpaths $b_1$, $b_2$, and $b_3$ on a wavelength. The backup lightpaths $b_1$ and $b_2$ share the channel on link $5 \rightarrow 2$, as $p_1$ and $p_2$ are link disjoint. The channel on link $2 \rightarrow 3$ is shared by $p_3$ and $b_1$ and the channel on link $6 \rightarrow 3$ is shared by $p_1$ and $b_3$. Therefore, both $p_1$ and $p_3$ are non-recoverable. This approach has been explored in [16].

## 14.2 *L+1* Fault Tolerance Routing Strategy

The *L+1* routing strategy attempts to provide a passive form of redundancy to optical networks in the event of a single-link failure. It is passive in that,

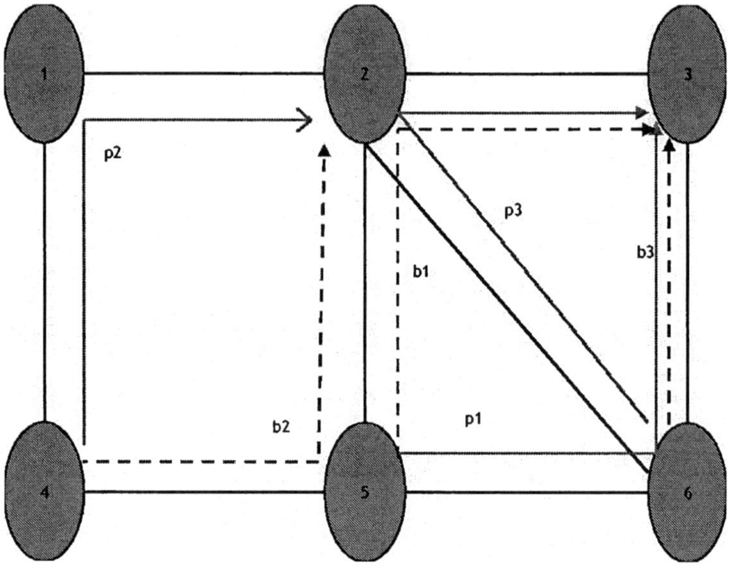


*Figure 14.3.* Example of primary-backup multiplexing technique.

before a connection is established, it is subjected to the constraints of *L+1* routing and is thus guaranteed in the event of a single link failure. The end user experiences nominal interruption in service due to network state restoration and it is characterized in [26]. The key characteristics of the *L+1* strategy are as follows:

1 No additional system transmission resources are used to provide connection redundancy.

2 Fault recovery network states are maintained throughout the operation of the network.

3 *L+1* fault tolerance provides a 100% guarantee that any single link fault can be recovered from.

4 *L+1* is a path-based fault tolerance strategy.

The first characteristic highlights one of the most important aspects of the strategy; it does not require the allocation of system transmission resources to ensure recoverability after the detection and location of a link fault. Simply put, there is no link capacity lost due to the routing of backup connections because no backup connections exist in this strategy. Instead, all connections have a different set of routes to take depending on the actual location of the fault. The second characteristic is important because, upon the occurrence of a fault, the network restores itself to a state that eliminates the defective link from consideration, and the network operates as if it never existed. Third, the strategy provides a 100% guarantee that any single link failure is recoverable. This becomes important when comparing *L+1* to other fault tolerant strategies. Fourth, *L+1* is a path based recovery strategy because it does not guarantee that any of the same links are used to reroute a connection upon the occurrence of a fault.

A disadvantage of *L+1* fault tolerance is that it can potentially require a complete reconfiguration of the network to a predetermined new state. Not all connections may be affected by a network reconfiguration, but no connection is guaranteed to be unaffected by a link fault recovery. *L+1* fault tolerance assumes that the probability of suffering a double link failure during network operation is very low. The strategy guarantees the recovery of the network from any single link failure at any given time. It is also assumed that each node knows the entire network state at any given time. This is key because all nodes need to know when and where a fault has occurred so that they can appropriately retask themselves to adopt the backup network state. Each node is required to maintain all subgraph network state information. Once a link fault occurs and is detected and located, all nodes are informed of the location to start the recovery process. Of course, if there is a centralized recovery station, then all such information is part of the centralized recovery server.

## 14.2.1 *L+1* Fault Tolerance Model

Networks consist of a set of nodes and links that correspond to the various servers, routers, switches and cables that make up its physical implementation. These nodes and links can be viewed as a set of vertices and edges in a graph. Each graph, $G$, is defined as a set of $V$ vertices and $E$ edges or, $G = (V, E)$. There exists a set of subgraphs of $G$, denoted as $G_i$, where $e_i$ is removed from the graph $G$, or mathematically, $\forall 1 \leq i \leq L,\ G_i | G_i = (V,\ E - \{e_i\})$, where $L$ is the cardinality, $|E|$, of the set of edges in graph $G$. Therefore there exist $L$ subgraphs of graph $G$, each one missing one of the $|E|$ edges. The set of $L$ subgraphs of $G$ represents all possible single-link failures in the network. The original full link graph is called the *base network*. The base network's constituent subgraphs are treated as virtual networks because only there state of utilization is maintain after the current set of active connections have been admitted. When a fault occurs on link $e_i$, the current set of connections are routed on the routes chosen in subgraph $G_i$.

A graph with five nodes and six edges, as shown in Figure 14.4, there will be six subgraphs. For the purposes of this example, we assume that each edge in the base network (and its constituent subgraphs) has a capacity of one and that the distance between any pair of adjacent vertices is one.

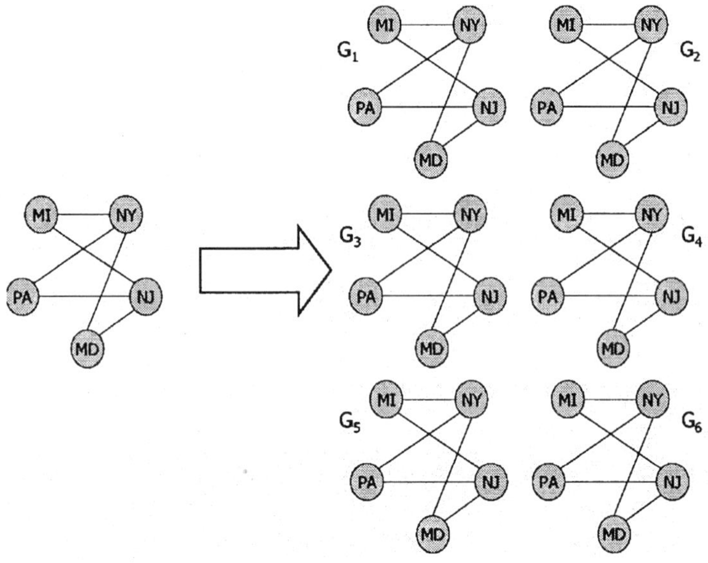


*Figure 14.4.* Example of subgraphs.

**Connection Request Servicing in an $L + 1$ Network.** Consider the northeast part of the NSF network subgraph as shown in Figure 14.5. It has six nodes and six bi-directional links. Let there be a request issued by vertex MI to connect with vertex NY. This connection attempts to find a path from MI to NY on all of the L subgraphs of the base network as shown in Figure 14.5. The connection request from MI to NY is accepted as a path is available in all subgraphs. Another request from vertex NJ to PA is routed in the same way as the path from NJ to PA is available in all subgraphs (although it may be a different path in different subgraphs). When a third request from NY to MD arrives, it can only find paths in subgraphs $G_2$, $G_3$, and $G_6$. It blocks in subgraphs $G_1$, $G_4$, and $G_5$. Thus the connection request between from node NY to node MD now fails due to non-availability of resources, and consequently not routed in the base network.

## 14.2.2 Fault Tolerance in an $L + 1$ Network

In the event of a fault, the $L + 1$ based routed network can fully recover by accepting the subgraph network state corresponding to the located edge

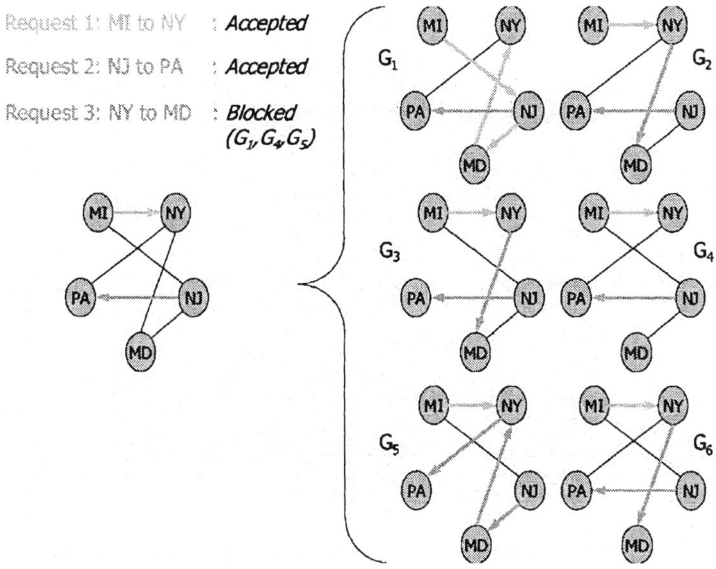


*Figure 14.5.* $L + 1$ model of northeast part of NSF network.

failure. For example, assume that there is an arbitrary failure of edge NJ to PA. We assume that the failure is in both directions and for some reason the edge is left non-operational. To recover, the network reroutes all current connections to reflect the network state depicted by subgraph $G_5$. The fault occurrence and recovery cycle is such that connections from the base network are rerouted to the paths in the selected subgraph (corresponding to the failed link). For the example, the request from MI to NY need not be rerouted as the path taken in the bases network as well in subgraph $G_5$ are the same. However the path for the request node NJ to PA has to be now rerouted through the links (NJ, MD), (MD, NY), and (NY, PA) as shown in $G_5$ of Figure 14.5.

## 14.3 Performance Evaluation

In this section, we compare the performance of the backup multiplexing and subgraph based $L + 1$ routing strategy. For comparison purpose, the shortest path length (in terms of the number of hops) routing strategy is utilized to evaluate the effectiveness of these schemes. Shortest path length routing attempts to dynamically route connections along the path with the least number of links between source and destination nodes. Each link is known as a hop. Wavelength assignment within a link is done at random. Connections are routed dynamically in that each request is routed based on the network state at the time it enters the network. Dynamic routing and wavelength assignment typically performs better than fixed path routing, although it requires higher control overhead because each node must maintain network state information. The no-backup and backup multiplexing routing strategies are compared to provide a reference to measure the effectiveness of $L + 1$ fault tolerance. No-backup routing uses the shortest path in terms of hop strategy, and makes no provision for fault tolerance. Backup multiplexing uses shortest cycle routing. The results laid forth for backup multiplexing are based on the selection of the shortest path for the primary connection. Shortest cycle routing [8] is used to increase the performance of backup multiplexing by guaranteeing both a potential primary and backup path are found while attempting to establish a connection and that the primary/backup path pair is the shortest pair of paths from source to destination.

For comparison purposes, the connection requests are generated using a Poison distribution with arrival rate of $\lambda$ where the arrival rates are varied to compare the different schemes. The request hold time follows negative exponential probability distribution with a parameter value of $\mu = 1.0$. Source and destination nodes for any request are uniformly distributed.

### 14.3.1 Performance Metrics

Several metrics are used to evaluate the effectiveness of the various schemes. These metrics are designed to measure both the efficiency and the feasibility of such a scheme and are only measured in the base network. They include, blocking probability, average path length, average shortest path length, effective network capacity used, and probability of path re-assignment

**Blocking probability.** Blocking probability is the most common indicator used to assess network routing and fault tolerance strategies. It is the probability that a request entering the network is rejected. Blocking probability is a ratio of $B$ and $R$ where $B$ is defined as the total number of blocked requests and $R$ is the total number of requests, i.e.

$$BP = B/R.$$

**Average path length.** Average path length and average shortest path length are intended to compare how metrics perform within a network. As the ultimate goal in routing a connection is usually to use the shortest path between two points, average shortest path length provides a way to compare how effectively a routing strategy performs in a given network configuration. Both average path length and average shortest path length are calculated in terms of number of "hops" or number of links along the path.

These averages are calculated for the accepted requests only. The average shortest path is computed by using the shortest possible path possible in the network, even if it is not available. The average path length is calculated using the path length of the path taken (or available for the request) when the request arrives.

**Network utilization.** Network utilization metrics are characterized by their inclusion of the ideas of connection and link capacity. They are an indication of how much of the network is being used over the course of operation and whether there are enough resources available to handle the request load demands.

Effective utilization refers to the minimum amount of system resources needed to service all accepted connections if they were to have been routed along the shortest path. In order for the effective utilization metric to be useful, it first has to be normalized. The first step is to normalize it to the time duration of the simulation so that data obtained at different arrival rates can be compared. Dividing by the time duration of the simulation normalizes utilization. The simulation time is known only to the network, and can either be obtained by knowing the time that the last request enters the network, or by calculating it as a function of $R/\lambda$ where $R$ is the number of arriving requests and $\lambda$ is the arrival rate. Normalizing the utilization with respect to time yields a value that is bounded on the low side by 0 and on the high side by the total available capacity in the network. The second step to normalization is to normalize it by dividing it by the total available capacity of the network, given by $L \times C$, where $L$ is the total number of links in the network, $C$ is the total available capacity per link. Thus for utilization computation, $0 \leq \lambda/R \times U \leq L \times C$, where $U$ is the utilization. In other words, $0 \leq \lambda \times U/(R \times L \times C) \leq 1$.

## 14.3.2 Probability of Path Reassignment

The effectiveness of a fault tolerance scheme depends on how the networks recover from a failure. This potentially may require all connections in the network to be reconfigured to different paths. In order to quantify the amount of path reassignment taking place, path reassignment probability must be measured, which is the probability that a connection's path on the base network needs to be changed upon the occurrence of a link failure. Let $P_j(R_i)$ be the probability that the path for request $R_i$ remains the same if link $j$ of the network fails. Then probability of reassignment for $L+1$ routing strategy is given by $P(\text{L+1 } reassignment) = 1 - \sum_{i=0}^{i=R-B} \sum_{j=0}^{j=L} P_j(R_i)/(L \times (R-B))$. For the backup multiplexing a path is reassigned only if the failed link is used by a request. So if $L_i$ is the path length of request $i$ then the probability of reassignment is given by $P(\text{L+1 } reassignment) = \sum_{i=0}^{i=R-B} L_i/(L \times (R-B))$.

**Link load.** Link load is a measure of the load placed on each node in the networks at any given time. It is useful in providing a baseline for the comparison of the effectiveness of routing strategies across different network topologies. Link load, or $\gamma$, is calculated using the following equation, where each duplex link is treated as 2 links, N is the total number of nodes in the network and $\lambda_n$ is the arrival rate per node.

$$\gamma = N \times \lambda_n \times \overline{H}/L.$$

$\overline{H}$ is the expected length of a primary connection in the topology in hops. Link load, expressed in units of Erlangs and is used to compare results among different networks.

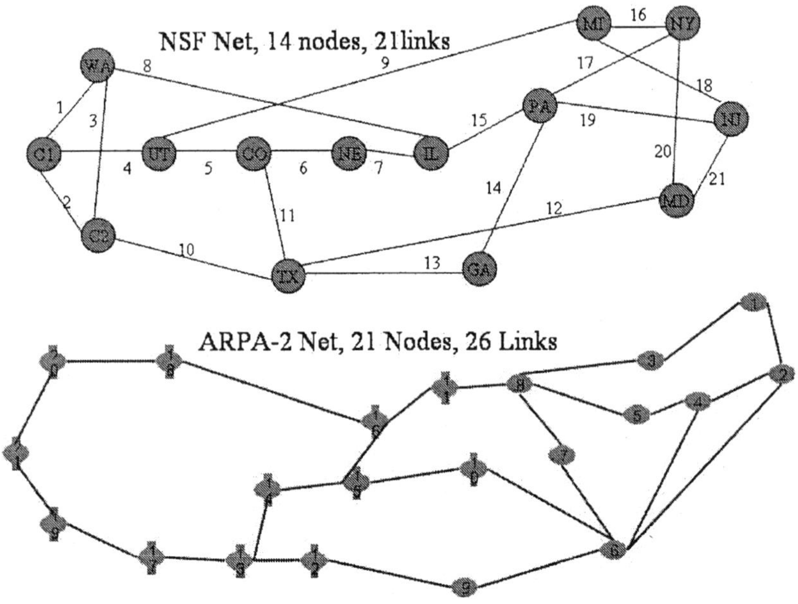


*Figure 14.6.* NSF network and ARPA2 network.

### 14.3.3 Network Structures

Three standard network structures are used to assess and compare the performance of different restoration techniques. The networks considered are a 14-node 21-link NSF net topology, a 4 by 4 mesh-torus network, and a 21-node 26-link ARPA-2 network topology. The first and the last topologies are shown in Figure 14.6. The NSFNET and APRA-2 networks are real world topologies. The 4x4 mesh torus network possesses a high level of connectivity. Each node in a 4x4 mesh torus has a degree of 4, resulting in many potential paths between a node pair.

All links in network have one fiber and the number of wavelengths per fiber is 16. In general, the capacity of a link is a direct function of the number of wavelengths on each fiber in the link. The wavelength continuity constraint has to be followed for each request. Each link is a duplex link (as all of the tested networks do) and each link is considered as two simplex links operating in opposite directions. In the event of a link failure such as a fiber optic cable being severed, both simplex links are severed.

## 14.4 Results

This section presents the performance evaluation.

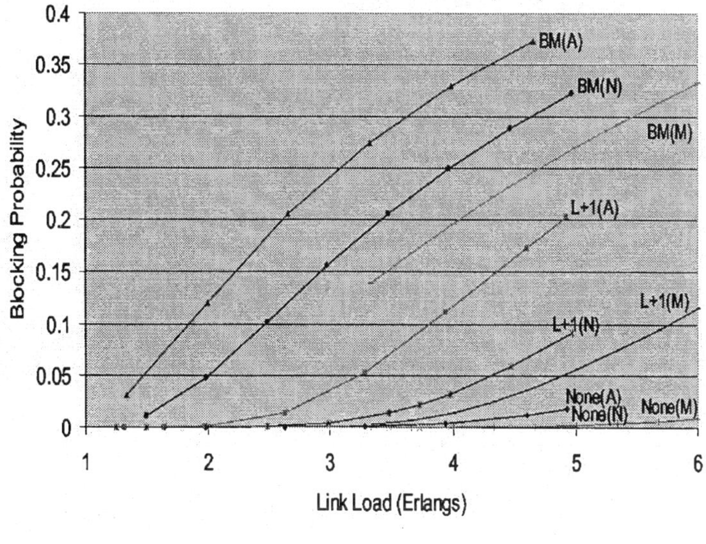


*Figure 14.7.* Blocking probability of three topologies.

### 14.4.1 Blocking Probability

The results for $L+1$ fault tolerance and backup multiplexing and no backup strategy for the NSFNET, ARPA-2 and 4x4 mesh torus topologies are shown in Figure 14.7. In the Figure 14.7, note that $BM$ stands for backup multiplexing, $L+1$ stands for $L+1$ method and $None$ is for no- backup. Also, $A$ is for ARPA-2 network, $N$ is for NSFNET and M is 4x4 Mesh network. It is noticed that the $L + 1$ fault tolerance performs much better than backup multiplexing under the same parameters. In the best case, the NSFNET, the blocking probability of backup multiplexing is roughly 3.5 times that of the $L + 1$ strategy, and is approximately 3 times higher than $L + 1$ in the ARPA-2 and mesh torus topologies.

One factor in the increasing the blocking probability of $L + 1$ fault tolerance is the presence of nodes with degree of two present two of the topologies. For example, there are two nodes of degree two in the NSFNET topology, and when subgraphs are formed; these two nodes become nodes of degree one in four of the 21 subgraphs. These isolated nodes are much more difficult to route connections because of the severely limited capacity in and out of the nodes. A node with degree of two in the base network is referred to as a dead-end node. Dead-end nodes account for 14 of the total 21 nodes in the ARPA-2 topology, and consequently the performance of $L + 1$ in that topology is slightly worse when compared to the NSFNET and mesh torus.

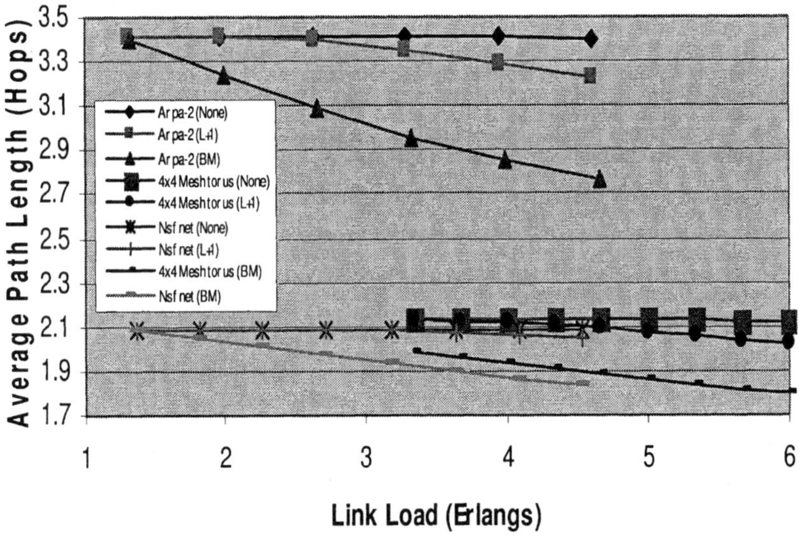


*Figure 14.8.* Average path length of three topologies.

## 14.4.2 Average Path Length

The results for $L + 1$ fault tolerance and backup multiplexing and no backup strategy for the NSFNET, ARPA-2 and 4x4 mesh torus topologies are shown in Figure 14.8. Average path length indicates the average number of hops a connection must contain. The figure shows that average path length decreases as arrival rate increases. The large decrease in path length can be attributed to higher blocking probability and the consequent lower number of connections in the network for a request to have to route around. In general the average path length decreases as the arrival rate increases. As more requests enter the network at a time, the network becomes more congested and more requests are blocked. The result is that the requests that are accepted as connections are only those that are able to find shorter paths to route on.

The backup multiplexing path lengths are higher across all topologies because the paths are based on shortest cycle routing, and the primary paths are not necessarily the shortest paths between two nodes. Shortest cycle routing actually improves performance because, although primary path lengths are longer, a primary-backup pair is almost always found and the total length of the primary-backup pair is the shortest possible.

### 14.4.3 Effective Utilization

Effective utilization measures the minimum amount of network resources needed to accept the connections in the network at any given time for any given link load. In other words, in order to accept the requests that the network did, the network had to provide a minimum amount of resources to the establishment of the connections. The effective utilization figures are shown in Figure 14.9. For the most part, the effective utilizations of each strategy mirror each other, the exception being under high arrival rates. The difference is again attributed to fewer requests being accepted as arrival rate increases. Utilization is directly proportional to the sum of the products of the capacity and path length of each accepted connection, it consequently decreases as fewer requests are accepted. Again effective utilization is used primarily to compare the performance of both fault tolerance strategies to a no tolerance strategy.

### 14.4.4 Comparison Between Network Topologies

An indication of the connectivity of a topology is the number of dead-end nodes each of the subgraphs of that topology contains. Along that reasoning, the ARPA-2 topology has the worst connectivity with 14 dead-end nodes out of 21, the NSFNET next with 2 of 14 and the mesh torus best with no dead-end nodes. The blocking probabilities confirm this, as the blocking probability is higher per Erlang of link load for the ARPA-2 topology, followed by the NSFNET and lastly by the mesh torus. The comparatively lower blocking probability per link load of the mesh torus indicates that $L + 1$ fault tolerance performs much better in topologies with higher connectivity.

### 14.4.5 Probability of Reassignment

In a network recovery situation, $L + 1$ fault tolerance requires the network to reconfigure to take the state given by the subgraph corresponding to the link failure. This could potentially require all connections in the network to

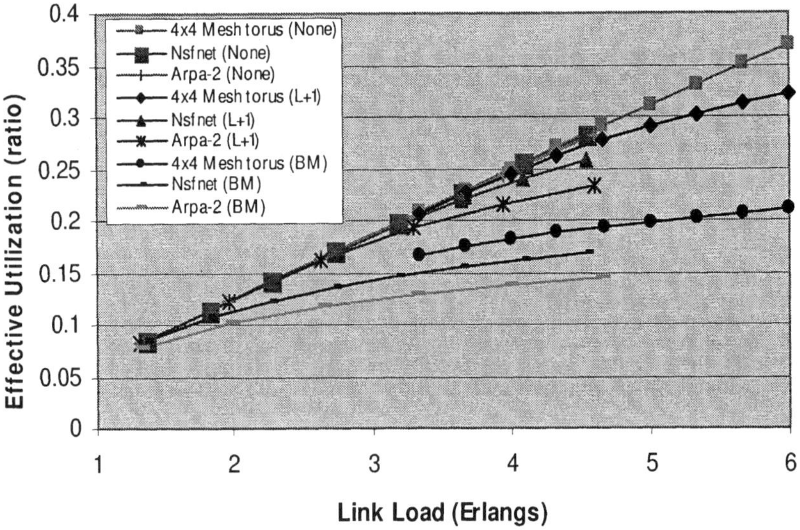


*Figure 14.9.* Average effective utilization of three topologies.

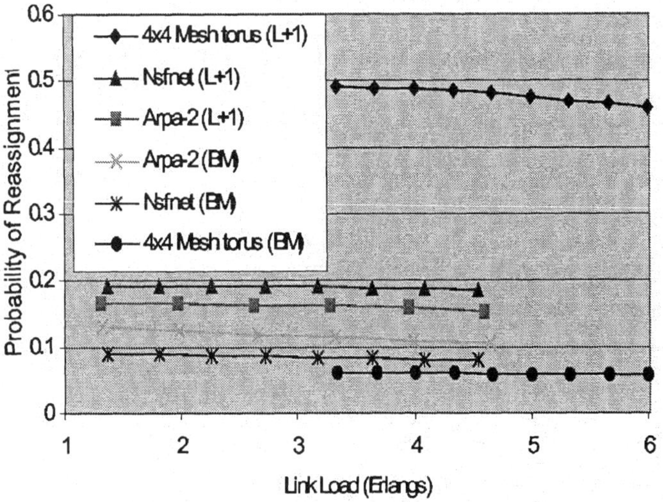


*Figure 14.10.* Average probability of path reassignment for the three topologies.

change how they are routed. The probability of reassignment indicates how likely a connection's path will change during recovery. In all three topologies, as the arrival rate increases, the probability of reassignment decreases. This observed decrease is due to the higher probability of a request being blocked as arrival rate increases. As requests enter the network at a higher rate, fewer connections are established. This makes routing the connections that do get accepted on each subgraph easier. Thus there is a much higher probability that a subgraph routes the connection exactly the same as the base network does.

Figure 14.10 shows the probability of reassignment for the NSFNET, ARPA-2 and 4x4 mesh topologies, respectively. As mass connection reassignment is unique to $L+1$ fault tolerance, it is important to show how much more reassignment it requires than backup multiplexing. As the link load increases, the probability of reassignment decreases, indicating that there is less of a chance of a connection having to be rerouted during a network recovery.

The probability of reassignment for the ARPA-2 and NSFNET topologies remains fairly constant and the probability of reassignment for the mesh torus topology decreases slightly more as the link load increases. The mesh torus also has a much higher probability of reassignment overall. This is probably due to the higher connectivity of the mesh torus. More links means that there are more options to route a path on, and the shortest hop length routing metric uses this to the fullest. Backup multiplexing requires far less reassignment in the event of a fault occurrence.

## 14.5 Multi-Link and Other Failures

$L+1$ routing strategy can be easily used to handle multiple link failures, node failures, and specific failure scenarios. Such efforts have been reported in [27, 28, 29, 30, 31]. Other approaches to fiber failure and multipath routing has been addressed in [32, 33, 34, 35, 36, 37, 38, 39, 40]. Other researchers have considered a combination of resources like shared risk link group (SRLG) [41] or other resources combinations [42]. Back up multiplexing can also be used in tolerating multiple faults and efforts reporting such results are [43, 44]. Essentially for multiple failure, the network must have capability to find alternate path that can account for failure of general or specific multi-link or node failures. We do not consider them here further.

## 14.6 Conclusions

Fault tolerance in optical networks has become increasingly important as reliance on these networks has increased. The ability to provide guaranteed connections in the event of a link failure without adversely affecting the operation of the network is imperative. In this chapter, we addressed the problem of routing dependable connections in WDM networks with dynamic traffic demands. We developed several algorithms to efficiently utilize the wavelength channels. These algorithms use pro-active, path based, failure independent restoration approach.

Backup multiplexing is a strategy designed to allocate capacity on a network to enable 100% recovery from a single-link failure. The major drawback of this strategy is that the capacity allocated to backup connections remains unused if there is no link failure.

We also presented an alternate technique for recovering from single-link failures known as $L+1$ fault tolerance. $L+1$ fault tolerance has the advantage of not requiring a network to allocate capacity for backup connections. As has been shown, the performance of the $L+1$ strategy was very good in comparison to that of backup multiplexing. This being said, there are still several ways to possibly improve the performance of the strategy.

## Acknowledgments

The author would like to thank his students, in particular Michael Frederick, Wensheng He, Murari Sridharan and Pallab Datta. The results reported here are based on author's research with them and are funded in part by NSF Grant ANI-9973102, ANI-0323374 and Defense Advance Research Projects Agency (co-funded by the National Security Agency) under Contract N66001-00-1-8949.

## References

[1] Ling Li and Arun Somani, "Fiber Requirement in Multifiber WDM Networks with Alternate-Path Routing," in Proc. ICCCN'99, Boston, MA, October 1999.

[2] I. Chlamtac, A. Ganz, and G. Karmi, "Lightpath Communications: An Approach to High Bandwidth Optical WANs", *IEEE Transactions on Communications*, vol. 40, no. 7, pp. 1171-1182, July 1992.

[3] R. Ramaswami and K. N. Sivarajan, "Routing and Wavelength Assignment in All-Optical Networks", IEEE/ACM Transactions on Networking, vol. 3. no. 5, pp. 489-500, October 1995.

[4] D. Banerjee and B. Mukherjee, "A Practical Approach for Routing and Wavelength Assignment in Large Wavelength Routed Optical Networks", IEEE Journal on Selected Areas in Communications, vol. 14, no. 5, pp. 903-908, June 1996.

[5] A. Mokhtar and M. Azizoglu, "Adaptive Wavelength Routing in All-Optical Networks", IEEE/ACM Transactions on Networking, vol. 6, no. 2, pp. 197-206, April 1998.

[6] S. Subramaniam, M. Azizoglu, and A. K. Somani, "All-Optical Networks With Sparse Wavelength Conversion", IEEE/ACM Transactions on Networking, vol. 4, no. 4, pp. 544-557, August 1996.

[7] H. Harai, M. Murata, and H. Miyahara, "Performance of Alternate Routing Methods in All-Optical Switching Networks", In Proc. of INFOCOM'97, 1997.

[8] B. T. Doshi, S. Dravida, P. Harshavardhana, O. Hauser, and Y. Wang, "Optical Network Design and Restoration", Bell Labs Technical Journal, pp. 58-84, January-March 1999.

[9] E. Karasan and E. Ayanoglu, "Performance of WDM Transport Networks", *IEEE Journal on Selected Areas in Communications*, vol. 16, no. 7, pp. 1081-1096, September 1998.

[10] P. F. Falcao, "Pan-european multi-wavelength transport networks: Network: design, architecture, survivability and SDH networking," Proceedings of the 1st International Work-shop on Reliable Communication Networks, Brugge, Belgium, May 17-20, 1998.

[11] C. S. Li and R. S. Ramaswami, "Automatic fault detection, isolation and recovery in transparent all-optical networks," IEEE/OSA Journal of Lightwave Technology, Volume 15, Issue 10, October 1997, pp. 1784-1793.

[12] S. Ramamurthy and B. Mukherjee, "Survivable WDM Mesh Networks, Part I - Protection", In Proceedings of IEEE INFOCOM '99, pp. 744-751, 1999.

[13] S. Ramamurthy and B. Mukherjee, "Survivable WDM Mesh Networks, Part II - Restoration", In Proceedings of ICC '99, 1999.

[14] M. Alanyali and E. Ayanoglu, "Provisioning Algorithms for WDM Optical Networks", IEEE/ACM Transactions on Networking, vol. 7, no. 5, pp. 767-778, October 1999.

[15] Sahasrabuddhe, L, Ramamurthy, S. and Mukherjee, B. "Fault management in IP-over-WDM networks: WDM protection versus IP restoration," IEEE Journal on Selected Areas in Communications," Volume 20, Issue 1, January 2002, pp. 21-33.

[16] Mohan, G, Siva Ram Murthy, C. and Somani, A.K. "Efficient algorithms for routing dependable connections in WDM optical networks," IEEE/ACM Transactions on Networking, Volume 9, Issue 5, October 2001, pp. 553-566.

[17] N. Nagatsu, S. Okamoto, and K. Sato, "Optical Path Cross-Connect System Scale Evaluation Using Path Accommodation Design for Restricted Wavelength Multiplexing", IEEE Journal on Selected Areas in Communications, vol. 14, no. 5, pp. 893-902, June 1996.

[18] S. Baroni, P. Bayvel, R. J. Gibbens, and S. K. Korotky, "Analysis and Design of Resilient Multifiber Wavelength-Routed Optical Transport Networks", IEEE/OSA Journal of Lightwave Technology, vol. 17, no. 5, pp. 743-758, May 1999.

[19] M. Sridharan, M. V. Salapaka, and A. K. Somani, "Operating mesh-survivable WDM transport networks," SPIE International Symposium on SPIE Terabit Optical Networking: Terabit Optical Networking, pp. 113–123, November 2000.

[20] M. Sridharan, A. K. Somani, and M. V. Salapaka, "Approaches for capacity and revenue optimization in survivable WDM networks," Journal of High Speed Networks, vol. 10, no. 2, pp. 109 – 125, August 2001.

[21] M. Clouqueur, and W.D.Grove, " Computational and design studies on the unavailability of mesh-restorable networks," Proc. IEEE/VDE Design of Reliable Communication Networks 2000, pp. 181-186, April 2000.

[22] M. Clouqueur and W. D. Grover, "Availability analysis of span restorable mesh networks," IEEE Journal on Selected Areas in Communication, vol.20, issue.4, pp.810-821, 2002.

[23] M. Clouqueur and W. D. Grover, "Mesh-restorable networks with complete dual failure restorability and with selectively enhanced dual-failure restorability properties," SPIE Optical Networking and Communications Conference (Opticomm 2002), Boston, MA, July-Aug 2002.

[24] J. Doucette and W. D. Grover, "Capacity design studies of span-restorable mesh transport networks with shared-risk link group (SRLG) effects," SPIE Optical Networking and Communications Conference (Opticomm 2002), Boston, MA, July-Aug 2002.

[25] G. Mohan and A. K. Somani, "Routing dependable connections with specified failure restoration guarantees in WDM networks," in Proc. IEEE INFOCOM 2000, pp. 1761–1770, March 2000.

[26] N. Jose and A. K. Somani, "Reconfiguring Connections in Optical Networks", in Proc. of DRCN, October 2003.

[27] Y. Liu, D. Tipper and P. Siripongwutikorn, "Approximating optimal spare capacity allocation by successive survivable routing," in Proc. of INFOCOM '01, pp.699-708, 2001.

[28] W. He, M. Sridharan and A. K. Somani, "Capacity optimization for surviving double-link failures in mesh-restorable optical networks", OPTICOMM 2002: Optical Networking and Communications, vol.4874, pp.13-24, June 2002.

[29] D. Schupke, A. Autenrieth and T. Fischer, "Survivability of Multiple Fiber Duct Failures", in Proc. of Design of Reliable Communication Networks (DRCN) Workshop, 2001.

[30] S. Lumetta and M. Medard, "Classification of two link failures for all-optical networks" Optical Fiber Communication Conference and Exhibit, vol.2, pp.TU03 1-3, 2001.

[31] S. Kim and S. S. Lumetta, "Evaluation of Protection Reconfiguration for Multiple failures in Optical Networks", in Proceedings of the Optical Fiber Communication Conference, Atlanta Georgia, March 2003.

[32] R. Doverspike and B. Wilson, "Comparison of capacity efficiency of DCS network restoration routing techniques," Journal of Network and System Management, vol. 2, no.2, pp. 95-123, 1994.

[33] D. Xu, Y. Xiong and C. Qiao, "Protection with multi-segments (PROMISE) in Networks with Shared Risk Link Groups (SRLG's)," IEEE/ACM Transactions on Networking, vol. 11, no.2, pp.248-258, April-2003.

[34] E. Modiano and A. Narula-Tam, "Survivable lightpath routing: a new approach to the design of WDM-based networks," IEEE Journal of Selected Areas in Communication, May 2002.

[35] H. Zhang and B. Mukherjee, "Path-protection routing and wavelength assignment(RWA) in WDM mesh networks under duct-layer constraints," IEEE/ACM Transactions on Networking, vol.11, no.2, pp.248-258, April-2003.

[36] E. Bouillet, J. Labourdette and et.al, "Stochastic approaches to compute shared mesh restored ligthpaths in optical network architectures," in Proc. of INFOCOM '02, pp.801-807, 2002.

[37] O. Gerstel and R. Ramaswami, "Optical layer survivability-an implementation perspective," IEEE Journal on Selected Areas in Communications, vol. 18, no.10, pp. 1885–1889, Oct 2000.

[38] R. Doverspike and J. Yates, "Challenges for MPLS in optical network restoration," IEEE Communications Magazine, Feb 2001.

[39] G. Li, B. Doverspike and C. Kalmanek, "Fiber span failure protection in mesh optical networks," SPIE Optical Networking and Communications Conference (Opticomm 2001), 2001.

[40] G .Z. Li, B. Doverspike and C .Kalmanek, "Fiber span failure protection in mesh optical networks", SPIE Optical Networking and Communications Conference (Opticomm 2001), vol.4599, pp.130-142, 2001.

[41] E. Oki, N. Matsuura, K. Shiomoto and N. Yamanaka, "A Disjoint Path Selection Scheme with Shared Risk Link Groups in GMPLS Networks", IEEE Communication Letter, Fall 2002.

[42] P. Datta, M. T. Frederick and A. K. Somani, "Sub-Graph Routing: A Novel Fault-tolerant Architecture for Shared-Risk Link Groups in WDM Optical Networks," 4th International Workshop on the Design of Reliable Communication Networks (DRCN 2003), Banff, Canada, October 2003.

[43] H. Choi, S. Subramaniam, and H. A. Choi, "On double-link failure recovery in WDM optical networks," in Proc. IEEE INFOCOM 2002, pp. 808–816, June 2002.

[44] W. He, M. Sridharan and A. K. Somani, "Capacity optimization for tolerating double link failures in WDM mesh optical networks", Proc. SPIE vol. 4874, pp. 13-25, July 2002.

# Chapter 15

# FACILITATING SERVICE LEVEL AGREEMENTS WITH RESTORATION SPEED REQUIREMENTS

Gokhan Sahin and Suresh Subramaniam[1]

[1] *George Washington University, Washington, DC 20052*

Email: suresh@gwu.edu

**Abstract** Optical networks with wavelength-division-multiplexing are expected to serve client networks with different optical layer protection requirements. This chapter considers the problem of providing Quality of Protection (QoP) classes based on restoration speed, and improving the failure-recovery time performance in such networks. We first present an approach which uses a different restoration method for each service class in order to meet the restoration speed requirement. We then focus on the signaling process required in mesh restoration for reconfiguring the nodes along the pre-planned restoration paths, and propose a novel approach for reducing the restoration time and meeting the QoP requirements by coordinating the set-up procedures for the backup paths through scheduling. We present priority-based online scheduling algorithms that are amenable to distributed implementation for the problems of *(i)* minimizing the worst-case restoration time, and *(ii)* maximizing the number of connections that meet their QoP-class-specific restoration time deadlines. The online scheduling methods that we propose use simple connection and/or class-specific information and can be easily implemented with minor modifications to the currently proposed signaling protocols. We apply these methods to signaling protocols that require cross-connect configurations at different nodes to be done in sequence as in the current GMPLS specification, as well as signaling protocols that allow cross-connect configurations to be done in parallel. It is shown that in both cases, significant performance improvements are achievable through scheduling in terms of both the QoP grades that can be supported and the restoration times, with both the heuristics, and two related mixed-integer-linear-program formulations which we use for comparison purposes. The improvement in restoration time and restorability through our heuristics can be quite high (e.g., increase from a network restorability performance of 40% to a network restorability of 88%, and a 17% reduction in restoration time).

**Keywords:** Optical networks, Quality of Protection, mesh restoration, SLA, restorability, signaling, message scheduling, MILP, GMPLS, OXCs.

## 15.1 Introduction

Optical networks with wavelength-division-multiplexing are considered to be among the most prominent candidates for the transport backbone. These networks provide static *lightpaths*, which serve as high-capacity virtual links for the higher layer client networks, such as IP, SONET, or ATM. It is widely recognized that optical networks will likely serve various classes of traffic with different Quality of Protection (QoP) requirements from the optical layer [1, 2] due to the variety of client networks running on top of the optical layer, and their restoration mechanisms.

There has been a significant amount of work on routing and capacity assignment in optical networks with multiple protection classes, with various goals such as meeting a pre-specified probability of restoration, minimizing blocking probability, and maximizing revenue [1, 2, 3, 4, 5]. In this chapter, we consider the problem of providing differentiated restoration services in terms of the restoration time in mesh WDM networks with capacity sharing, and introduce the concept of using control-message scheduling for this purpose. The problem of providing differentiated services in terms of the restoration times was also considered in [6], where each connection belonged to one of the following classes depending on the maximum restoration time allowed: *platinum* for fastest restoration (50msec), *gold* (50-100msec) and *silver* (1-10sec), excluding propagation delays. The authors proposed using dedicated protection for connections in the platinum QoP class, shared protection using a logical ring for gold QoP, and shared mesh protection for silver QoP. Our proposal differs from [6] in that we use only shared mesh protection schemes, which have better capacity efficiency, and provide the QoP classes using *control message scheduling*. It is worth pointing out that our proposed control message scheduling strategy may be used with *any* shared protection mechanism, for example, the mechanism used in [6] for the silver class to further improve the performance.

### 15.1.1 Motivation

Connections may have different restoration speed requirements, which are either explicitly or implicitly specified as part of an Service Level Agreement (SLA) between the network operator and service provider. Such requirements may arise simply because a higher layer network completely relies on the optical layer for restoration and wishes to provide guaranteed levels of availability to its customers. On the other extreme, they may also arise in order to prevent race conditions, e.g., when a client network also has a protection mechanism and the optical layer needs to restore the lightpath in time before the higher layer restoration mechanisms are triggered. In these cases, it is important for the client network to prevent any possibility of race conditions, where protection mechanisms in two layers are activated simultaneously. To avoid this phenomenon, the client network may either request the lightpath to be *unprotected* at the optical layer, or it may require extremely fast restoration guarantees at the optical layer so that optical layer restoration is completed before the higher layer mechanisms are triggered. Finally, restoration speed is a very natural service differentiation metric based on the criticality of the data being carried over the lightpaths. It is reasonable to assume that connections which are served on a best-effort basis would not require the same restoration speed as high-priority connections which require protection against any single-failure scenario. Accordingly, several approaches have been proposed in the literature to improve the restoration speed, or to facilitate providing service level agreements based on the restoration speed. We will discuss two of these approaches in the following.

## 15.2 Using a Different Restoration Method for each Service Class

Restoration methods, in general, offer a trade-off between capacity efficiency, restoration speed, and restoration success rate. For example, restoration methods with dedicated capacity can transmit the optical signal on both the service and the back-up paths simultaneously, thus providing very rapid recovery in the event of a failure, while using at least 100% protection capacity. On the other hand, restoration methods with shared capacity allow multiple connections that are not expected to fail simultaneously to access the same restoration capacity to improve efficiency, while providing a slower recovery from failures due to the need to reconfigure the optical cross connects (OXCs) after the failure occurs. In [6], the authors use the trade-off between capacity efficiency and restoration speed to support three different grades of service: *platinum* connections with a restoration deadline of 50ms, *gold* connections with a restoration deadline of 100ms, and *silver* connections with a restoration deadline of 1sec. In all cases, the restoration deadline is assumed to exclude the propagation delay. The authors propose using a different restoration method for each grade of service: dedicated protection for platinum connections, shared protection using logical rings for gold connections, and shared mesh protection for silver connections.

### 15.2.1 Dedicated Protection

In restoration with dedicated capacity, each source-destination pair is connected by two paths, each with a sufficient number of wavelengths: an active path, and a protection path that is diversely routed with the active path. Failure recovery simply involves, **(i)** detection of the failure by the end nodes of each affected connection, and **(ii)** switching to dedicated protection wavelength using an operation identical to Automatic Protection Switching (APS) in SONET. Thus, failure recovery can be completed within a few tens of milliseconds with this method, making it a feasible option for Platinum connections.

### 15.2.2 Shared Protection Using Logical Rings

This method groups source-destination pairs into logical WDM rings each of which carries no more than a pre-defined number of wavelengths per fiber for active as well as protection paths. The source-destination pairs on a given ring essentially use ring protection: each connection is assigned an active path, and protection wavelengths are reserved in the complementary routes. Non-overlapping connections within the same logical ring can share protection wavelengths, improving the capacity efficiency. The authors estimate the restoration time to be approximately 50ms, allowing about 40ms for cross-connect remapping in the intermediate nodes. We note, however, that this estimate does not include the potential queueing delays that can arise during the control-messaging to reconfigure the OXCs. It is suggested that this approach may be appropriate for Gold connections.

### 15.2.3 Shared Mesh Protection

In shared mesh protection, all source-destination pairs can use the same restoration capacity provided that they are not expected to fail simultaneously. Since the back-up path does not have to be on a given logical ring, this method may be more complicated than shared protection using logical rings, and possibly slower. The authors estimate the reconfiguration time for this method to be approximately 80msec.

### 15.2.4 Numerical Results

Table 15.1 gives a comparison of the 3 restoration methods when applied to an example core network of 29 nodes and 53 candidate links, with 138 node-pairs with positive connections. The trade-off between the capacity efficiency and restoration speed is clearly observed in the results: Dedicated restoration provides the fastest recovery, followed by shared restoration over logical rings and shared mesh restoration, where as shared mesh restoration provides the best capacity efficiency, followed by the shared restoration over logical rings, and dedicated restoration.

## 15.3 Motivation for Control-Message Scheduling

In restoration methods with capacity sharing, the backup paths need to be set up in the event of a failure through a signaling procedure, even when the backup paths are completely pre-planned. While there are some differences

*Table 15.1.* Comparison of Various Restoration Methods.

| Algorithm | Cost saving | Recovery Speed | Operation |
|---|---|---|---|
| Dedicated | 0% | Fast (20ms) | Easy |
| Logical Rings | 10% | Medium (50ms) | Moderate |
| Shared Mesh | 20% | Slow (100ms) | Complicated |

among the various signaling protocols proposed in the literature [7, 8, 9, 10, 11], the rerouting procedure generally involves sending a restoration message along the restoration path and reconfiguring the optical cross-connects (OXCs) along the path. Since a DWDM link may carry up to hundreds of lightpaths, a single failure may affect a large number of connections, and many restoration messages and OXC configuration commands may need to be processed at the network nodes. This introduces queueing delays, which may be a significant component of the restoration time [12, 13]. More importantly, these queueing delays may make it difficult to provide restoration time guarantees to connections which have pre-determined restoration time requirements. We propose the coordination of the rerouting process among connections that fail simultaneously, in order to optimize the failure recovery time and provide QoP classes. Significantly, the current signaling protocols do not consider scheduling the control messages, and assume that these messages are processed according to FIFO (first-in-first-out) service discipline. In the following, we first show that the restoration times for various signaling protocols can be reduced significantly through *control-message scheduling*, i.e., by coordinating the rerouting process among connections that fail simultaneously. For this purpose, we first present MRTF, a scheduling algorithm designed with the specific goal of optimizing the worst-case restoration time performance. We then consider the problem of supporting different Quality of Protection classes with different restoration speed requirements. We propose a new control-message scheduling algorithm, FIFO-CI, for supporting QoP classes with different restoration deadlines.

## 15.4 Restoration Architecture and Signaling

We assume that the network has an OAM (operations, administration, and maintenance) channel dedicated for exchanging control messages for signaling. The OAM channel is terminated at every node and the control messages are processed electronically. We assume a generic node model that includes the units shown in Figure 15.1, based on the model presented in [7]: a message receiver, a message processor, a cross-connect controller, and the cross-connect (switch) fabric. The message receiver unit is responsible for receiving messages, determining the message type, and forwarding them to the message processor. The message processor is the algorithm logic of the node. It can access/change the records regarding the availability of the wavelengths on the links originating from the node. It also has an interface to the cross-connect controller, which controls the cross-connect fabric. When the cross-connect configuration needs to be changed, a *cross-connect* command is issued to the cross-connect controller. Each cross-connect controller, message processor, and message receiver acts as a queue with a deterministic service time, and has a buffer to store multiple messages/commands if it receives new messages while still handling other messages.

We note that the architecture of an arbitrary OXC is not restricted to the model used here. The various components in our node architecture are generic ones, and most node architectures can be made to "fit" our model by choosing the component parameters (such as the message processor service time) appropriately.

Several signaling protocols have been proposed for path restoration in mesh networks in the literature [14, 7, 8, 9, 12], as well as in the Internet Engineering Task Force (IETF) drafts [10, 11]. We consider two broad categories of signaling protocols. In *signaling with OXC confirmation* (WXC), as in the current Generalized MPLS (GMPLS) specifications, a node may not send the connection set-up messages to the next node along the path before the completion of the OXC reconfiguration [8]. Alternatively, *signaling without OXC confirmation* (WOXC) allows a node to forward the control message to the next node without waiting for OXC configuration. Such a method was proposed in [7] to expedite restoration, and has been implemented with extensions to GMPLS [15]. Similar approaches have also been considered in [11, 12, 8]. We describe a generic protocol for signaling WOXC below, and note that signaling WXC can be similarly done. Note that these signaling protocols are

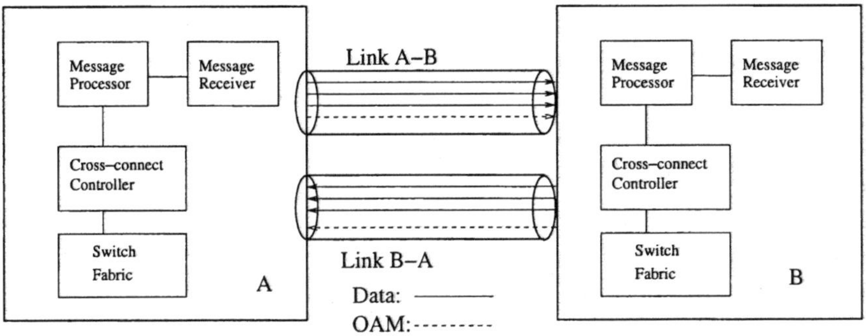


*Figure 15.1.* Node and link model for restoration signaling.

similar to the extensions to RSVP-TE proposed in [11], as well as the methods used in [7]. We emphasize that the signaling protocols described here are generic. Different signaling protocols will have slight performance differences, and it is not our objective here to investigate the performance of any single proposed standard such as [14, 10, 8]. Our main objective is to show that control-message scheduling is an effective mechanism to facilitate QoP in mesh networks that use path protection.

We assume that the pre-determined restoration path and wavelength for each connection are stored at the end-points of the connection. When a link on the service path of a connection fails, the source node detects the failure when the loss of signal (LOS) propagates to the source. After retrieving the stored restoration path and wavelength, the source issues a cross-connect command to its OXC controller to reconfigure its switch fabric. Without waiting for the completion of this command, it also sends a control message (*set-path*) to the next node on the restoration path. Similarly, each intermediate node that receives the set-path message issues an OXC command, and simultaneously sends the set-path message to the next node. The source switches to the restoration path when it first detects the LOS. The destination switches to the restoration path as soon as it receives the set-path message. Restoration is completed when the last OXC configuration command along the restoration path is executed.

## 15.5 Optimizing Restoration Time Performance Through Scheduling

In this section we focus on the problem of optimizing the restoration time performance of the signaling protocols that we described in Section 15.4. It is possible to formulate the problem as a mixed-integer-linear-program (MILP), which can be solved through commercial optimization software. We present such an MILP formulation of the problem for signaling without OXC confirmation in [16]. The formulation can easily be adjusted for other signaling protocols as well. However, the MILP turns out to require a substantial amount of time to solve for large networks and/or when the number of connections to be restored is large, and may thus not be practical under dynamic traffic. Accordingly, we propose an online scheduling heuristic, maximum-remaining-time-first (MRTF), to coordinate the restoration signaling process without requiring any pre-planning. This algorithm can be easily implemented with slight modifications to the current GMPLS proposals, or other signaling protocols suggested in the literature. Moreover, it can also be used for failure-independent restoration methods and for dynamic traffic conditions, which is not the case for the MILP formulation given in [16]. Note that the MRTF heuristic is not derived from the MILP formulation, (e.g., by solving an integer-constraint-relaxed version of the MILP) but is an independent algorithm that schedules control messages with the goal of minimizing the worst-case restoration time. However, we will use example MILP solutions in our numerical results for comparison purposes.

### 15.5.1 The Maximum-Remaining-Time First Algorithm (MRTF)

As mentioned earlier, existing signaling protocols do not consider control message scheduling and are assumed to use first-in-first-out (FIFO) scheduling to process the tasks at each unit. However, this may be inefficient in terms of minimizing the overall restoration time. In the following sections, we will demonstrate that the worst-case restoration time can be reduced significantly by allowing network nodes to use other scheduling strategies. To this end, we propose the Maximum-Remaining-Time-First (MRTF) algorithm to improve the worst-case restoration time performance. MRTF is an online scheduling algorithm which aims to optimize the restoration time performance by giving priority to the tasks belonging to connections that are further away from completing the rerouting process. We describe the algorithm for failure-dependent restoration methods. Extension to methods with failure-independent restoration paths is straightforward.

In MRTF, as in ordinary failure-dependent restoration, the restoration path is stored at the source node of the connection. When the failure-notification message arrives at the source node, the source node can compute the remaining processing time before the rerouting for that connection can be completed, *excluding the possible queueing delays*. For methods that do not wait for cross-connect confirmation before forwarding the set-path message to the next node, this will be equal to the sum of the message processing times at each node on the path, message transmission times and propagation delays on each link of the path, message receiving times at each node except the source, and the cross-connect configuration time at the destination node. Note that all of these times are deterministic because the backup route is completely pre-planned. They also do not depend on the processing times of other connections, as queueing delays are not included. Therefore, MRTF can also be used in a dynamic traffic scenario. For methods that wait for cross-connect confirmation, the cross-connect configuration times at the source and the intermediate nodes need to be included as well. Similarly, when an intermediate node receives a set-path message from the previous node, it also can compute the remaining processing time for that particular connection. One way to achieve this would be to include the *remaining time* information in the set-path message, and allow each node receiving the message to update this information by subtracting from it the amount of processing that has already been completed. The nodes may also compute this information by simply looking at the restoration path information, and calculating the time it would take to go through the remaining part of the path.

In MRTF signaling, each unit at a node gives priority to connections that have the highest remaining-time, rather than serving the tasks waiting in the queue in a FIFO fashion. This can be done without requiring any major additions to the node/restoration architecture. The remaining time information for the given connection can be updated at the completion of each task. Thus, MRTF can be easily implemented with minimal modification to the currently proposed restoration signaling methods, such as the GMPLS extensions for restoration [10, 11]. Note that we propose MRTF only to improve a particular performance measure, worst-case restoration time. Optimizing the worst-case restoration time is similar to optimizing the *makespan* in a traditional job scheduling problem, and MRTF is inspired by a commonly used scheduling algorithm to improve makespan. Other scheduling algorithms can be used for other performance measures. For example, when the goal is to optimize the average restoration time, a scheduling algorithm which processes messages belonging to connections with the *minimum* (as opposed to maximum) remaining time first could be used. Note also that MRTF is not unresponsive to the queueing delays. Since the remaining processing time for each connection is updated as the message goes through each intermediate node, a connection which initially had a low remaining time, but was delayed excessively due to queueing may be processed earlier in the remaining intermediate nodes.

## 15.5.2 Numerical Results

We have applied the MRTF algorithm to signaling protocols that wait for cross-connect confirmation, as well as those that do not wait for cross-connect confirmation. In this section we present numerical results demonstrating the failure-recovery time performance using these signaling protocols. We first present the solutions obtained through the MILP formulation in representative problem instances for comparison with the simulation results.

We have developed a network of queues model to simulate the signaling protocols for different restoration methods [12]. The node and link models shown in Figure 15.1 form the basis of our simulation model. Each message processor, message receiver, cross-connect controller unit, and link is represented as a queue with a deterministic service time, as in [7, 12]. The service rates at which jobs are processed at each of these queues, as well as the physical length of each link in the network are input parameters to the model. These are the: message processing rate ($P$ messages/sec), message receiving rate ($R$ messages/sec), cross-connect configuration rate ($X$ cross-connects/sec), message transmission rate ($T$ messages/sec), and link length ($L$ km). Simulations for signaling protocols that do not use scheduling are performed by treating each unit as a FIFO queue. For signaling protocols using the MRTF algorithm, we assume that each unit (except the message receiver) along the restoration path has access to the *remaining processing time* corresponding to each task in its queue, and processes the message belonging to the connection with the highest remaining time first. In our simulations, we assume that the control network has the same topology as the optical network, although other control network topologies are possible.

We have simulated these algorithms with randomly generated connection sets, where each source-destination pair is equally likely to be in the connection set. For a given set of connections, we compute the restoration time for each connection for each link failure, and record the worst-case restoration time over all failures. For a given number of connections $N$, the experiment is repeated over 100 connection sets with the given $N$, and the worst-case restoration time for each of these sets is averaged. We use two test topologies: the reasonably well-connected Arpanet (20 nodes, 32 links) shown in Figure 15.2, and a 20-node ring topology as an extreme example of a sparsely-connected network. Note that we are not necessarily suggesting the use of these signaling protocols for ring networks, but only using the ring topology as an example to evaluate the performance of the algorithms in sparsely connected networks. We use two different sets of parameters in our simulations. The parameters used in the first set are: $L = 200$, $P = 1,000$, $R = 10,000$, $X = 1,000$, and $T = 200,000$. This assumes a relatively small OXC configuration time (1ms) and message processing time. In the second set of parameters, we set $X = 100$, corresponding to a configuration time of 10ms, and $P = 500$. These parameter sets are in line with the values reported in the literature, as well as the expected range of values that can be achieved with today's technology [7, 15, 8]. We have not varied the transmission rate, since the restoration time is dominated by the cross-connect configuration and electronic processing of the restoration messages, which are much slower. A detailed study of the effects of these parameters on the restoration time can be found in [13]. In this section we only consider failure-dependent restoration, since the signaling for failure-independent restoration is identical except for the starting times for restoration. For simplicity, we use minimum hop routing for service path selection, and edge-disjoint minimum hop routing for restoration path selection as in [12].

**Example MILP Solutions.** We have used ILOG CPLEX 7.1 to solve the MILP problem. Table 15.2 summarizes the restoration time with and without MRTF, as well as the optimal restoration time obtained by CPLEX for our test networks. For these results, we used signaling without cross-connect confirmation with parameter set 1, and averaged the experiements over 5 different set of 60 connections.

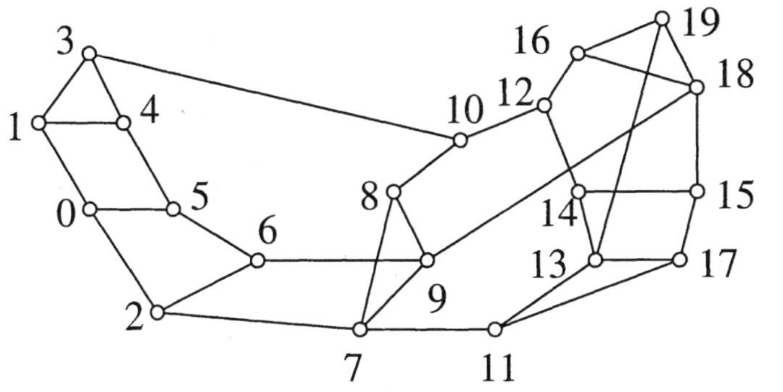


*Figure 15.2.* The Arpanet topology that is used as a test network.

We observe that the MILP solution provides significantly faster restoration than FIFO. In ring, the worst-case restoration time can be reduced from an average of 89.51ms to 65.49ms, whereas in Arpanet it can be reduced from 41.77ms to 33.91ms. This demonstrates the potential importance of control-message scheduling in optimizing restoration times. We further note that MRTF is also useful in reducing the restoration times. In particular, for the ring network, MRTF can provide about 14% reduction in restoration time. Given the simplicity and implementation advantages of MRTF over the MILP solution, these reductions are significant.

*Table 15.2.* Restoration time with MILP solution, and simulations in a 20-node ring and in Arpanet with 60 connections, with parameter set 1 and OXC confirmation.

| Test Network | MILP solution | MRTF simulation | FIFO simulation |
|---|---|---|---|
| Ring | 65.495 | 76.495 | 89.510 |
| Arpanet | 33.919 | 38.204 | 41.772 |

**Performance Evaluation of MRTF.** We first look at the worst-case restoration time performances for methods without cross-connect confirmation (WOXC). Table 15.3 shows the restoration time performances of FIFO and MRTF with various parameter sets and signaling options. We observe that MRTF provides significant reductions in the restoration time, over a wide range of $N$. The reduction varies between 11.8% when $N$=60 and about 13.8% when $N$=220. As expected, MRTF is more useful when the number of connections is larger, because then the number of connections affected by a failure is larger as well. In this case, queueing delays constitute a significant part of the restoration time, increasing the benefits of using MRTF. It is worth noting that savings remain high over a wide range of $N$: the percentage reduction in worst-case restoration time through MRTF is more than 10% for all $N$, reaching up to 17.8% for $N$=100.

It is known that cross-connect configuration times may be as high as 5-10ms for MEMS based switches [15]. We next use parameter set 2, corresponding to the extreme case of 10ms OXC configuration time, and $P$=500, to see the performance of the algorithms under very slow OXC reconfiguration. Table 15.3 shows that the performance with and without MRTF is very similar for parameter set 2 when there is no waiting for XC confirmation. Note that when we allow forwarding the set-up messages along the restoration path without waiting for cross-connect confirmation, cross-connect reconfigurations at different nodes can proceed in parallel, reducing their contribution to the end-to-end queueing delay. Moreover, when the cross-connect times are significantly higher than the message forwarding and processing times, the restoration time is strongly dominated by the cross-connect time [13], reducing the potential benefits of giving priority to connections that need to go through a larger number of hops as in MRTF. This however does not hold for signaling methods with cross-connect confirmation. We will also see in the following sections that this observation will not hold when we consider the QoP requirements for individual connections. In that case, control-message scheduling will turn out to be useful even with slow OXC configuration.

We next consider signaling methods with cross-connect confirmation (WXC). Table 15.3 shows that reduction in restoration time through MRTF remains high for the ring topology for such signaling protocols as well. We observe that the MRTF algorithm reduces the restoration time significantly for all values of $N$. For example, when $N$=80, MRTF reduces the restoration time by about 17.2%. Similar observations can be made for the Arpanet topology, where the savings reach up to 9.5% (not shown). Table 15.3 shows similar results for parameter set 2 in the ring. Unlike the case for signaling without OXC confirmation with the same parameter set, we observe that MRTF provides significant restoration time reductions in the ring topology. In this case, the savings in restoration time vary between 8.5% and 15.4%. It is interesting to note that the reduction in restoration time through MRTF remains more than about 10% in the ring in all scenarios, except for the case of extremely slow cross-connection time and parallel OXC reconfiguration.

We have shown in this section that control-message scheduling can provide significant improvements in the restoration time performance of shared path-protected mesh networks. We will next focus on the problem of maximizing the restorability performance in networks providing QoP classes. We next proceed to show that control-message scheduling can be an effective means of providing QoP classes.

*Table 15.3.* Restoration time performance with FIFO and MRTF for various parameter sets and signaling options in the ring topology.

| num. of cons | MRTF WOXC set 1 | FIFO WOXC set 1 | MRTF WXC set 1 | FIFO WXC set 1 | MRTF WOXC set 2 | FIFO WOXC set 2 | MRTF WXC set 2 | FIFO WXC set 2 |
|---|---|---|---|---|---|---|---|---|
| 60 | 49.4 | 56.0 | 76.9 | 89.6 | 235.8 | 235.8 | 368.5 | 436.4 |
| 80 | 51.5 | 61.0 | 83.6 | 101.0 | 295.8 | 295.8 | 426.0 | 504.5 |
| 100 | 55.4 | 67.4 | 94.0 | 112.7 | 360.7 | 360.7 | 495.0 | 574.8 |
| 140 | 65.4 | 78.3 | 115.0 | 134.3 | 474.6 | 474.6 | 624.3 | 706.0 |
| 180 | 75.6 | 89.2 | 136.2 | 156.7 | 590.7 | 590.7 | 755.5 | 840.6 |
| 220 | 86.9 | 100.8 | 159.3 | 179.6 | 710.7 | 710.7 | 895.52 | 979.6 |

## 15.6 Providing QoP Classes Through Control Message Scheduling

In this section, we focus on the problem of satisfying the restorability requirements for various QoP classes. We describe the potential performance metrics for supporting QoP classes, and propose an online scheduling heuristic based on class information, which is amenable to distributed implementation.

### 15.6.1 Optimizing Restorability

We assume that each connection is given a pre-specified deadline to complete the restoration. This deadline may be either determined on a connection-by-connection basis, or it may be a class-specific parameter. When QoP classes are assumed, we define the following two performance metrics for each class: *connection restorability* and *network restorability.* Connection restorability for a given class is the probability that a randomly chosen connection in that class can be restored within its deadline for *any* failure. Let us say that a particular connection is restorable if it can be restored within its deadline for all failures. Then, if the connection restorability for a class is $p$, then a fraction $p$ of the total number of connections in that class are restorable. On the other hand, network restorability for a given class is defined as the probability that a randomly chosen connection in that class can be restored within its deadline when a random link failure occurs. Let us define $U_{c,f}$ to be 1 if a connection $c$ cannot be restored within its deadline when link $f$ fails, and $N(\alpha)$, where $\alpha \in \{Platinum, Gold, Silver\}$, to be the number of connections in QoP Class $\alpha$. Let $N_f(\alpha)$ be the number of connections in Class $\alpha$ that fail when failure $f$ occurs. Then, connection restorability (C.R.) and network restorability (N.R.) for QoP Class $\alpha$ can be defined as:

$$\text{Connection Restorability}(\alpha) = \frac{\sum_{c \in \alpha}\{1 - \max_f U_{c,f}\}}{N(\alpha)},$$

$$\text{Network Restorability}(\alpha) = \frac{\sum_{c \in \alpha}\sum_f\{1 - U_{c,f}\}}{\sum_f N_f(\alpha)}.$$

In [17], we present an MILP formulation, which minimizes the weighted number of failed connections that cannot be restored within their deadlines. This is equivalent to maximizing the average network restorability (over all classes) if the weights are set to 1. However, the MILP formulation is not amenable to distributed implementation. Hence, we propose a simple online scheduling algorithm for supporting QoP classes in the following, and use the MILPs only for comparison purposes.

## 15.6.2 Online QoP Scheduling Algorithms

We earlier described the maximum-remaining-time-first (MRTF) heuristic as a means of reducing the restoration time performance, ignoring the QoP requirements. However, in a more practical scenario, it would be more important to ensure that connections meet their individual QoP deadlines, rather than optimizing the worst-case restoration time among all connections. We now describe FIFO-CI (FIFO with class-information) as a means of providing QoP classes.

FIFO-CI attempts to improve the restorability performances by allowing each node along the restoration path to serve higher priority connection messages first. The control message (set-path) sent from the source node to the destination along the restoration path carries the deadline information for the connection either explicitly or implicitly as part of the class-information. For example, the header of the packet may carry the class information that the connection belongs to. Each unit along the restoration path that receives a set-path message first checks the remaining processing time (excluding queueing time as in MRTF) to complete the restoration for that connection. Note that this processing time can be easily computed since the set-path message contains the restoration path information. If the node determines that it is not possible to restore the connection within its deadline based on the arrival time of the message to the node and the remaining processing time required, then the message is discarded in order not to delay the connections that do have a chance of meeting their deadlines. All the other messages are processed according to their class information, with higher class connections receiving higher priority in service. Note that messages within a class are processed in FIFO order; hence the name FIFO-CI. Thus, FIFO-CI can be implemented easily in a distributed fashion with various signaling protocols. We will refer to signaling schemes that do not use such deadline or class information as non-CI (NCI).

### 15.6.3 Numerical Results on QoP Performance

We now evaluate the QoP restorability performances of the various algorithms. We first give example MILP solutions for a small number of connections, and then present simulation results for various $N$ and parameter sets. Throughout this section, we assume that each connection belongs to one of the three restoration classes used in [6]: platinum, gold, and silver, with restoration times (excluding the propagation delays) equal to 50ms, 100ms and 1sec. Connection sets were generated randomly as in the previous sections, with each connection likely to be in Platinum, Gold, and Silver Classes with probabilities of 0.25, 0.25, and 0.5, respectively. We use the same parameter sets (parameter set 1 and parameter set 2) as in Section 15.5.2.

**Example MILP Solutions.** We first present the solutions obtained through the MILP formulation over representative problem instances, and compare them with the simulation results. We set the weights for the penalty function to 1 in our objective function, so the goal is to maximize the number of connections that meet their deadline (network restorability). For each topology, the results presented in this section are the averages over the solutions obtained for 5 different connection sets with $N$=60.

Table 15.4 summarizes the connection and the network restorability performances of various algorithms as well as the MILP solution for the ring network, with parameter set 1, and OXC confirmation. Throughout this section, the results for Silver connections will not be shown, since they have 100% restorability for all methods. We observe that the MILP solution can increase both the connection and the network restorability for Platinum to 100%, from the corresponding values of 29.68% and 62.42% for FIFO-NCI. We also note that significant improvements over FIFO-NCI are also achievable by the online scheduling heuristic that uses class information (FIFO-CI). Similar observations can be made in Table 15.5 for the Arpanet topology, with parameter set 2 and no OXC confirmation.

**QoP Restorability Performance Evaluation.** Figures 15.3-15.5 summarize the network and connection restorability performances of FIFO-NCI and FIFO-CI in the ring network. We observed that FIFO-CI provides major increases in the QoP grades that can be supported. In particular, for the case of $N$=220, connection and network restorability for Platinum connections can be increased from 2.28% to 100%, and from 44.32% to 100%, respectively by employing FIFO-CI. For this parameter set, all Gold connections already meet their deadlines using FIFO-NCI, so there is no room for additional performance

*Table 15.4.* Restorability with MILP solution and simulations in ring with 60 connections, parameter set 1 and OXC confirmation.

| Metric | MILP soln. | FIFO-NCI | FIFO-CI |
|---|---|---|---|
| Platinum (conn. resto.) | 100 | 29 .68 | 72.08 |
| Platinum (netw. resto.) | 100 | 62.42 | 92.07 |
| Gold (conn. resto.) | 100 | 100 | 100 |
| Gold (netw. resto.) | 100 | 100 | 100 |

*Table 15.5.* Restorability with MILP solution and simulations in Arpanet with 60 connections, parameter set 2 and NXC.

| Metric | MILP soln. | FIFO-NCI | FIFO-CI |
|---|---|---|---|
| Platinum (conn. resto.) | 100 | 46.41 | 92.51 |
| Platinum (netw. resto.) | 100 | 65.86 | 96.43 |
| Gold (conn. resto.) | 100 | 96.78 | 100 |
| Gold (netw. resto.) | 100 | 98.73 | 100 |

improvement through FIFO-CI (not shown). However, QoP improvements are significant for Gold when parameter set 2 is used. In this case, Gold network and connection restorability with FIFO-CI are 92.4% and 78.69% for $N$=60, as opposed to 23.22% and 1.22% for FIFO-NCI (Figure 15.5).

Figures 15.6 and 15.7 show similar results for the Arpanet topology with parameter set 2 (all values are 100 with parameter set 1). In this case, FIFO-CI provides major restorability improvements for both Platinum and Gold. For $N$=60, network restorability for Platinum increases from 70.5% to 97.2%, and connection restorability increases from 53.6% to 93.8%. For $N$=220, network restorability for Platinum increases from 16.1% to 66.9%, while network restorability for Gold increases from 55.14% to 76.86%. For parameter set 1, restorability performances with FIFO-CI were already 100% leaving no room for improvement.

We observed significant performance through class-based scheduling in the case of signaling with XC as well. For example, in the ring topology, for $N$=80 and parameter set 1, platinum network restorability performance is 78.6% for FIFO-CI, as opposed to 38.8% for FIFO-NCI. However, we observed that under parameter set 2, which has very slow OXC configuration (10ms), Platinum and Gold restoration time deadlines could not be achieved even with the use of FIFO-CI. This is mainly due to the sequential OXC configuration requirement,

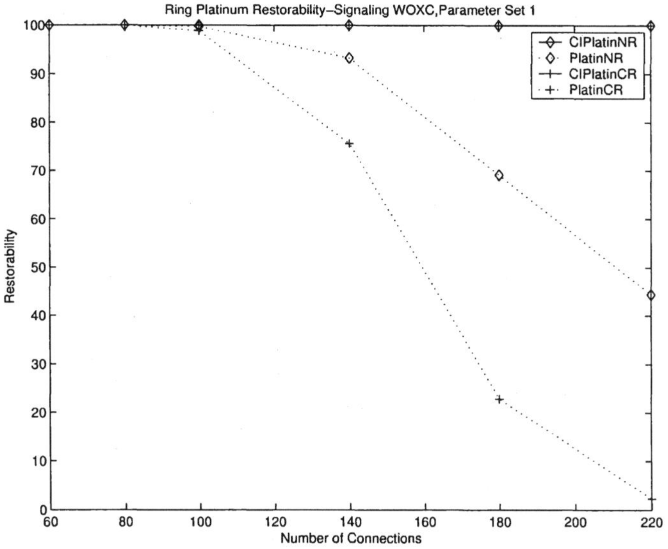


*Figure 15.3.* Platinum Restorability in Ring for Signaling without OXC confirmation with parameter set 1.

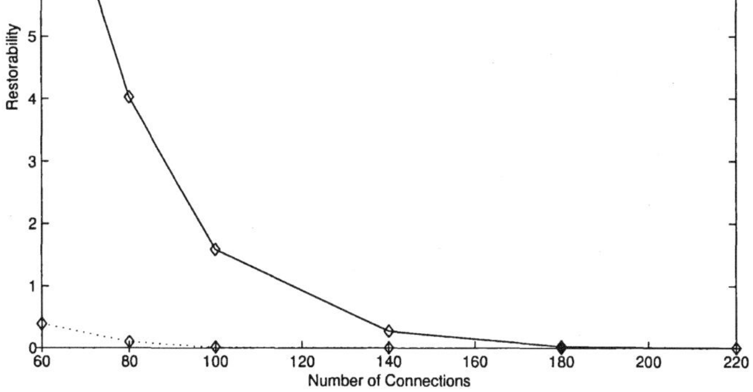


*Figure 15.4.* Platinum Restorability in Ring for Signaling without OXC confirmation with parameter set 2.

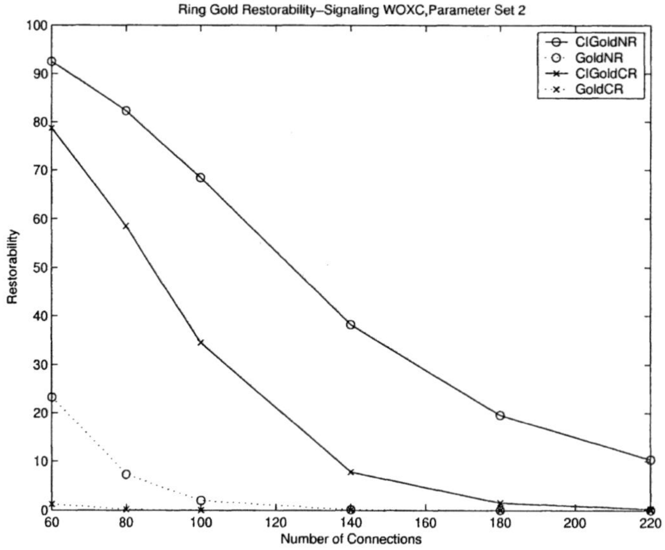


*Figure 15.5.* Gold Restorability in Ring for Signaling without OXC confirmation with parameter set 2.

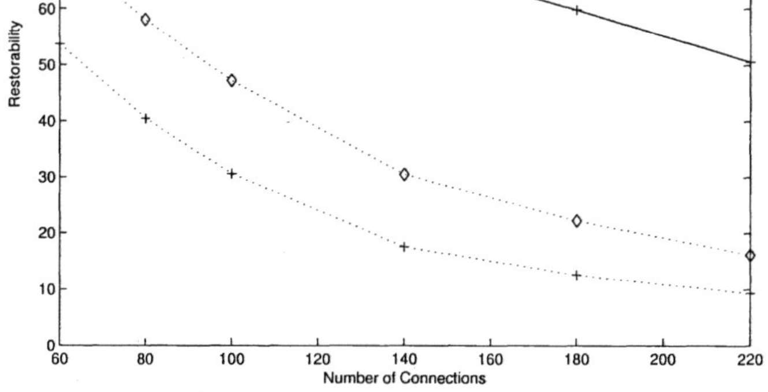


*Figure 15.6.* Platinum Restorability in Arpanet for Signaling without OXC confirmation with parameter set 2.

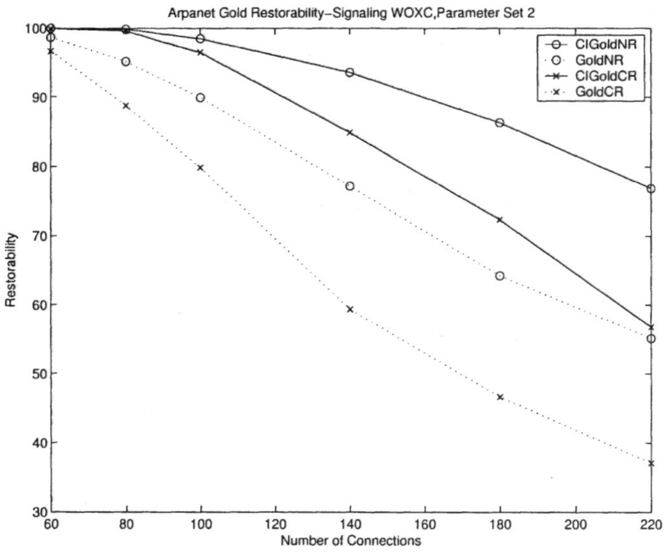


*Figure 15.7.* Gold Restorability in Arpanet for Signaling without OXC confirmation with parameter set 2.

which makes 50 or 100ms restoration targets unrealistic under slow OXC configuration.

## 15.7 Conclusions

Optical networks will likely serve many client networks with different restoration/protection requirements from the optical layer. In this chapter, we have considered the problem of providing Quality of Service Classes based on the restoration speed at the optical layer. We first presented an approach which aims to meet the varying speed requirements by employing different restoration methods for each class. We then proposed a novel approach for providing quality of protection (QoP) classes by coordinating the reconfiguration procedures for the backup paths through scheduling. While it is possible to formulate these scheduling problems as Mixed-Integer-Linear-Programs (MILP), MILPs take a significant amount of time to solve and are certainly impractical in dynamic traffic scenarios. Therefore, we have proposed online scheduling algorithms, maximum-remaining-time-first (MRTF), and FIFO with class information (FIFO-CI) which do not require any off-line schedule computation, and can be implemented with minor modifications to the restoration signaling protocols proposed in the literature and the IETF drafts. We have applied MRTF and FIFO-CI to different signaling protocols similar to the existing proposals, and observed that the performance improvements through these algorithms can be very significant. FIFO-CI can increase the restorability performances from 40% to 88%, and MRTF can reduce the restoration time by up to 17.8%. These performance improvements are significant considering the implementation simplicity of these algorithms.

## References


[1] O. Gerstel and G. Sasaki, "Quality of protection (QoP): a unifying paradigm to protection service grades," *Optical Networks Magazine*, vol. 3, pp. 40–49, May 2002.

[2] G. Mohan and Arun K. Somani, "Routing dependable connections with specified failure restoration guarantees in WDM networks," in *Proceedings of INFOCOM '2000*, April 2000, pp. 1761–70.

[3] G. Sahin and M. Azizoglu, "Optical layer survivability for single and multiple service classes," *Journal of High Speed Networks*, vol. 10, pp. 91–108, 2001,

[4] M. Sridharan, A. Somani, and M. Salapaka, "Approaches for capacity and revenue optimization in survivable WDM networks," *Journal of High Speed Networks*, vol. 10, pp. 109–127, 2001,

[5] A. Fumagalli and M. Tacca, "Differentiated reliability (DiR) in WDM rings without wavelength converters," in *ICC '2001 Proceedings*, June 2001, vol. 9, pp. 2887–2891.

[6] E. Bouillet, K. Kumaran, G. Liu, and I. Saniee, "Wavelength usage efficiency versus recovery time in path-protected DWDM mesh networks," in *Optical Communications Conference, OFC '01*, March 2001, vol. 2, pp. TuG1–T1–3.

[7] R. Doverspike, G. Sahin, J. Strand, and R. Tkach, "Fast restoration in a mesh network of optical cross-connects," in *OFC '99*, February 1999, vol. 1, pp. 170–172.

[8] G. Li, J. Yates, D. Wang, and C. Kalmanek, "Control plane design for reliable optical networks," *IEEE Communications Magazine*, pp. 90–96, February 2002.

[9] A. Banerjee, J. Drake, J. Lang, B. Turner, D. Awduche, L. Berger, K. Kompella, and Y. Rekhter, "Generalized multiprotocol label switching: An overview of signaling enhancements and recovery techniques," *IEEE Communications Magazine*, pp. 144–151, July 2001.

[10] B. Rajagopalan, D. Saha, G. Bernstein, V. Sharma, A. Banerjee, J. Drake, J. Lang, J. Yates, and G. Li, "Signaling for protection and restoration in optical mesh networks," *IETF Internet Draft*, November 2001.

[11] G. Li, C. Kalmanek, J. Yates, G. Bernstein, F. Liaw, and V. Sharma, "RSVP-TE extensions for shared-mesh restoration in transport networks," *IETF Internet Draft*, July 2002.

[12] G. Sahin, S. Subramaniam, and M. Azizoglu, "Signaling and capacity assignment for mesh-based restoration schemes in optical networks," *Journal of Optical Networking*, vol. 1, no. 5, pp. 188–205, May 2002.

[13] G. Sahin, *Service and Restoration Routing and Wavelength Assignment, and Restoration Signaling in Optical Networks*, Ph.D. thesis, University of Washington, 2001.

[14] S. Ramamurthy and B. Mukherjee, "Survivable WDM mesh networks, part II- Restoration," in *ICC '99 Proceedings*, pp. 2023–2030, 1999.

[15] G. Li, J. Yates, R. Doverspike, and D. Wang, "Experiments in fast restoration using GMPLS in optical/electronic mesh networks," in *OFC '2001*, March 2001.

[16] G. Sahin and S. Subramaniam, "Control-message scheduling for improving restoration times in optical mesh networks," in *Proceedings of 14th IASTED International Conference Parallel and Distributed Computing and Systems*, November 2002, pp. 833–838, Invited paper.

[17] G. Sahin and S. Subramaniam, "Quality of protection through control-message scheduling in optical mesh networks," in *Proceedings of 4th International Workshop on the Design of Reliable Communication Networks*, (Banff, Canada), October 2003.

# Chapter 16

# FAILURE LOCATION IN WDM NETWORKS

Carmen Mas,[1] Hung X. Nguyen[2] and Patrick Thiran[2]
[1]*AIT, GR-19002 Athens, Greece,*
[2]*LCA, EPFL, CH-1015 Lausanne, Switzerland*

**Abstract** Fault identification and location in optical networks must cope with a multitude of factors: (i) the redundancy and the lack of coordination (internetworking) of the managements at the different layers (WDM, SDH/SONET, ATM, IP); (ii) the large number of alarms a single failure can trigger; (iii) the difficulty in detecting some failures and the resulting need to cope with missing or false alarms.

This chapter first details the behavior of network components in transparent WDM networks when a failure occurs. Using this model, we then describe an efficient algorithm (Fault Location Algorithm, FLA) pointing out the element(s) which is (are) most likely to be the cause of the received alarms. Although the problem of multiple failure diagnosis is known to be NP-hard, the non-polynomial complexity of the algorithm is pushed ahead in a pre-computational phase, which can be done off-line, and not at the time of a failure. The diagnosis phase is therefore very rapid. We discuss the time and space complexity of the FLA.

**Keywords:** WDM network components, failure model, failure management, complexity.

## 16.1 Introduction

Because of the huge data rates that a single optical fiber can carry using WDM technology, a ribbon break yields the interruption of hundreds of thousands of flows, and the loss of thousands of megabits of data [1]. Survivability of optical networks, which includes fault identification and location, has thus become a crucial problem.

When a failure occurs at the physical layer, the lightpaths that are affected have to be restored as soon as possible so that higher layers do not see the failure and do not start their own restoration mechanisms. In the meantime, the failure has to be located and repaired.

Failures are located from the alarms received by the management system. When there are two or more simultaneous failures, the number of alarms considerably increases, the alarms arrive intermingled at the management system and the problem of locating the failures becomes even more difficult. The problem of locating multiple failures has been shown to be NP-hard by Rao [2].

The location of the failure(s) must be fast and accurate, so that a small set of faulty candidates can be rapidly identified, before expensive repair actions are undertaken. Failures are less rare than one might expect; [3] has recently reported a failure rate of 1 per year per 300km of fiber. Submarine cables, which are vulnerable to damage from submarines, anchors and fishing gears, have to be repaired once every five weeks [4].

Network management is essential to ensure the good functioning of these networks. Every network management performs several functions, which have been classified into five different functional areas by OSI, and are briefly recalled here.

- **Configuration Management** deals with the initialization of the network components, the establishment, maintenance and updating of relationships among them. These relationships are based on the connections established and cleared down in the network. Configuration management must also reconfigure the network whenever necessary and include routines that are able to inform about any change in the configuration (for example, when a protection switch changes its position, the manager should be informed about the new paths of the established channels).

- **Performance Management** monitors and controls the components in the network. *Monitoring* is the function that tracks activities in the network, whereas the *control* function enables adjustments in the components to improve network performance. The main performance issues are: network capacity utilization, existence of excessive traffic and of bottlenecks, and increase of response time. Performance management collects information from the network and analyzes it so that the management system can recognize situations of performance degradation.

- **Security Management** deals with the generation, distribution and storage of encryption keys. It also monitors and controls access to computer networks and to management information.

- **Accounting Management.** Many network services are charged to the users. Network management performs not only the internal accounting but also other tasks, such as checking allowed access privileges.

- **Fault Management** deals with [5]:

- fault detection, to know whether there is a failure or not in the network,
- fault location, to know which is/are the component(s) that has/have failed and caused the received alarms,
- fault isolation (so-called protection) in order for the network to continue to operate, which is the fast and automated way to restore interrupted connections. In general, it is implemented with protection switches that change positions when the optical powers drop below a certain threshold.
- network (re-)configuration (so-called restoration) that minimizes the impact of a fault by restoring the interrupted connections using spare equipments. This involves some processing to discover the best paths to re-route the connections.
- replacement of the failed component(s).

Protection and restoration mechanisms in optical networks is an active field of research. This chapter focuses on the fault location problem. This problem is not specific to optical networks, but is encountered in many fields, such as electrical power plants monitoring, medical diagnosis, electric circuit analysis, nuclear power station maintenance and management of communication networks at large. Here we consider only its application to optical networks, although many methods are actually valid or have been developed in more general settings. Fault location is particularly important for optical networks, where the quantity of information carried is much larger than over other physical media, making the effects of a failure much more severe.

## 16.2 Fault Location Problem definition

Optical communication networks, and all networks in general, need a fault management system that performs fault diagnosis, that is, able to identify the faults that occur from the information given by the network components. A *fault* can be defined as an unpermitted deviation of at least one characteristic parameter or variable of a network element from acceptable/usual/standard values, whereas a *failure* can be defined as the manifestation of the fault. For example, if the ventilator in a laser stops, and if the temperature increases and overpasses an accepted temperature limit, the fault is the stopped ventilator and the failure is the temperature of the laser, which is too high. Because both terms are closely related, we use them indistinctly.

Fault detection relies on the monitoring of the state of the network components. Simple fault detection mechanisms are often based on locally monitored variables. The faulty values reached by these variables are logged as errors.

Critical errors are sent to the network manager as alarms. However, it is not always possible to detect complex faults on the sole basis of locally monitored variables: it is then necessary to have a global knowledge of the network and to do some processing to diagnose the presence (or absence), the nature and the location of the fault. Also, and because the fault can propagate to components that depend on the failed component, the influences of faulty components on other components have to be taken in account to perform an efficient fault management.

Communication networks are built on several layers, each performing fault management functions independently. When a failure occurs, several symptoms or event indications are issued to the network manager from different management layers, and fault management functions start in parallel. Research is carried out to allow interoperability between different layers, to avoid task duplication and increase efficiency.

The fault location problem is solved by Fault Management Systems that take the events generated by the network elements as input (these events can be alarms, warnings or parameters of the network elements), and produce an output, which is the set of network elements whose failures explain the input events.

These fault management systems differ in:

1 the way they solve the problem: using neural networks [6, 7], Finite State Machines [8, 9, 10], the history of previous cases [11], a (detailed) modeling of the network [12]. A classification of these methods into model-based methods and Black box learning-based approaches is made in [13]. The first one relies on an abstract model of the network, capturing the dependency relations between the elements and capable of pointing out the element(s) which is (are) most likely to be the cause of the received alarms. The second one does not attempt to model the network in detail, but leaves it as a black box. An abnormal situation is then diagnosed from a set of rules obtained by learning or thanks to expertise of the human manager.

2 the information they need: failure propagation probabilities [14], timestamps, set of established channels [12], etc.

3 the assumptions on which they rely: existence of only single failures [15], existence of multiple failures [12, 16, 17], etc.

4 the quantity of memory they use: large memory requirements [11, 12] to store failure history, modest memory requirements or even absence of memory requirements [15, 14, 9, 8, 10], etc.

5 intolerance [11, 18] or tolerance [6, 12] to false and missing alarms.

## 16.3 A failure model of optical networks

We distinguish two classes of network components: (i) Optical components, which take care of the optical signal transmission and are not able to send alarms, except, in some cases, for their own failures (but never when the received optical signals are not as expected); and (ii) Monitoring equipments that are able to send alarms and notifications when the monitored optical signals are not the expected one. The alarms sent by monitoring equipments depend on the equipment type and characteristics. Failure of a monitoring equipment does not interrupt/modify the data transmission, it may only result in the loss of an alarm. It is not as relevant as the failure of an optical component, and therefore is not considered in the present study.

### 16.3.1 Optical equipment

The optical equipment in a transparent optical network can be listed as follows.

- Transmitters (Txs), which are located at the beginning of an optical channel, are lasers or laser arrays converting electrical signals into optical ones at a certain wavelength. New lasers used in advanced WDM networks are tunable and can change the emission wavelength within a prescribed range. Some lasers do include a wavelength locker so that when the emitted wavelength deviates from the expected value due, for example, to temperature changes, it resets the transmitter to the original wavelength.

- Receivers (Rxs), which are located at the end of an optical channel, convert the received optical signal of a certain wavelength, into an electrical one.

- Optical switches: There are different switches architectures, each of them having different crosstalk characteristics: crossbar, Clos, Spanke, Benes, and Spanke-Benes. Different technologies can be used for their implementation (except MEMS, all these technologies are used in crossbar architectures: Micro-Electro-Mechanical System (MEMS), Bulk mechanical, Bubble-based waveguide, Liquid crystal, Thermo-optical, SOA).

- Amplifiers, which output a signal at a higher power level than the input signal, usually add distortion to the signal. A fault may occur when the pump laser (in the case of EDFA and Raman amplifiers) fails or when the fiber or a passive component within the amplifiers fails. Other faults may involve the failure of the gain monitoring system causing gain variations. They send alarms for example when the pump laser does not work properly, or when the incoming power falls below a threshold value.

- Optical regenerators and wavelength converters: These two types of elements are included in the same category since they are based upon similar physical principles and technologies. There are three techniques to perform optical wavelength conversion and regeneration: optical gating, interferometric, and wave mixing based.
- Couplers (Splitters/combiners): These elements are included in some demultiplexers/multiplexer architectures. Their key performance parameter is their insertion loss that should be kept low so that when included in serial architectures the overall loss is still acceptable.
- Optical filters: These components have two important applications: to be used to multiplex and demultiplex wavelengths in a WDM system, and to provide equalization of the gain and filtering of noise in optical amplifiers. The most important characteristics of optical filters are: insertion loss, temperature coefficient, flat passband, and sharp passband skirts.
- Protection switches, which receive multiple optical signals, and select the one with an acceptable power level. Note that these elements could also be considered as monitoring equipments, since they send alarms when they change the switch positions due to unacceptable incoming optical powers.

### 16.3.2 Monitoring equipment

Different types of monitoring equipment exist in the market and are used in transparent optical networks [19, 20]. They monitor the optical signals using tapping couplers. We assume that for monitoring purposes, the optical signal can be converted to the electrical domain. We distinguish six different types of monitoring equipment:

- Optical Power Meter: detects any change in the power of the optical signal over a wide band. It may be able to send an alarm when the measured power is different from the expected one. When the power decrease is very slow, it takes a longer time to be detected and an alarm takes longer to be generated. In some cases, the BER can be severely degraded without a corresponding decrease in the optical power. For example, amplifiers with automatic gain control deliver signals with the expected power even if they contain too much noise.
- Optical Spectrum Analyzer [21] performs analog optical signal monitoring by measuring the spectrum of the optical signal. The parameters

that can be measured are channel power, channel center wavelength and optical signal to noise ratio (OSNR), which is useful and provides important information on the health and quality of the optical signal. For example, it is able to detect OSNR changes (even if they do not cause optical power variations) and the unexpected out-of-band signals. However, this equipment suffers from slow responses and possible sampling errors.

- Eye Monitoring derives from the eye diagram information on the time distortion and interferences. The resulting histogram is used to study the statistical characteristics of the optical signal [22, 23].

- BER Monitoring: After converting the signal to the electrical domain, this equipment is able to calculate the Bit Error Rate which is sensitive to noise and time distortion. BER Testers compare the expected with the received bit pattern and the differences that are found, give the estimated BER. This equipment is sensitive to impairments such as crosstalk, chromatic and polarization mode dispersion, and optical nonlinearities. Most of the BER techniques are based on the synchronous [24, 25] or asynchronous [23, 26] sampling of the optical signal.

- Wavemeter is an accurate monitoring equipment able to detect any variation in the used wavelength.

- Pilot tones are signals that travel with the data but can be retrieved easily. For example, pilot tones may use different carrier frequencies from the transmitted signals (in WDM systems), different time slots (in TDM systems), or different codes (in CDMA systems). Pilot tones can detect transmission disruptions, but not in-band jamming problems unless they affect the pilot tone frequency (which in this case may not affect the transmitted signals).

- Optical Time Domain Reflectometry (OTDR) techniques are based on pilot tones. However, the difference is that OTDR analyzes the pilot tone's echo. This method is used to detect fiber cut and bending that causes data signal loss. OTDR may be able to detect some faults that pilot tones cannot. For example, a jamming signal may not be detected by a pilot tone if it does not cover its frequency, but the pilot tone's echo may contain remainders of the jamming signal that have reflected.

Table 16.1 summarizes the monitoring properties of the components listed above, for a transparent WDM network.

A WDM network is formed by a number of optical channels, which are composed of the different components listed above. The classification of this

*Table 16.1.* Failure detection capabilities of monitoring equipment (Under Columns 3 and 4, *ST* stands for sometimes).

| *Monitoring Component* | *Power* | *In-band Jamming* | *Out-band Jamming* | *Wavelength misalignment* | *Time distortion* |
|---|---|---|---|---|---|
| Opt. Power Meter | yes | no | no | no | no |
| Opt. Sp. Anal. | yes | no | yes | no | no |
| Eye Monitoring | yes | no | yes | no | yes |
| BER Monitoring | yes | yes | yes | no | yes |
| Wavemeter | yes | no | no | yes | no |
| Pilot Tones | yes | no | no | no | yes |
| OTDR | yes | *ST* | *ST* | no | yes |

section enables us to abstract the network components in two categories: optical (passive) components, and monitoring components. To keep the exposition of the Fault Location Algorithm sufficiently simple in the next section, we assume that only power is monitored, which is the only variable that will be detected by all monitoring components. In reality, the algorithm is much more complex, as each monitoring component belongs to a different sub-category, as defined by the alarms that can be generated according to Table 16.1 (see also [27, 28]). The classification becomes even richer, when we include opaque networks.

Figure 16.1 shows an example of two channels in a transparent network, abstracted using the models elaborated in this section.

## 16.4 Fault location algorithm (FLA)

Time to locate failure(s) is critical, any good fault location algorithm must be able to locate fault as quick as possible. Unfortunately, the fault localization problem has been shown to be NP-Hard by Rao in [2] even in the ideal scenarios where there are no erroneous alarms. Nevertheless, the computation that has to be carried out when alarms reach the manager can be kept short despite the potentially large size of the network, if we follow the fault location algorithm (FLA) proposed in [12] to pre-compute, as much as possible, the functions that can be executed independently of the received alarms. This phase is called the *pre-computation phase* ($PCP$). The pre-computation phase is executed only when relationships between fault sources and alarm elements change, for example when a channel is set up or torn down, not when the alarms are received. Once the manager starts receiving alarms from the network, the algorithm does not have to perform complex computation but simply traverses a binary tree. This minimizes the time the algorithm needs to deliver results to the manager when failures occur.

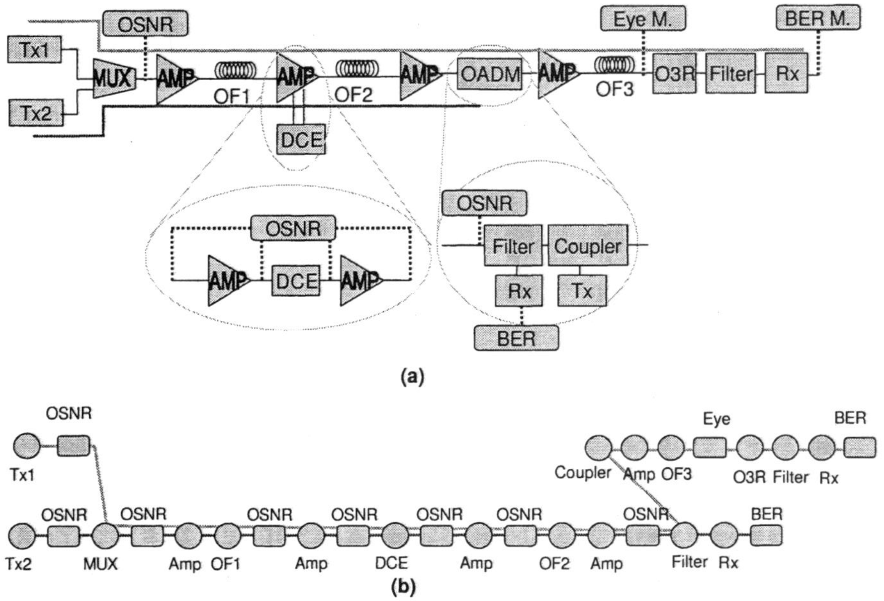


*Figure 16.1.* Example of a transparent ultra long haul WDM network including some monitoring equipments at some relevant components such as an amplifier with a Dynamic Spectrum Equalizer (DSE) or an Optical Add-Drop Multiplexer with Wavelength Selective architecture. The figure also shows two established channels that have been modeled: the optical components are represented by circles, and the monitoring equipments by rectangles.

## 16.4.1 FLA to localize single failure in the ideal case

We first start by solving the problem of localizing single failure. In this case, only a single network component can fail at a time and when a network component fails, the monitoring components that follow it on a channel will issue alarms.

Let us illustrate the steps of the algorithm on the network in Figure 16.1. For the sake of simplicity in our description, we will only consider a fraction of the network as in Figure 16.2, this is equivalent to assume that all optical components preceding $OF2$ in both channel 1 and channel 2 never fail. There are two established channels in the network ($\mathcal{CH} = \{CH_1, CH_2\}$). We label the network elements by $p$ for optical (passive) components and $e$ for the monitoring components. In this example, there are 10 passive elements $p_1, ..., p_{10}$ and 4 monitoring elements $e_1, ..., e_4$.

The $PCP$ consists in the following steps.

1 Compute the *domain* for each optical component. The *domain* of an optical component is defined as the set of monitoring equipments that

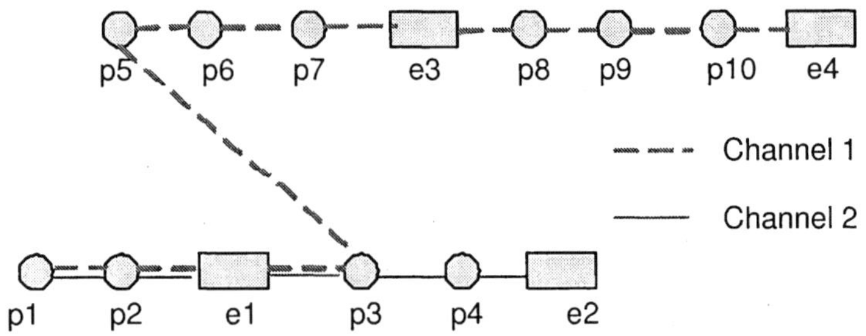


*Figure 16.2.* Reduced network example.

will trigger alarms when that optical component fails. In the example, we obtain

$$\begin{aligned}
Domain(p_1) &= \{e_1, e_2, e_3, e_4\}, Domain(p_2) = \{e_1, e_2, e_3, e_4\}, \\
Domain(p_3) &= \{e_2, e_3, e_4\}, Domain(p_4) = \{e_2\}, \\
Domain(p_5) &= \{e_3, e_4\}, Domain(p_6) = \{e_3, e_4\}, \\
Domain(p_7) &= \{e_3, e_4\}, Domain(p_8) = \{e_4\}, \\
Domain(p_9) &= \{e_4\}, Domain(p_{10}) = \{e4\}.
\end{aligned}$$

2 Group all identical domains into *fault classes*

$$\begin{aligned}
C_1 &= Domain(p_1) = Domain(p_2) = \{e_1, e_2, e_3, e_4\}, \\
C_2 &= Domain(p_3) = \{e_2, e_3, e_4\}, \\
C_3 &= Domain(p_5) = Domain(p_6) = Domain(p_7) = \{e_3, e_4\}, \\
C_4 &= Domain(p_4) = \{e_2\}, \\
C_5 &= Domain(p_8) = Domain(p_9) = Domain(p_{10}) = \{e_4\}.
\end{aligned}$$

3 Compute the set $P(C_i)$ of elements that belong to the same fault class $C_i$, in other words, the optical components whose failures cannot be distinguished by the network manager from the monitoring information. In this example, these sets are:

$P(C_1) = \{p_1, p_2\}$ , $P(C_2) = \{p_3\}$ , $P(C_3) = \{p_5, p_6, p_7\}$ , $P(C_4) = \{p_4\}$ and $P(C_5) = \{p_8, p_9, p_{10}\}$.

4 Construct the dependency matrix where each column of the dependency matrix is a binary vector $Bin(C_i)$ with as many elements as the number of monitoring components in the established channels (4 in this example). The $j^{th}$ component of $Bin(C_i)$ is equal to 1 if the $nth$ monitoring equipment fires alarm when elements in the fault class $C_i$ fail, and 0

otherwise. In the example, let us denote by $G$ the dependency matrix, then

$$G = \begin{bmatrix} 1\,0\,0\,0\,0 \\ 1\,1\,0\,1\,0 \\ 1\,1\,1\,0\,0 \\ 1\,1\,1\,0\,1 \end{bmatrix}$$

5 Build a binary tree for single failures: The binary tree has a depth equal to the number of alarm elements and leaves correspond to different binary combinations. Occupied leaves point to the fault class $C_i$ whose corresponding column in the dependency matrix, $Bin(C_i)$, is the path from the root of the tree to the leaves. The binary tree constructed for this example is shown in Figure 16.3.

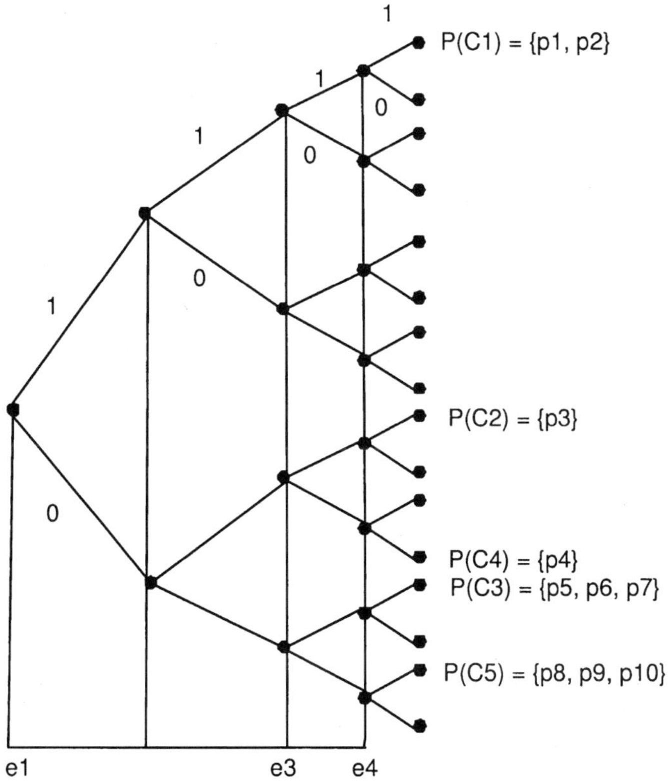


*Figure 16.3.* Binary tree for single failures.

### 16.4.2 FLA to localize multiple failures in the ideal case

Let us now consider the case where several failures may happen in a short interval of time so that the alarms reaching the manager are intermingled. The algorithm has to be able to distinguish the failures based on the received alarms. To solve this problem, the binary tree is extended by adding leaves which correspond to multiple failures. This amounts to computing the domains of simultaneous failures which are the union of the domains of single failures. We begin with double failures. The construction are as follows. Let $C_i = Domain(p_i)$, $C_j = Domain(p_j)$ and $C_k$ be the domain for double failures of $p_i$ and $p_j$ , then

$$Bin(C_k) = Bin(C_i \cup C_j) = Bin(C_i) \vee Bin(C_j),$$

where $\vee$ stands for the point-wise $OR$ operation between $Bin(C_i)$ and $Bin(C_j)$.

If $Bin(C_k)$ is equal to $Bin(C_l)$ for an existing failure class $C_l$, the leaf is already occupied by the domain of a single failure. We can reasonably assume that single failure is more likely to occur than multiple failures, hence the occupied leaf points to the more likely single failure. Conversely, if $Bin(C_k)$ is different from any of the existing leaves, a new leaf is then occupied and pointed to the double failures. Once all the new leaves corresponding to double failures are filled, we proceed likewise for triple failures, etc. Note that if at some point of this procedure, there is a $C_k$ corresponding to a single failure which is such that for all the already computed $C_i$'s,

$$Bin(C_i) \vee Bin(C_k) = Bin(C_i) \text{ or } Bin(C_i) \vee Bin(C_k) = Bin(C_k),$$

then $C_k$ will not contribute to any new leaf anymore. Therefore, it needs not be considered for further steps with more failures. This property allows us to decrease the number of binary vectors corresponding to single failures needed for computing the domain of multiple ones. The process finishes when the set of single failures becomes empty.

Let us consider the example shown in Figure 16.2. The output of this part of the algorithm is the following:

$Bin(C_4) \vee Bin(C_5) = Bin(C_6)$.

One new failure class $C_i$ has been found. It is defined by the vector $Bin(C_6) = (0, 1, 0, 1)$. The resulting binary tree is shown in Figure 16.4.

### 16.4.3 FLA to localize multiple failures in the non-ideal case

Alarm errors are unavoidable in network diagnosis. There are two kinds of erroneous alarms: missing alarms and false alarms. Missing alarms occur when some alarms are lost, or arrive with such a delay that they cannot be

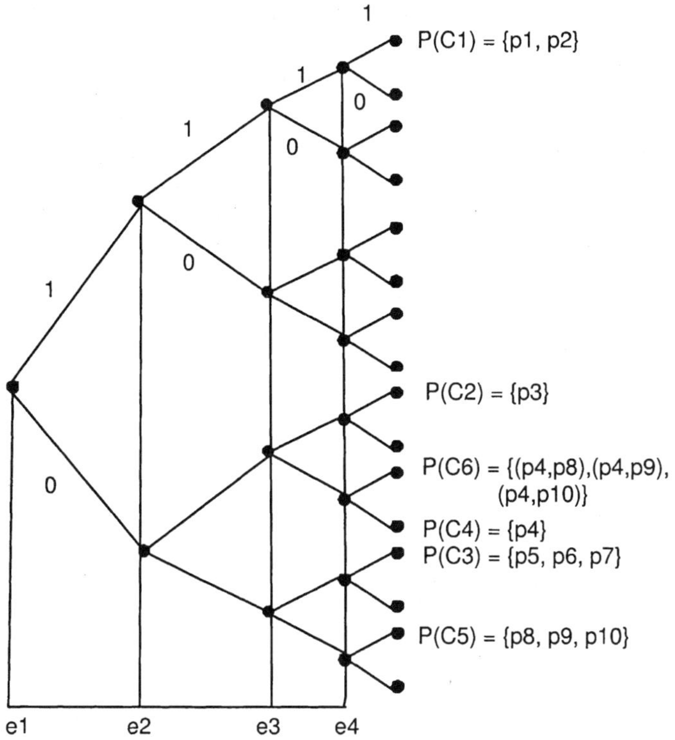


*Figure 16.4.* Binary tree for multiple failures.

considered during the ongoing computation of the algorithm or if a monitoring component is itself defective. False alarms can occur when, due to abnormal situations, a monitoring component sends an alarm although there is no real failure. The empty leaves in the binary tree for multiple failures in Figure 16.4 are those corresponding to missing or false alarms.

The tree can be viewed as a particular block error-correcting code, whose codewords have the property that the logical OR of any two codewords is another codeword. One empty leaf of the tree corresponds to an erroneous word, and the error correction task would be to replace the erroneous word by the correct codeword whose Hamming distance with the received word is minimal. The problem of finding the nearest codeword to an arbitrary word can be shown to be NP-complete, but however, there are special instances where this problem is easy to solve. We will discuss these cases in the next section when we discuss the computational complexity of the FLA. In the rest of this section, we will show how the binary tree can be extended to deal with erroneous alarms.

Contrary to the use of error-correcting codes for data transmission, the manager of a network does not require unique decoding. Indeed, he/she will prefer to get the set of all faulty candidaιes whose domains are close to the received alarms. In fact, we can use the bounded distance decoding approach by giving all the codewords that realize a given alarm mismatching threshold as possible answers. For example, if we tolerate a maximum of $m_1$ missing alarms and $m_2$ false alarms, for a set of alarms $\mathcal{R}$, the codewords that fall within this margin from the binary vector $Bin(\mathcal{R})$ are the codewords that have a '1' when a $Bin(\mathcal{R})$ has a '0' in at most $m_1$ positions, and have a '0' when $Bin(\mathcal{R})$ has a '1' in at most $m_2$ positions.

The error correction part of the algorithm can be computed off-line, in the $PCP$ module, or on-line, when the alarms are received. In the first case, one computes for each occupied leaf (i.e., for each codeword), the binary vectors whose number of 0's and 1's differ to this codeword respectively by at most $m_1$ and $m_2$ positions. The corresponding fault classes $C_i$ are added to the list pointed by the leaf (see for example in Figure 16.5 the new binary tree when $m_1 = 1$ and $m_2 = 0$ and in Figure 16.6 the new binary tree when $m_1 = 0$ and $m_2 = 1$). The on-line approach is discussed in the next section when trade-offs between storage requirement and time requirement for the FLA are considered.

Let us illustrate the algorithm with different scenarios in the example of Figure 16.2 when the set of received alarms is $\mathcal{R} = (a_2, a_3)$ ($a_2$ is issued by $e_2$, and $a_3$ is issued by $e_3$). Hence, $Bin(\mathcal{R}) = (0110)$, which corresponds to an empty leaf of the tree in Figure 16.4. Let us check the following scenarios:

- $m_1 = 1, m_2 = 0$: One missing alarm and no false alarm are tolerated. In this case (Figure 16.5), the output of the algorithm is the leaf $Bin(C_2) = (0111)$ with one mismatch which corresponds to the following solution:

  Failure of $p_3$ with one mismatch.

- $m_1 = 0, m_2 = 1$: One false alarm and no missing alarm are tolerated. In this case (Figure 16.6) the output of the algorithm is the leaf $Bin(C_4) = (0100)$ with one mismatch which corresponds to the following solution:

  Failure of $p_4$ with one mismatch.

In the considered example, there are only four monitoring elements, the tolerance $m_1 = m_2 = 1$ is too loose because it amounts to accepting 50% erroneous alarms. For this reason, we will not consider this tolerance or other tolerances with higher values of $m_1$ and $m_2$.

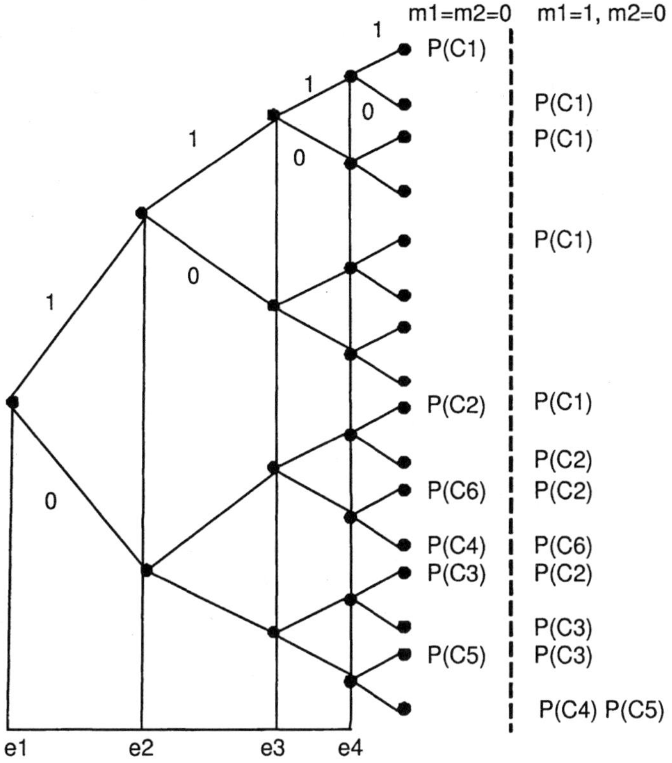


*Figure 16.5.* Binary tree when $m_1 = 1$ and $m_2 = 0$.

## 16.5 Algorithmic complexity

In this section, we present the complexity of the FLA and show how the algorithm can be modified to meet certain complexity requirements.

Let us denote by $n$ the number of different components in the established channels, by $n_a$ the number of alarming elements and by $n_{na} = n - n_a$ the passive (non-alarming) elements. $n_a$ is also the depth of the binary tree and therefore the binary vector size. Let us also denote by $t$ the number of single fault classes and by $c$ the number of established channels.

### 16.5.1 Pre-Computing Phase (PCP)

**Time requirements.**

1 Computation of *Domains* : The computational complexity for this step is at most of $O(\frac{n(n-1)c}{2}) = O(n^2)$ because the number of channels $c$ is much less than the number of network components.

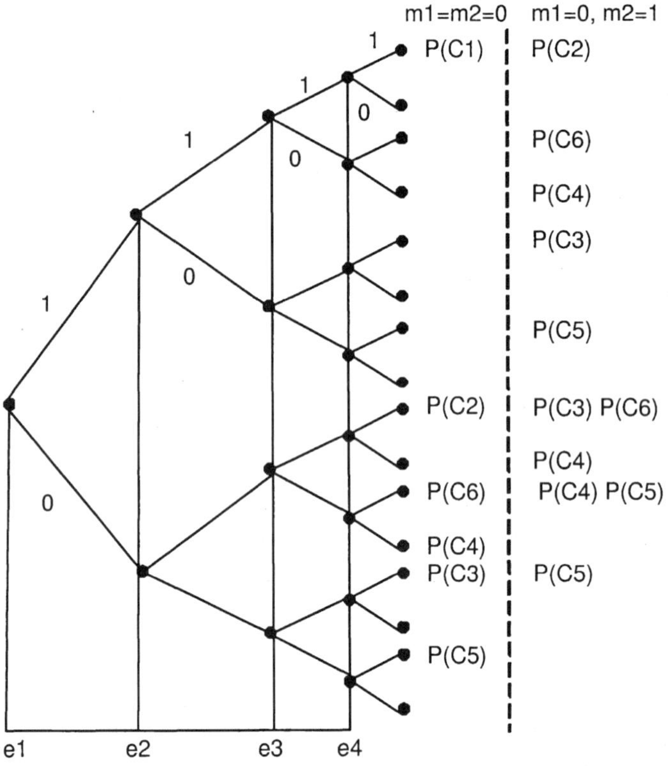


*Figure 16.6.* Binary tree when $m_1 = 0$ and $m_2 = 1$.

2 Grouping of *Domains* into *fault classes* and production of the dependency matrix: each domain is associated to one $n_a$-dimensional binary vector. The computation time needed to group identical domains, and to compute the codeword $Bin(C_i)$ is at most $O(n^2 n_a)$.

3 Computation of the domains of multiple failures and of the corresponding codewords. If $d$ is the total number of different classes $C_i$ ( or equivalently of different codewords), the computational time of this step is $O(\frac{d(d-1)}{2} n_a)$. Since $d \leq 2^t$, the (very) worst case bound is $O(4^t n_a)$.

4 Computation of codewords with mismatching thresholds $m_1$ and $m_2$. This step can be implemented as follows. For each codeword $\overrightarrow{x} = Bin(C_i)$, it finds all the vectors $\overrightarrow{y}$ such that $\sum_{k=1}^{n_a} [x_{ik} - y_k]^+ \leq m_1$ and $\sum_{k=1}^{n_a} [y_k - x_{ik}]^+ \leq m_2$ where $[z]^+ = z$ if $z > 0$ and 0 otherwise. The worst case computation time is $O(2^{n_a} \sum_{k=1}^{m_1+m_2} C_k^{n_a})$, where $C_k^{n_a} = \frac{n_a!}{(n_a-k)!k!}$ is the combinatorial coefficient $(n_a, k)$.

**Storage requirements.** The storage memory for a binary tree which has $2^{n_a}$ leaves is $O(2^{n_a})$.

### 16.5.2 FLA Main Part

1 Compute $Bin(\mathcal{R})$: This module only requires a computational time of $O(n_a)$.

2 Find $P(C_i)$'s of the $Bin(\mathcal{R})$ leaf: This module also requires a computational time of $O(n_a)$.

The total computational time of the on-line part of the algorithm is $O(n_a)$. This low complexity comes at a cost of a high computation time in the $PCP$ which is $O(4^t n_a)$ and an exponential storage memory of $O(2^{n_a})$.

### 16.5.3 The Error Correction Problem

Let us consider a received alarm vector $\vec{r} = Bin(\mathcal{R})$, which is a binary vector of dimension $n_a$. The problem of correcting the erroneous alarms amounts to finding a codeword $Bin(C_i)$, for some fault class $C_i$, with the minimum Hamming distance to the received word; that is, such that the number of 1's by which $\vec{r}$ and $Bin(C_i)$ differ is minimal for all possible fault classes $C_i$.

Note at this point that the code is not linear, and that it will have very poor performances in terms of minimal distance. Nevertheless, the following results are important for managing monitoring equipments in a transparent optical network. When there is no missing alarm, i.e., $m_1 = 0$, finding the nearest codeword to a received word can be achieved in polynomial time. The polynomial time algorithm first finds all the codewords that do not have 1s at the positions where the received word has 0s, and then sequentially remove all the common '1' entries between the received word and each of the codewords identified above. The '1' entries left in the received word after this operation correspond to the false alarms. On the other hand, when there are missing alarms, i.e. $m_1 > 0$, finding the nearest codeword to a received word is NP-complete. This is due to the fact that when $m_1 > 0$ and $m_2 = 0$, the problem of finding the nearest codeword becomes the red-blue set cover problem as defined in [29]. The red-blue set cover problem is an extension of the set cover problem which is an NP-complete problem, hence it is also NP-complete.

From the above result, we see that for the error correction problem, it is more desirable to have false alarms than missing alarms. This has a significant practical influence as the most common source of alarm errors comes from setting wrong threshold values for monitoring equipments. Our analysis above indicates that one should set the threshold values high rather than low in order to restrict all missing alarms (with the possibility of producing more false alarms).

### 16.5.4 Trading time for storage

In this section, we look at the problem of reducing the exponential storage requirement for the binary tree. Since the storage memory is proportional to the number of leaves, the way to reduce the storage memory is to reduce the number of leaves in the tree. The leaf reduction can be achieved in two steps.

In the first step, we remove all the leaves that correspond to erroneous alarms (that is, those obtained for $m_1 > 0$ and/or $m_2 > 0$) and carry out the error correction module of the algorithm on-line. There are two reasons for carrying out this module in the on-line part of the algorithm : first, it helps speeding up the $PCP$ phase of the algorithm; second, the overall complexity of the algorithm is lower, as the received alarm vector $\overrightarrow{r} = Bin(\mathcal{R})$ is compared only with the codewords corresponding to single or multiple failures in the absence of erroneous alarms (i.e., with $m_1 = m_2 = 0$), not with all the leaves. This is particularly true in cases where the problem of finding the nearest codeword can be solved in polynomial time (i.e., when $m_1 = 0$).

After this first step, we are left with a tree whose leaves point to at most one set $P(C_i)$ each. Each of these sets $P(C_i)$ is made of subsets of network components, the union of the domains of all elements within a subset is $C_i$. Recall that in the tree construction process for multiple faults, among all the subsets of faulty elements which satisfy the condition that the union of the domains of all elements within a subset is $C_i$, we only keep in $P(C_i)$ subsets with the smallest cardinality. Let us define a failure scenario associated to a set $P(C_i)$ as a set of fault classes for single failure which satisfies the condition that the unions of the fault classes in that set is $C_i$. In the second step of the leaf reduction process, we further reduce the leaves in the binary tree by exploiting the fact that the online decoding time for a leaf depends on the number of failure scenarios associated to the corresponding set $P(C_i)$. We proceed as follows. Pick a number $m \geq 1$. If the number of failure scenarios associated to $P(C_i)$ is less than $m$, then $P(C_i)$ is removed from the tree. Otherwise, it is maintained in the tree.

With such a trimmed tree, whenever the received codeword $\overrightarrow{r}$ corresponds to a set $P(C_i)$ that is associated to at least $m$ failure scenarios, it is retrieved as before, since it is pointed by the corresponding leaves. Otherwise, if the received codeword $\overrightarrow{r}$ corresponds to a set $P(C_i)$ that is associated to less than $m$ failure scenarios, we find these on-line, using an exhaustive search. This exhaustive search only needs to examine at most $m$ different sets of single fault classes that correspond to $P(C_i)$ to find the optimal subsets. Let us denote by $t$ the number of single failure codewords. Since there is a maximum of $2^t$ different possible subsets of single fault classes, there are less than $2^t/m$ codewords whose corresponding set $P(C_i)$ is associated to more than $m$ different failure scenarios. Furthermore, the computational time to identify one subset of fault classes that corresponds to a codeword is $O(n_a t)$. Hence, by storing in the binary tree only the codewords whose corresponding set $P(C_i)$ that has more than $m$ failure scenarios, one can store less than $2^t/m$ leaves in the binary tree and still guarantee a worst case decoding time of $O(mn_a t)$. By varying $m$, one can actually choose different levels of time and storage complexities for the FLA.

The on-line decoding strategy when a word $\overrightarrow{r} = Bin(\mathcal{R})$ is received is as follows. The algorithm iteratively identifies all codewords within a mismatching threshold $m_1$ and $m_2$ from $\overrightarrow{r}$. This step has a time complexity of $O(\sum_{k=1}^{m_1+m_2} C_k^{n_a})$. For each codeword, the algorithm first traverses the binary tree in $O(n_a)$ time, to find the optimal failure scenario. If the codeword is not in the binary tree, the tree search will fail, the algorithm will then carry out an exhaustive search in time $O(mn_a t)$ to find the optimal failure scenario, where $m$ is the number of failure scenarios that are associated with $\overrightarrow{r}$. The worst case bound for the on-line decoding module is $O(mn_a t \sum_{k=1}^{m_1+m_2} C_k^{n_a})$.

## 16.6 Conclusion

Ideally, methods to locate faults in a WDM network should be sensitive to hard and soft failures, should rely on the analog signals or cope with lost and false alarms, should avoid the use of probabilities since they are time-varying parameters that are difficult to estimate, should be aware of the possible failure scenarios and how they propagated under correlation rules [30], or an appropriate taxonomy of the network elements as developed in this chapter and [12], and should rapidly identify an accurate set of faulty candidates. Despite the NP-hard complexity of the multiple-failure problem, an efficient algorithm that meets these requirements has been described in this chapter. We have also discussed the complexity of the various modules of the algorithm, and shown how the NP-hard part of the algorithm can be done off-line.

Without monitoring equipments along the optical physical layer, optical networks provide few information about failures. These monitoring components detect different signals and thus failures, and the Fault Location Algorithm should be extended in order to take this into account.

Instead of deploying additional monitoring equipment, or in complement to this deployment, a second solution is to exploit the interoperability between the layers so that a single fault management system using information from the different layers could become more accurate and locate a larger number of failure scenarios.

## Acknowledgments

This work was financially supported by grant DICS 1830 of the Hasler Foundation, Bern, Switzerland.

# References


[1] C. Metz. IP Protection and Restoration. *IEEE Internet Computing*, pages 97–102, March-April 2000.

[2] N. S. V. Rao. Computational Complexity Issues in Operative Diagnosis of graph-based Systems. *IEEE Transactions on Computers*, 42(4):447–457, April 1993.

[3] P. Arijs, P. van Caenegem, P. Demeester, P. Lagasse, W. Van Parys, and P. Achten. Design of ring and mesh based WDM transport networks. *Optical Networks Magazine*, pages 20–45, 2000.

[4] Submarine Networks. *FibreSytems*, 3(5):26, June 1999.

[5] S. Abek H. Hegerin and B. Neumair. *Integrated Management of Networked Systems*. Morgan Kaufmann Publishers, 1998.

[6] R. Gardner and D. Harle. Alarm Correlation and Network Fault Resolution using Kohonen Self-Organising Map. *Proc. of IEEE GLOBECOM*, pages 1398–1402, 1997.

[7] C. Rodriguez, S. Rementeria, J. I. Martin, A. Lafuente, J. Muguerza, and J.Perez. A Modular Neural Network approach to Fault Diagnosis. *IEEE Transactions on Neural Networks*, 7(2):326–340, March 1996.

[8] C. S. Li and R. Ramaswami. Fault Detection and Isolation in transparent All-Optical Networks. *IBM Research Report*, RC-20028, April 1995.

[9] C. Wang and M. Schwartz. Fault Detection with Multiple Observers. *Proc. IEEE INFOCOM*, pages 2187–2196, 1992.

[10] A.T. Bouloutas, G. W. Hart, and M. Schwartz. Fault Identification Using a FSM model with Unreliable Partially Observed Data Sequences. *IEEE Transactions on Communications*, 41(7):1074–1083, July 1993.

[11] H.-G. Hegering and Y. Yemini (Editors). *Integrated Network Management*. Elsevier Science Publishers B.V., 1993.

[12] Carmen Mas and P. Thiran. An efficient algorithm for locating soft and hard failures in WDM network. *IEEE Journal on Selected Areas in Communications*, 18(10):1900–1911, October 2000.

[13] Carmen Mas and P. Thiran. A review on fault location methods and their application to optical networks. *Optical Networks Magazine*, 2(4), July/August 2001.

[14] R. H. Deng, A. A. Lazar, and W. Wang. A probabilistic Approach to Fault Diagnosis in Linear Lightwave Networks. *IEEE Journal on Selected Areas in Communications*, 11(9):1438–1448, December 1993.

[15] N. S. V. Rao. On Parallel Algorithms for Single-Fault Diagnosis in Fault Propagation Graph Systems. *IEEE Transactions on Parallel and Distributed Systems*, 7(12):1217–1223, December 1996.

[16] J. Kleer and B.C. Williams. *Diagnosing multiple faults-Artificial Intelligence*, volume 32. Elsevier Science Publishers, 1987.

[17] Y. Y. Yang and R. Sankar. Automatic failure isolation and reconfiguration. *IEEE Network*, pages 44–53, September 1993.

[18] I. Katzela and M. Schwartz. Schemes for Fault Identification in Communication Networks. *IEEE/ACM Transactions on Networking*, 3(6), December 1995.

[19] M. Medard, S. R. Chinn, and P. Saengudomlert. Node wrappers for QoS monitoring in transparent optical nodes. *Journal of High Speed Networks*, 10:247–268, 2001.

[20] R. Habel, K. Roberts, A. Solheim, and J. Harley. Optical domain performance monitoring, *Proc. IEEE Optical Fiber Communication conference (OFC)*, (Baltimore, MD), pp. 174–175, 2000.

[21] K. J. Park S. K. Shin and Y. C. Chung. A novel optical signal-to-noise ratio monitoring technique for WDM networks, *Proc. IEEE Optical Fiber Communication conference (OFC)*, (Baltimore, MD), pp. 182–184, 2000.

[22] K. Mueller et al. Application of amplitude histograms for quality of service measurements of optical channels and fault identification, *Proc. ECOC 1999*, (Nice, France), Sept. 1999. Madrid, 1998.

[23] I. Shake, H. Takara, S. Kawanishi, and Y. Yamabayashi. Optical signal quality monitoring method based on optical sampling. *Electronic Letters*, 34(22):2152, October 1998.

[24] S. Ohteru and N. Takachio. Optical signal quality monitor using direct Q-factor measurement. *IEEE Photonics Technology Letters*, 11:1307, 1999.

[25] R. Wiesmann, O. Bleck, and H. Heppne. Cost-effective performance monitoring in WDM systems, *Proc. IEEE Optical Fiber Communication conference (OFC)* (Baltimore, MD), pp. 171–173, 2000.

[26] N. Hanik, A. Gladisch, C. Caspar, and B. Strebel. Application of amplitude histograms to monitor performance of optical channels, *Electronic Letters*, 35:403, 1999.

[27] Carmen Mas and Ioannis Tomkos. Failure detection for secure optical networks. *International Conference on Transparent Optical Networks*, (Warsaw, Poland), Volume 1, pp.70–75, June 2003.

[28] Carmen Mas, Ioannis Tomkos, and Ozan Tonguz. Optical Networks Security: A Failure Management Framework. *Information technologies and Communications ITCom 2003*, pp. 230-241, 2003.

[29] R. D. Carr, S. Doddi, G. Konjevod, and M. Marathe. On the red-blue set cover problem. *Proc. ACM-SIAM Symposium on Discrete Algorithms*, pp. 345–353, 2000.

[30] A.T. Bouloutas, S. Calo, and A. Finkel. Alarm Correlation and Fault Identification in Communication Networks. *IEEE Transactions on Communications*, 42(2/3/4):523–533, Feb/March/April 1994.

VI

# TESTBEDS

Chapter 17

# A MULTI-LAYER SWITCHED GMPLS OPTICAL NETWORK

Aihua Guo, Zhonghua Zhu and Yung J. (Ray) Chen
*University of Maryland, Baltimore County, Baltimore, MD 21250*
Email: { aihguo1, zzhu1, ychen } @umbc.edu

**Abstract** In this chapter, a new multi-layer switched optical network scheme is proposed based on the GMPLS paradigm. In the new scheme, we introduce an integrated reconfigurable electronic- optical switching layer, that enables traffic grooming and wavelength conversion in the core region of the GMPLS network. Benefits of the proposed network architecture are analyzed and a brief introduction of the UMBC GMPLS testbed, based on this scheme, follows.

**Keywords:** Generalized MPLS, routing, QoS, traffic grooming, label switching.

## 17.1 Introduction

The Internet traffic has been increasing rapidly in recent years. In order to satisfy the surging traffic demands, large amount of bandwidth has been deployed utilizing wavelength-division multiplexing (WDM) technology. In a WDM network, multiple signals can be carried over a single fiber via different wavelengths, therefore extending the capacity of fiber links. The initial employment of WDM technology is point-to-point, where WDM links are established as high-capacity transport channels. As optical network evolves, the concept of wavelength switched optical network has emerged as an effort to introduce networking functionality to the optical transport plane. One of such concepts is generalized-multiple protocols label switching (GMPLS), or MP$\lambda$S. GMPLS extends the MPLS concept from packet- switching capable interfaces to non-packet-switching capable interfaces, such $\lambda$ interfaces.

Although GMPLS paradigm contains various network scenarios, for the sake of simplicity here we only consider a two-layer GMPLS network case: consisting of an IP/MPLS layer, whose network elements are label switching routers (LSRs), and an optical layer, whose network elements are optical cross-connects (OXCs). The network nodes consist of two main components in this scenario: the OXC and the LSR. A typical architecture of the two-layer GMPLS network as well as a depiction of the core node structure is presented in Figure 17.1. In such networks the IP/MPLS routers are inter-connected through a set of lightpaths crossing multiple underlying OXCs and fiber spans. Incoming traffic flows are groomed at ingress LSRs and then transmitted all-optically (through designated lightpaths) to the destinations. In this process, traffic engineering (TE) is used to manage the establishment, maintenance, teardown of the lightpaths, and grooming of traffic flows to optimize resource usage and to keep the network performance above a certain level.

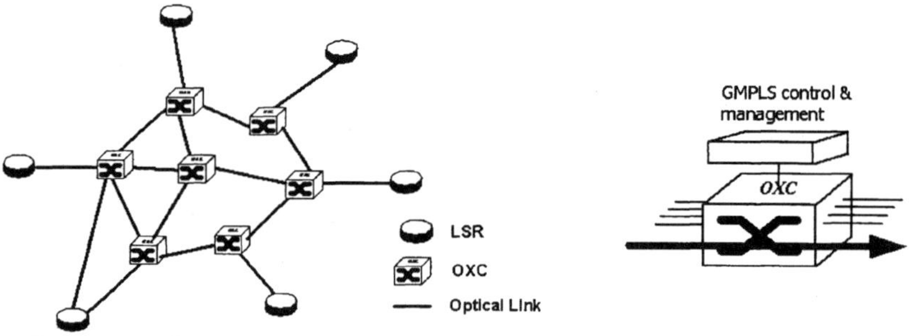


*Figure 17.1.* GMPLS network architecture and core node structure.

Traffics carried on the lightpaths in this scheme have high end-to-end signal and service qualities, e.g., low delay and low jitter. Because of the elimination of O/E conversions at intermediate core nodes, the costs of these IP based networks are expected to be lower than the traditional SONET/SDH-over-WDM networks. However, due to the nature of all-optical routing, certain constraints are imposed against the network operations and, as a result, can hinder the efficiency of the network. One typical constraint is the wavelength continuity constraint, in which two lightpaths sharing a common fiber link cannot be assigned to the same wavelength. This can result to a wavelength blocking. A partial solution to this problem is to introduce wavelength converters to the network. Using a wavelength converter can improve the connectivity at the node it is present. However the cost would be prohibitively high if wavelength converters are placed on every node for every wavelength channel. In addition, normally a wavelength converter, due to technology limitation, does not have full wavelength coverage of the network, but operates only at a subset of wavelengths. The elimination of electronic layer involvement at core nodes also results in the removal of electronic buffers, making the network less flexible to accommodate traffic flows especially in case of congestion and optical path reconfiguration.

Moreover, in real networks not all traffic flows require the same level of quality of service (QoS). If in the core optical region only one unified lightpath service is provided to all types of traffic requests, differentiated QoS cannot be offered in the core region.

In order to improve the network efficiency, it is therefore desirable for GMPLS networks to have the capability of provisioning more fine-grained services to different traffic flows. More specifically, the GMPLS core network region should be capable of performing sub-wavelength traffic grooming, wavelength conversion, and differentiated QoS. If routers were used at core in the traditional fashion to handle all incoming traffics, even with GMPLS cut-through, it will call for ultra-high capacity high-performance routers at each core node (due to the high volume of traffic in a DWDM network). The line card for each router channel is also expensive, due to the complexity and high performance design of the router backbone. The challenging issue is how to provide differentiated QoS and achieve efficient resource utilization while preserving the cost-effectiveness of all-optical networks.

## 17.2 Proposed Network Architecture

Our approach to this problem is to utilize an integrated optical-electronic switching layer at core nodes in GMPLS networks. In fact, the concept of electronic switching has been well established by MPLS, in which packets are forwarded at intermediate nodes by simply looking up a fixed electronic "label". Networks, such as MPLS, that utilize this type of electronic switching can achieve higher performance than traditional store-and-forwarding networks, like traditional IP networks. The same concept can be applied to GMPLS. In our proposed network architecture, an integrated optical-electronic switching layer is introduced to core network nodes, as shown in Figure 17.2. The electronic switch is capable of executing both layer-2 switching and layer-3 store-and-forwarding functions. Differentiated QoS is provided by executing different forwarding schemes at core nodes, including all-optical bypass, layer-2 switching, and layer-3 store-and-forwarding. Each forwarding scheme corresponds to a specified type of QoS. The all-optical bypass offers the highest QoS with negligible delay and jitter, while the layer-2 switching scheme offers bounded QoS with predicted delay and controlled packet dropping rate. Traditional best-effort traffic can be routed utilizing the layer-3 store-and-forwarding scheme. In times of congestion these best-effort traffics can be shaped or reshaped according to TE arrangements and their routing paths adjusted dynamically. Flexible sub-wavelength traffic grooming at the core region is also achievable by directing traffics to electronic switches. At the same time wavelength conversion can be performed at electronic switches to avoid/reduce wavelength blocking. This greatly increases the routing efficiency of the network.

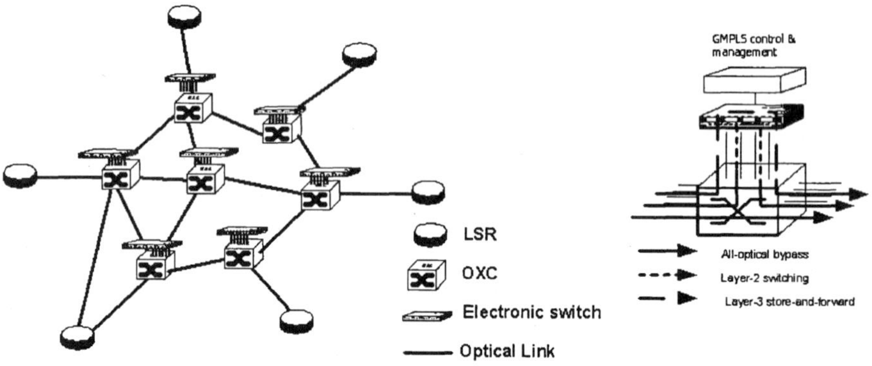


*Figure 17.2.* Proposed network architecture and core node structure.

The GMPLS "label" in our proposed system includes not only the wavelength (λ) and MPLS label, but also the layer-2 label, as illustrated in Figure 17.3. In this chapter we use the MAC address as a typical representation of layer-2 labels. It should be noted that the MAC label already exists in all GMPLS packets. In a traditional GMPLS system, the layer-2 is treated as a passive layer therefore the MAC label is not used for routing purposes.

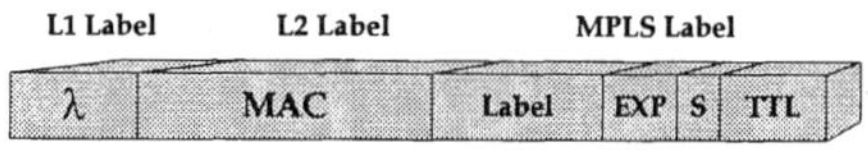


*Figure 17.3.* Proposed label structure.

Noted that in our proposed scheme, the optical and electronic switching layers are integrated as one. It is executed through the control of wavelength "label": an OXC directs (switches) the incoming optical signal either to the following OXC, ala all-optical switching, or to a local electronic switch port and thus allowing the incoming data packets to be processed by the electronic layer: to be groomed and, if needed, traffic to be directed to a new (wavelength) lightpath (layer-2 switching), or stored at the buffer and forwarded to a lightpath when bandwidth is available, based on the electronic "label" assigned through the TE process (layer-3 store-and-forwarding).

A salient feature of the proposed node structure is its ability to seamlessly execute path reconfiguration within the node when the forwarding scheme is reconfigured from layer-2 switching to layer-3 store-and-forwarding or vice versa. This is accomplished using the buffering capability of electronic switches, a feature that is not presented for all-optical networks.

The cost issue of such proposed network is not a show-stopper either. Although the cost will increase due to additional electronic switches and correspondent O/E transceivers/transponders, the increase is not formidable. This is because the layer-2 electronic switching function can be carried out by inexpensive legacy Ethernet switches, which are mass-produced and widely deployed in existing networks. Many current Ethernet switches are equipped with well-designed QoS functions and limited amount of buffer, so that layer-3 store-and-forwarding functions can also be built into the same switches. Furthermore, the OXC only direct the optical channels that require electronic processing to the local node, the line cards are shared in the network and the traffic to each node is controlled to a manageable size. Thus monster routers are not needed for most core nodes.

In summary, the key ideas of our proposed scheme are listed below:

- Implements an integrated reconfigurable optical-electronic switching layer; the management of switching layer is based on QoS-TE.
- Enables traffic grooming and O/E/O wavelength conversion in the core region.
- Wavelength information and MAC addresses are utilized collectively to perform the GMPLS "label" functions. It is fully compatible with the current IP packet formats.

## 17.3 Network Traffic and Performance Analysis

To analyze the benefits of our proposed network architecture, we have carried out the following simulations to compare the performance of an all-optical network and our proposed network, termed core O-E switching network. Although our network can support QoS, we did not include the QoS constrain in the network traffic engineering process, since QoS is not considered in traditional GMPLS all-optical routing. To make the proper comparison, we set the traffic type to be Internet best-effort traffic only. The rest of the system considerations are as follows:

- Network topology: we used a 32-node NSFNET topology; each link in the topology has 32 wavelength channels with a uniform line rate of 100Mbps per wavelength (noted that the line rate is not relevant to the conclusion of the simulations).
- Wavelength conversion: We assumed no all-optical wavelength conversion in the all-optical network and full wavelength conversion using electronic switch in core O-E switching network.
- Routing and wavelength assignment (RWA) algorithm: we used adaptive Dijkstra minimum-hop SPF and First-Fit (FF) wavelength assignment algorithm for all-optical networks, and Dijkstra minimum-hop SPF for core O-E switching networks.
- Traffic model: we assumed a Poisson distribution with mean value $\lambda$ for traffic inter-arrival rate and an exponential distribution with mean value $\mu$ for traffic holding time. The source, destination of individual traffic flow is uniformly distributed. This traffic model is appropriate for our case because we are doing flow-based traffic simulation and assuming that a resource reservation scheme is used.
- Requested bandwidth: We assumed each traffic flow request had the same unit bandwidth. We used bandwidth units (BW_Unit) varying from 2Mbps to 100Mbps in order to compare the performance under different traffic granularities.

We also applied the following metrics to measure the performance of the two networking schemes:

- **Normalized traffic load $T_L$** = $\lambda_L$ x (L/BW_Unit) x $\mu$,where L is the line rate (constant for all wavelengths), and $\lambda_L$ is the mean inter-arrival rate with bandwidth unit L.
- **Requested network capacity** = $\dfrac{\sum_{i,j} \frac{T_L}{N-1} \times Min\_Hop_{i,j} \times L}{\sum_{i,j,wl} L}$

  The requested network capacity is defined as a value normalized by the maximum capacity of the network. Figure 17.4 shows the relationship between total requested network capacity and normalized traffic load. It gives a correlation of the value of normalized traffic load to the actual network traffic load. Accordingly, we term that for values of normalized traffic load between (0,25) the network traffic load is "normal", while for values above 25 the network traffic load is "heavy".
- **Blocking probability** = $\dfrac{\sum_{i,j} Blocked\_requests_{i,j}}{\sum_{i,j} Total\_requests_{i,j}}$

  Blocking probability represents the percentage of blocked traffic over the bandwidth requirement of all traffic requests during the simulation.

- **Weighted blocking probability =**

$$\frac{\sum_{i,j} Blocked\_requests_{i,j} \times Min\_hop\_distance_{i,j}}{\sum_{i,j} Total\_requests_{i,j} \times Min\_hop\_distance_{i,j}}$$

Weighted blocking probability measures the lost revenue over all potential revenue during the simulation period. Each request carries revenue that is proportional to its source-destination minimum hop distance.

- **Hop-dependent blocking probability** =

$$\frac{\sum_{i,j,H} Blocked\_requests_{i,j,H}}{\sum_{i,j,H} Total\_requests_{i,j,H}}$$

Hop-dependent blocking probability calculates the blocking probability with respect to source-destination hop distances.

- **Average path hop =**

$$\frac{\sum_{i,j,H} Successful\_Request_{i,j,H} \times Path\_Hops_{i,j,H}}{H \times \sum_{i,j,H} Total\_Successful\_Requests_{i,j,H}}$$

Average path hop calculates the average number of hops for successfully allocated traffic requests and categorizes them according to their source-destination hop distances.

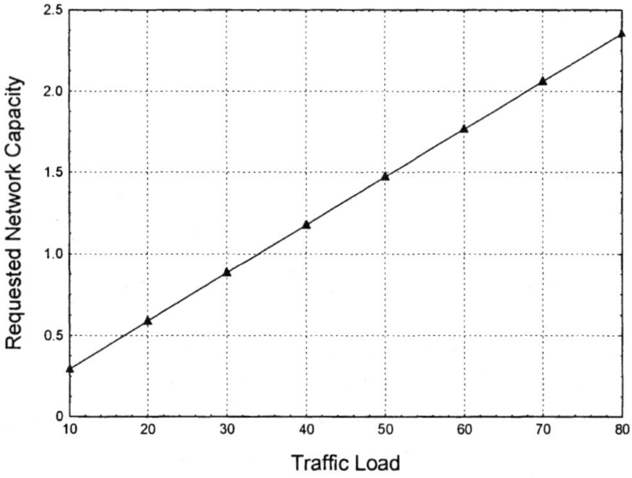


*Figure 17.4.* Requested network capacity vs. traffic load.

Figure 17.5a and 5b show the blocking probability vs. offered traffic load under different traffic granularities, and Figure 17.6a and 6b show the corresponding weighted blocking probability under the two network scenarios. We note that under normal traffic load condition (0~25), our proposed core O-E switching network that is equipped with electronic switches at the core nodes outperforms an all-optical network equipped with only OXCs, in both traffic blocking ratio and lost revenue comparisons. When traffic granularity becomes finer, the all-optical network becomes less efficient (more blocking). This is because traffic grooming can only be carried out at ingress routers and not at the core nodes. On the other hand, the core O-E switching network becomes more efficient, as the light-pipe can be more efficiently utilized by grooming at core nodes - the finer the traffic granularity is the more efficient grooming at the core becomes. This is consistent with the current "filling the fat light pipe at the edge" concept for all-optical networks. *The ability to fill the fat pipe at the core nodes makes it even better.*

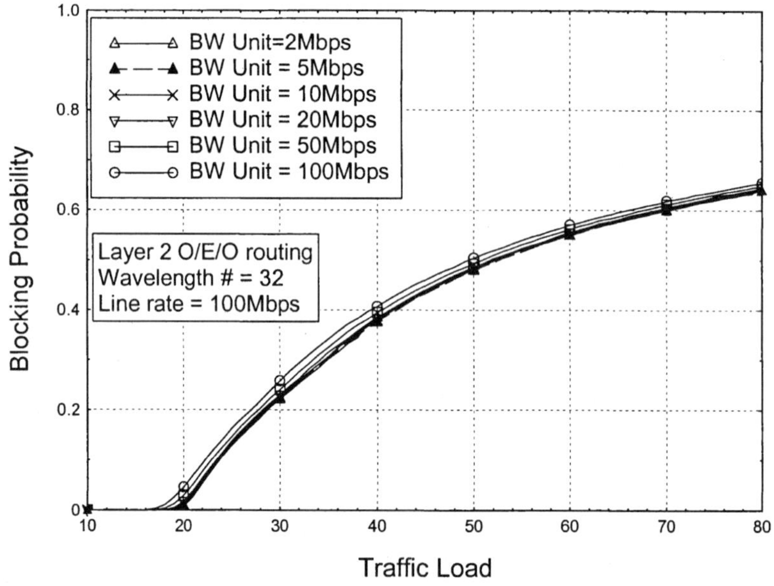


*Figure 17.5a.* Blocking probability vs. traffic load , layer 2 O/E/O.

We should point out that when the bandwidth unit equals to the line rate, in our case 100Mpbs, the curve in our core O-E switching network is also corresponding to the all-optical network case with wavelength conversion capability at the core nodes. Since each request occupies the whole lightpath, the OEO wavelength conversion function at the core nodes is equivalent to an all-optical wavelength conversion. Under this condition, the all-optical network shows some improvement over the corresponding all-optical case without wavelength conversion. However it is still inferior to the finer granularity cases in our core O-E switching network.

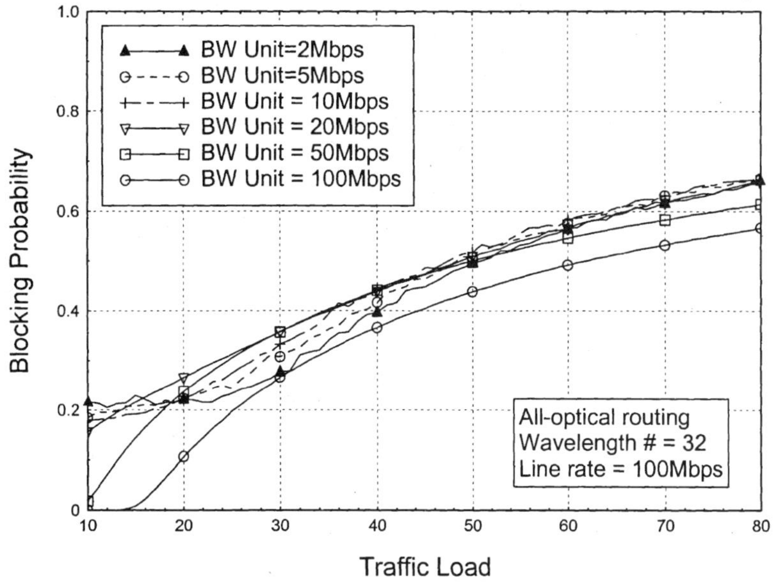


*Figure 17.5b.* Blocking probability vs. traffic load , all-optical routing.

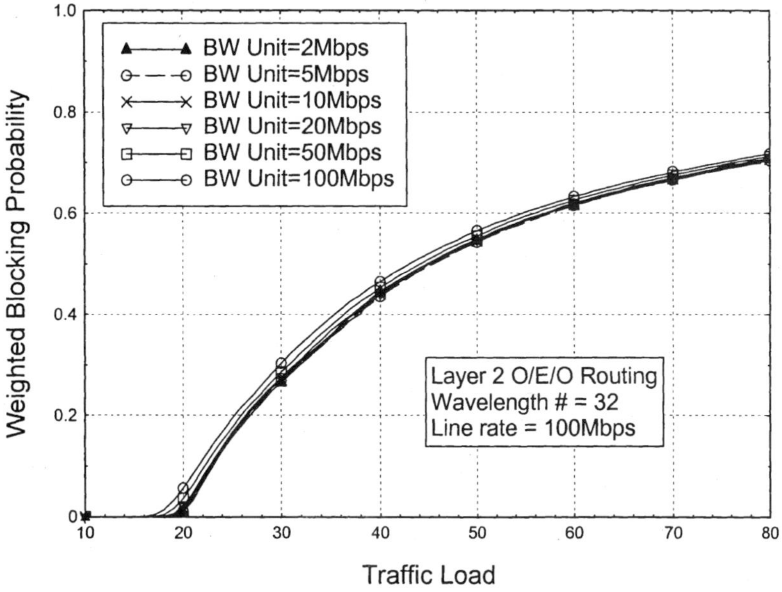


*Figure 17.6a.* Weighted blocking probability vs. traffic load, layer 2 O/E/O.

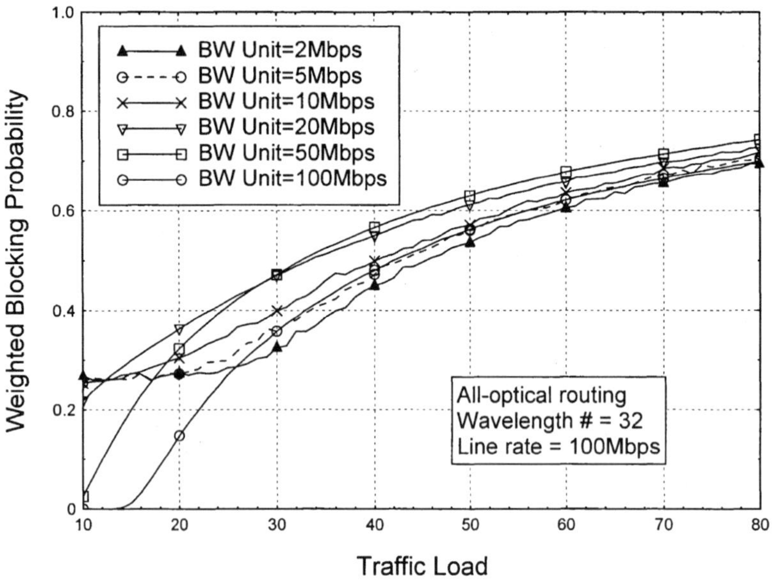


*Figure 17.6b.* Weighted blocking probability vs. traffic load, all-optical routing.

Under high traffic load (25 and up), we observe that by allowing unlimited hop distance that a path can traverse, the blocking probability in a core O-E switching network tends to increase faster than in the all-optical networks. Figures 17.7 and 17.8 illustrate this effect from another angle where average path hops are observed. It shows that the average path-hop in all-optical networks stays at a constant level insensitive to traffic load changes, while average path-hop scales with traffic load in core O-E switching networks. Average paths become longer under higher traffic load. Paths with longer hops occupy more resource and thus leave less resource for future traffic requests, resulting in a high traffic blocking rate. This suggests that one should limit hop distance of path requests when doing calculation in traffic engineering. To address this traffic path engineering issue, we devise a traffic engineering rule imposing dynamical restriction on path hops according to traffic load: for low traffic load (below 20), the maximum hop path allowed is the actual hop number required by the connection plus 2, for normal traffic load (20-25), the maximum hop path allowed is the actual hop number plus 1, and for heavy traffic load (above 25), the maximum hop path allowed is the actual hop number. Under the dynamic path restriction rule we can achieve a even better network performance, i.e., reduced blocking probability, reduced lost revenue, and shorter paths, as shown in Figure 17.9 and Figure 17.10, respectively. Restriction of path hops is also essential when considering traffic engineering with QoS constraints, where longer paths with more O/E/O conversion may incur accumulated high delay and jitter that exceed requested bounds.

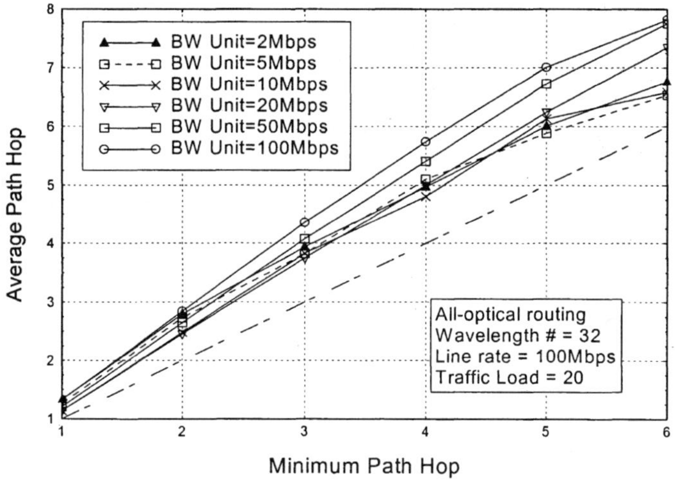


*Figure 17.7a.* Average path hop vs. minimum path hop, traffic load = 20, all-optical routing.

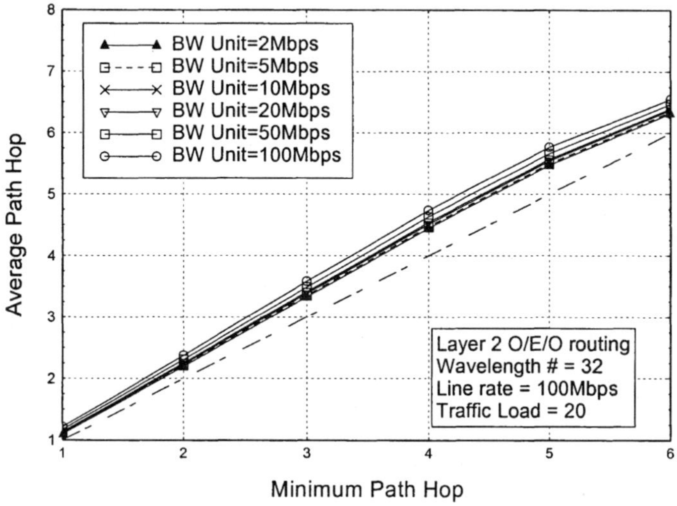


*Figure 17.7b.* Average path hop vs. minimum path hop, traffic load = 20, layer 2 O/E/O.

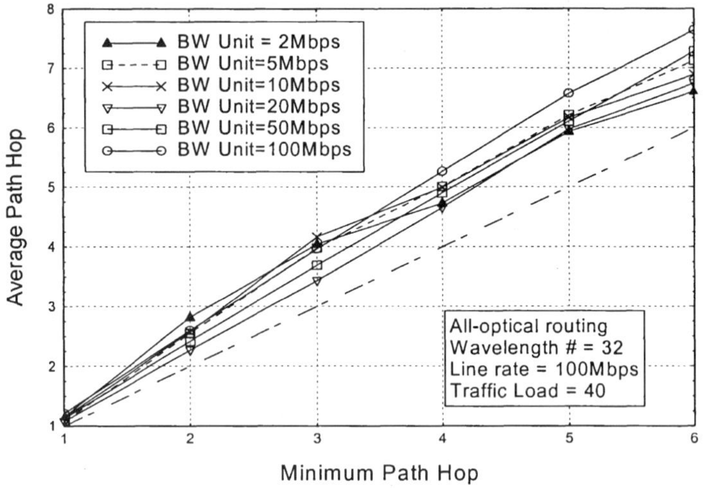


*Figure 17.8a.* Average path hop vs. minimum path hop, traffic load = 40, all-optical routing.

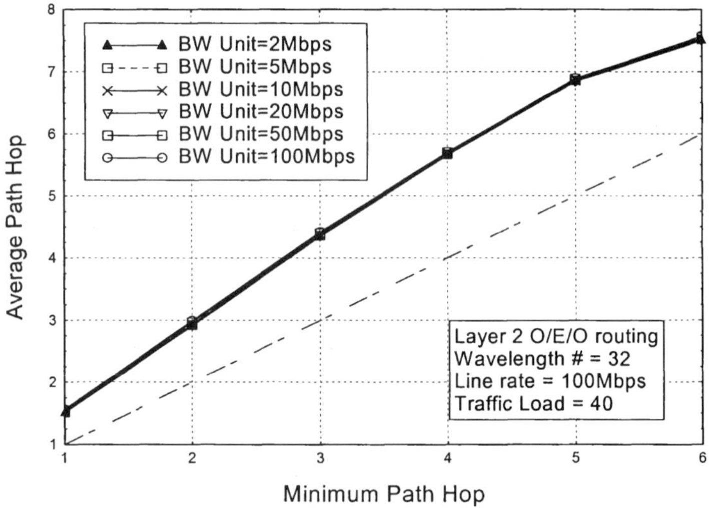


*Figure 17.8b.* Average path hop vs. minimum path hop, traffic load = 40, layer 2 O/E/O.

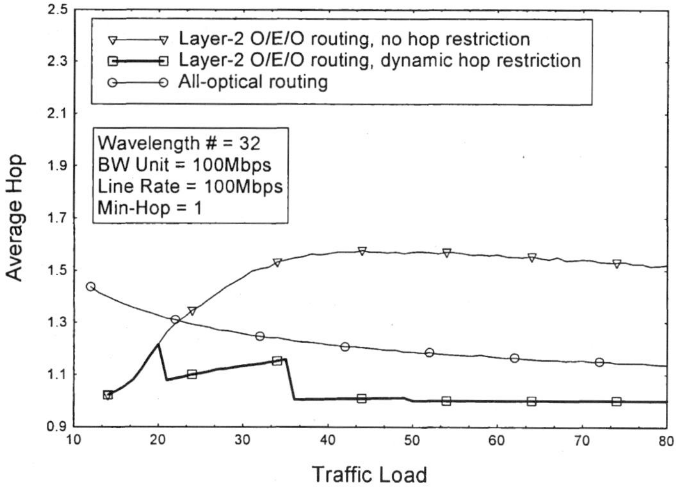


*Figure 17.9.* Improvement of average path hops with dynamic hop restriction. Type of traffic : min-hop = 1.

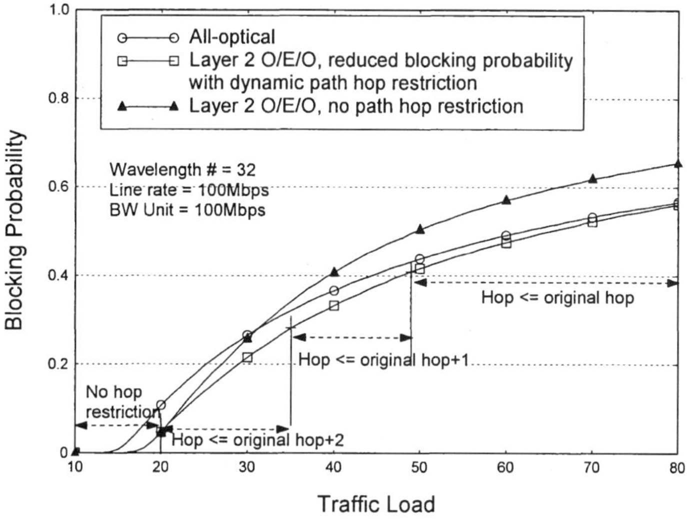


*Figure 17.10a.* Blocking probability vs. offered load, BW_Unit=100Mbps.

In another simulation, we studied the hop-distance-dependent blocking of the two network scenarios. In Figure 17.11, the hop-distance-dependent blocking probability vs. traffic load is shown. We observe that the blocking probability for longer hop distance requests is much higher than that of shorter hop distance requests in all-optical networks, while there is not much difference in core O-E switching networks. This is because, in all-optical networks, it is difficult for a long-hop distance request to successfully find a route due to high wavelength blocking. While in core O-E switching networks, the grooming at the core nodes eases this problem considerably. Hence our proposed scheme can achieve greater "fairness" with respect to the hop distance of traffic requests.

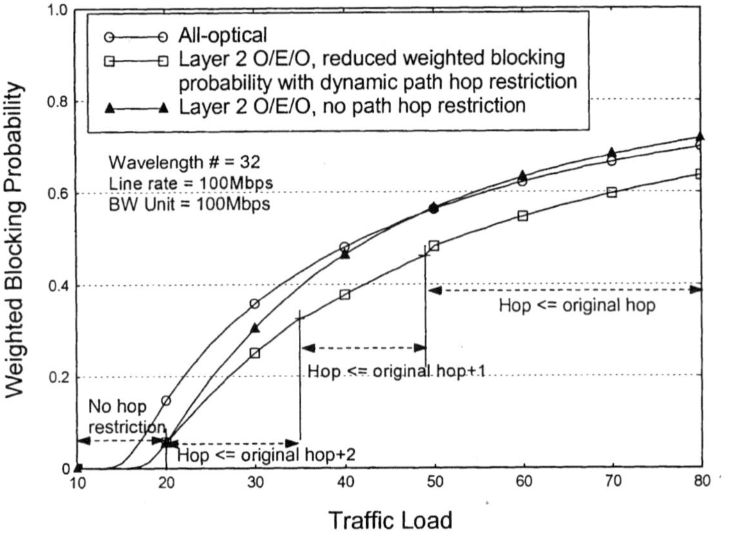


*Figure 17.10b.* Weighted locking probability vs. offered load, BW_Unit=100Mbps.

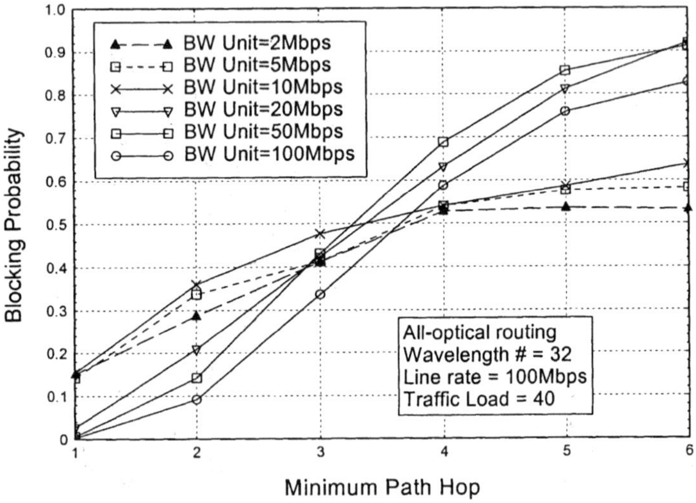


*Figure 17.11a.* Hop-dependent blocking probability, traffic load = 40, all-optical routing.

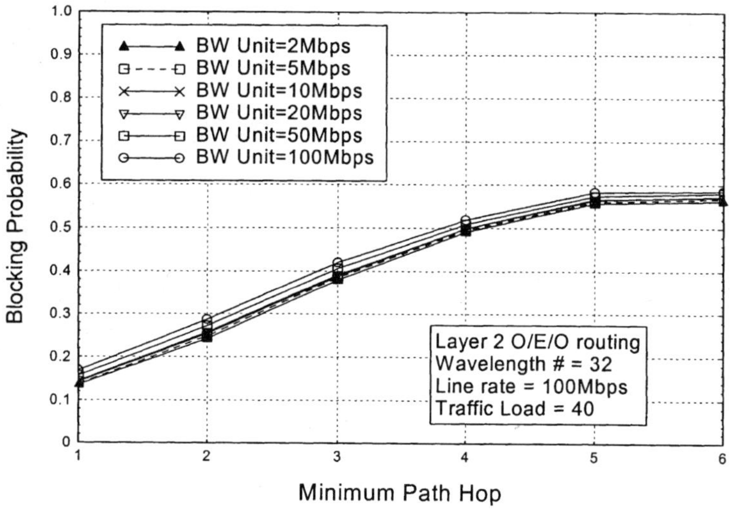


*Figure 17.11b.* Hop-dependent blocking probability, traffic load = 40, layer 2 O/E/O.

## 17.4 The UMBC GMPLS Testbed

In order to implement our core O-E switching concept in a real world environment, we have built a GMPLS testbed based on the proposed network architecture, as illustrated in Figure 17.12. The testbed consists of 3 core nodes and 6 edge nodes. Currently, the testbed is equipped with four operational wavelengths at a line rate of up to 1Gbps. It can be readily scaled to 16 or more wavelengths. Each core node is equipped with a 4x4 OXC for *each* operating wavelength. The 4x4 OXC connects the core node to the two adjacent core nodes and 2 edge nodes. The OXC is a MEMS-based device made by OMM that has built-in local add-drop capability, which allows us to provide all three switching schemes at each core node. The layer 2 switches are legacy Gigabit Ethernet switches by Dell that have build-in buffer and support QoS. The routers are based on standard high performance PCs, running a GMPLS network operating system developed by our group, based on NIST Free BSD MPLS engine [27]. QoS Traffic engineering software is developed in-house for our operating system as well as the control of various network elements (OXCs and Ethernet switches). A separate signaling channel is established to interchange control and management information among network nodes. The signaling channel is based on a legacy Ethernet network (the campus network).

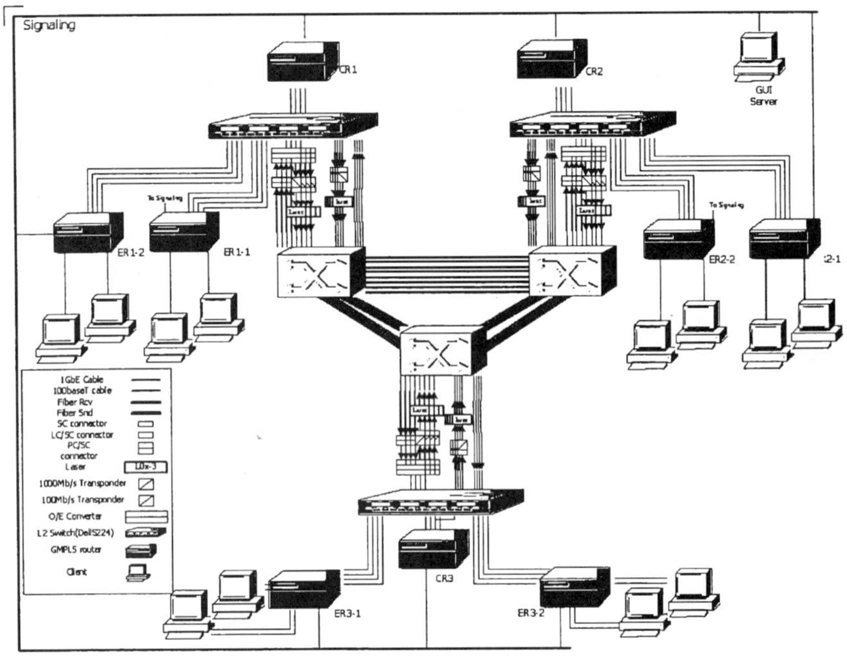


*Figure 17.12.* The UMBC GMPLS testbed.

To implement differentiated QoS, data traffic in our network is classified into three categories: premium, preferred, and best-effort, in accordance with current Differentiated Service (DiffServ) specifications. The premium traffic has fixed bandwidth, bounded delay and no packet drop, and is served with the highest priority in the testbed. The testbed will try the best effort to provide service to its request, including reconfiguration of lower priority traffic paths, and will guarantee its QoS conditions. The preferred traffic also allows bounded delay, but its requested bandwidth has a finite dropping rate: its traffic is divided into a premium portion that must be satisfied (like the premium service) and a best-effort portion that may be dropped when congestion occurs. The best-effort traffic has no specific QoS requirements and is implemented just as current Internet best-effort services do. Premium and preferred services are typically assigned to either all-optical bypass or layer-2-O/E/O paths, both of which have guaranteed QoS.

As mentioned earlier, we utilize MAC address as part of the electronic "label" so that the legacy Ethernet routers can be utilized in the testbed. Since MAC addresses do not carry any QoS information, using only MAC address would result in a coarse QoS granularity of one QoS per MAC address. Currently the more advanced Ethernet switches, deployed in our system, are capable of utilizing additional DiffServ bits in IP header to support QoS. Therefore, we have extended the MAC "label" to a modified version that includes not only MAC address but also additional DiffServ bits. The MAC address label cannot be swapped (label-swapping) at the Ethernet switch, due to the hardware limitation of our switch. However label swapping can be carried out by directing traffic to layer 2.5/layer 3. We can perform label swapping function by directing the packets to the PC based routers.

Each class of traffic has different processing disciplines. When a new client request comes through the User-network-interface (UNI), the traffic engineering process first determines the class of the new request and then executes route calculation. If a feasible route is successfully found, the traffic engineering procedure then tries to establish the route through the signal protocol, such as RSVP-TE [7]. The network status, including the status of network nodes, links and information databases, will be updated if the route is successfully established by the signaling protocol. The client request can be either rejected or rerouted according to carrier's policy if traffic engineering is not able to find a route for it or there are errors during the route establishment. Figure 17.13 shows an illustrative example of how traffic is routed and signaled in the traffic engineering process.

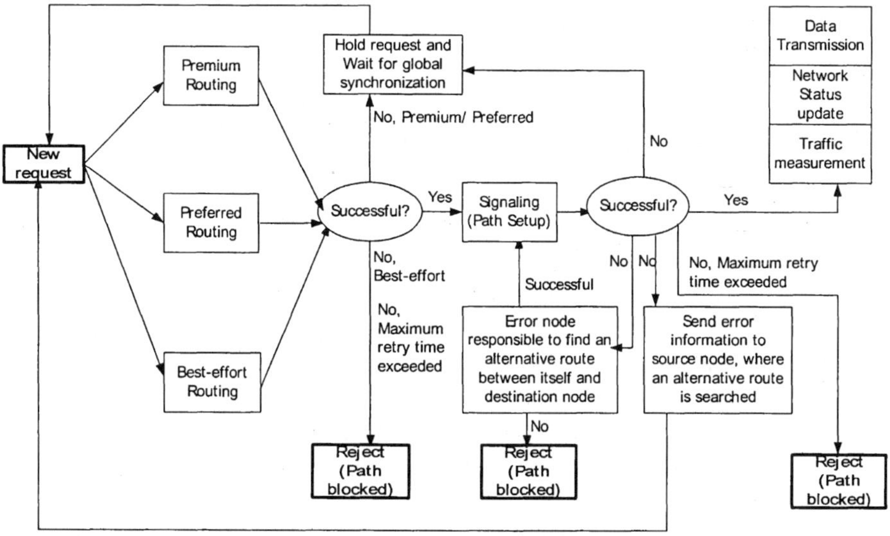


*Figure 17.13.* Illustration of routing and signaling process.

## 17.5 Conclusion

In this chapter we proposed a new scheme based on the GMPLS paradigm, in which we introduce an integrated reconfigurable electronic-optical switching layer, thus enabling traffic grooming and wavelength conversion in the core region of the GMPLS network. Benefits of the proposed network architecture are analyzed and a brief introduction of the UMBC GMPLS testbed follows.

## References

[1] D. Awduche *et al.*, "Overview and Principle of Internet Traffic Engineering," RFC 3272, May 2002.

[2] D. Awduche *et al.*, "Requirements for Traffric Engineering over MPLS," RFC 2702, Sept. 1999.

[3] E. Mannie *et al.*, "Generalized Multi-Protocol Label Switching (GMPLS) Architecture", draft-ietf-ccamp-gmpls-architecture-07.txt, Aug. 2002.

[4] S. Blake *et al.*, "An Architecture for Differentiated Services", RFC 2475, Dec. 1998.

[5] User Network Interface (UNI) 1.0 Proposal, Optical Internetworking Forum, July 2000.

[6] D. Katz, *et al.*, "Traffic Engineering Extensions to OSPF Version 2", Internet Draft, draft-katz-yeung-ospf-traffic-07.txt, August 2002.

[7] L. Berger, "Generalized MPLS Signaling – RSVP-TE Extensions", Internet Drafts, draft-ietf-mpls-generalized-rsvp-te-08.txt, August 2002.

[8] B. Rajagopalan *et al.*, "IP over Optical Networks: Architectural Aspects," IEEE Commun. Mag., vol. 38, no. 9, Sept. 2000, pp. 94–102

[9] Resilient Packet Ring, IEEE 802.17 Working Group.

[10] Optical Ethernet Feature Issue, ed. by Cedric F. Lam and Danny H. K. Tsang, OSA Journal of Optical Networking vol.1, 8/9 2002.

[11] A. Guo, Z. Zhu *et al.*, "Applying GMPLS On Optical Ethernet: A Hybrid Approach", Proceeding of OFC 2003.

[12] A. Guo, Z. Zhu *et al.*, "A Novel Packet-loss Free Reconfiguration Scheme for GMPLS Optical Networks", Proceeding of APOC 2002.

[13] K. Zhu *et al.*, "Traffic Engineering in Multigranularity Heterogeneous Optical WDM Mesh Networks through Dynamic Traffic Grooming", IEEE Network, Mar./Apr. 2003.

[14] T. Cinkler *et al.*, "Traffic and $\lambda$ Grooming", IEEE Network, Mar./Apr. 2003.

[15] E. Modiano and P. J. Lin, "Traffic Grooming in WDM Networks," IEEE Communication Magazine, Vol. 39, No. 7, July 2001, pp. 124–29.

[16] C. Xin *et al.*, "On an IP-centric Optical Control Plane," IEEE Communication Magazine, Vol. 39, No. 9, Sept. 2001, pp. 88–93.

[17] G. Papadimitriou *et al.*, "Optical Switching: Switch Fabrics, Techniques, and Architectures", IEEE J. of Lightwave Technology, Vol. 21, No. 2, Feb. 2003.

[18] I. Chlamtac *et al.*, "Lightpath (Wavelength) Routing in Large WDM Networks", IEEE. J. on Selected Areas in Communications, Vol. 14, No. 5, June 1996.

[19] M. Kodialam *et al.*, "Minimum Interference Routing with Applications to MPLS Traffic Engineering", Proceeding of IEEE INFOCOM 2000.

[20] G. Li *et al.*, "Control Plane Design for Reliable Optical Networks", IEEE Communication Magazine, Feb. 2002.

[21] S. Sengupta *et al.*, "From Network Design to Dynamic Provisioning and Restoration in Optical Cross-Connect Mesh Networks: An Architectural and Algorithmic Overview", IEEE Network, July/Aug. 2001.

[22] C. Assi *et al.*, "Optical Networking and Real-Time Provisioning : An Integrated Vision for the Next Generation Internet", IEEE Network, July/Aug. 2001.

[23] R. Dutta *et al.*, "Traffic Grooming in WDM Networks: Past and Future", IEEE Network, Nov./Dec. 2002.

[24] B. Mukherjee *et al.*, "Some principles for designing a wide-area WDM optical network," IEEE/ACM Trans. Networking, vol. 4, pp. 684–696, Oct. 1996.

[25] I. Chlamtac, A. Ganz, and G. Karmi, "Lightpath Communications: an Approach to High Bandwidth Optical WAN's," IEEE Trans. Commun., Vol. 40, July 1992, pp. 1171–82.

[26] H. Zang *et al.*, "A Review of Routing and Wavelength Assignment Approaches for Wavelength-Routed optical WDM Networks", SPIE Optical Network Magazine, vol. 1, No. 1, Jan 2002.

[27] Mark Carson, et al., "NIST Switch: A Platform for Research on Quality of Service Routing", SPIE Symposium on Voice, Video, and Data Communications, 1998.

[28] R. Stevens, "TCP-IP Illustrated", Addison-Wesley, 1994.

Chapter 18

# HORNET: A PACKET SWITCHED WDM METROPOLITAN NETWORK

Kapil Shrikhande[1], Ian White[1,2], Matt Rogge[1] and Leonid G. Kazovsky[1]

[1] *Photonics and Networking Research Laboratory, Stanford University, Stanford, CA 94305*
[2] *Sprint Advanced Technology Laboratories (ATL), Burlingame, CA*

Email: kapils@stanford.edu, iwhite@sprintlabs.com, mrogge@stanford.edu, kazovsky@stanford.edu

**Abstract** HORNET is a metro network designed to better face the challenges offered by next generation metropolitan area networks. HORNET uses a packet-over-WDM transport on a bi-directional ring network. Fast-tunable packet transmitters and wavelength routing enable it to scale cost-effectively to ultra-high capacities. A control-channel-based MAC protocol enables the nodes to share the bandwidth of the network while preventing collisions. The MAC protocol is designed to transport variable-sized IP packets and guaranteed bit-rate circuits while maintaining fairness in the network. Survivability protocols protect circuit and best effort traffic against a fiber cut or node failure. A burst mode receiver, in conjunction with the tunable transmitter and constant gain optical amplifiers make up the essential sub-systems inside a node. This article summarizes the accomplishments of the HORNET project, including the design, analysis, and demonstration of the architecture, protocols, sub-systems and testbed. As this work shows, the HORNET architecture is an excellent candidate for next-generation high-capacity metro networks.

**Keywords:** Metropolitan Area Networks, Tunable Lasers, Media Access Control, Ring Networks.

## 18.1 Introduction

The past five years have seen a significant deployment of high-capacity photonic systems in backbone networks. The increasing availability of high-speed access via DSL, cable, and wireless technologies, will continue to fuel the growth in traffic. It is well understood that the next frontier for photonic networks and technologies will be in the metro/access space. Besides increasing traffic, a noticeable trend has been the dominance of data, i.e. IP traffic (as opposed to voice); while future applications such as file-sharing may contribute to changing traffic patterns. For example, we may see a need for *any-to-any* communication in the metro instead of many-to-one (nodes talking to one another in addition to the Point of Presence or POP).

This dominance of *data-traffic* has been one of the prime motivating factors for the emergence of packet-switched solutions for ring topologies. Industry has proposed RPR (Resilient Packet Ring, IEEE 802.17) [1], a survivable packet-ring in which nodes uses a store-and-forward data-link layer on top of a point-to-point physical layer. The use of store-and-forward techniques allows nodes to tailor their bandwidth usage more dynamically (by simply transmitting more or less packets), drop packets during congestion, and in general benefit from statistical multiplexing. Some research groups have taken the packet-switching trend one step further by proposing architectures that switch small data-units (like packets) at the physical layer in addition to the data-link layer, in order to achieve added benefits and/or functionality. HORNET [2], amongst others [3, 4], is a prime example of the above trend. HORNET enables direct communication between nodes by using a tunable-transmitter-fixed receiver (TTFR) packet-based transport, eliminating costly regeneration, and creating a more scalable architecture.

In doing so, a number of challenges emerge at both the physical layer and the data-link layer of the architecture. Most of these problems have been investigated by using a combination of theoretical analysis, simulations and implementation. This article summarizes the accomplishments of the HORNET project: the architecture is described in Section 18.2; Section 18.3 describes the design and evaluation of the MAC and survivability protocols; Section 18.4 describes the design and implementation of the HORNET node and sub-systems; the testbed and recent results are presented in Section 18.5; and the work is summarized in Section 18.6.

## 18.2 HORNET Architecture

HORNET uses a 2-fiber ring topology to leverage currently deployed fiber infrastructure. Figure 18.1 shows the node architecture capable of bi-directional transmission. The physical layer is burst-mode, i.e. nodes transmit and receive optical packets and do not maintain a point to point link with their neighbors like in SONET or RPR networks. The node consists of four main functional components: the fast-tunable transmitter, packet receiver, the MAC and the data and control processor. Each node is assigned a unique wavelength. To transmit a packet, the node tunes the fast-tunable transmitter to the destination wavelength and transmits the optical packet. The packets are optically coupled into the ring using a wideband power coupler. At each node,

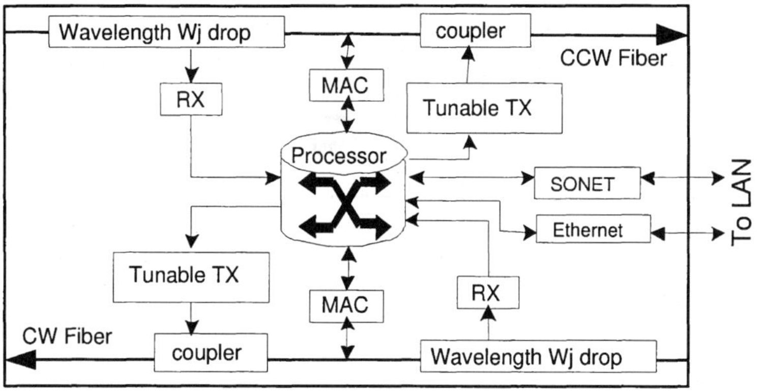


*Figure 18.1.* HORNET architecture.

an optical drop filter drops the assigned wavelength into the packet receiver. Thus, only the packets destined for the node are dropped into the node's receiver. All other wavelengths will pass through the drop filter transparently. The media access control (MAC) is required to avoid collisions on the network. It monitors all the wavelengths passing through the node and finds openings. It controls the transmitter's tuning and data transmission process such that the transmitted packet does not collide with an existing packet on the network. In addition the MAC performs the scheduling of the data queued inside the node. The data and control processor, shown in Figure 18.1 as a switch, handles all the *electrical* logic, processing and queuing functions of the node.

HORNET's packet switched architecture is well suited to handle the data-dominated traffic patterns that the metro will face. Moreover, all nodes can talk *directly* to all other nodes, a logical mesh, eliminating costly regeneration. In other words, the I/O and processing bandwidth of a HORNET node ($Rb_{man}$) is used exclusively to transport its local traffic ($Rb_{lan}$). By contrast, in ADM (add-drop-multiplex) based networks such as SONET and RPR, a large fraction of the node's bandwidth is wasted in regenerating upstream traffic. It can be shown that for an ADM-node, $Rb_{man}$ will have to be large enough to support a fraction of the *aggregate* traffic that the network expects to switch: $Rb_{man} = k * (\#ofnodes) * Rb_{lan}$, while for a HORNET node $Rb_{man} = k * Rb_{lan}$. In both cases, the exact value of *k* depends on a number of factors such as traffic patterns, protection needs, statistical multiplexing gain, etc. This scaling property is one of the *unique* advantages that HORNET offers over ADM-rings.

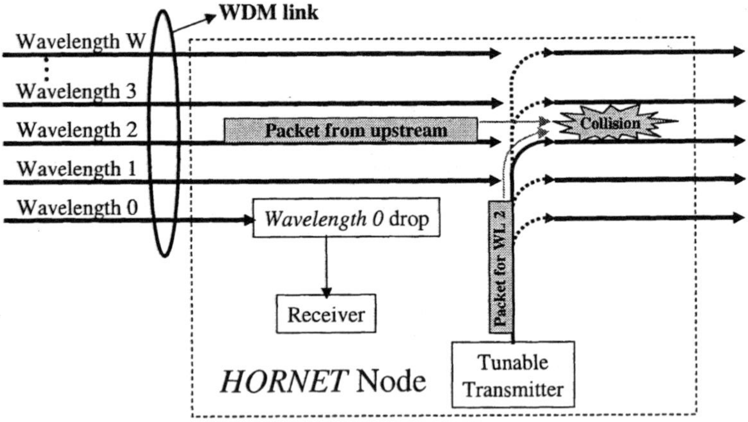


*Figure 18.2.* A collision occurs if a transmitter inserts a packet on a wavelength that is currently carrying a packet through the node.

## 18.3 HORNET Control Protocol and Mechanisms

### 18.3.1 Media Access Control (MAC)

The MAC has two inter-related parts. The first part is the physical mechanisms by which nodes are able to find openings and transmit packets without collisions: packet multiplexing. The second part is the mechanisms such as reservation protocols and fairness control protocols necessary to support circuit transport and best effort packet transport. It is important to realize that the data contained inside the HORNET packet could belong to a TCP/IP best effort service; or it could belong to a guaranteed bit-rate service such as a SONET TDM stream. In this article, a packet refers to the HORNET packet, while an IP packet is referred to as a best effort packet or simply IP packet.

**Packet Multiplexing.** Data-multiplexing in HORNET can suffer from collisions at the power-combiner inside the node, if the transmitter transmits on a wavelength that already has a packet on it (see Figure 18.2). To prevent collisions, the MAC mechanisms should be able to monitor the WDM traffic passing through the node, locate the wavelengths that are available, and inform the transmitter of which wavelengths it is allowed to use at a particular moment. As a result, the transmitter will not insert a packet on a wavelength that is currently carrying another packet through the node.

We have developed different solutions for monitoring and packet multiplexing. Our current solution of using a control wavelength has evolved from and improves upon our earlier work[5].

**Control channel design.** The control wavelength is dropped and added in every node. A point to point link is established between neighboring HORNET nodes using the control channel. The control channel is time slotted into fixed size frames. The frame (or slot) size is chosen to be 64 Bytes in order to minimize the overhead in transporting variable size IP packets. The ring therefore holds N circulating slots, where N depends on the size of the ring, the frame size and the bit-rate (a 100km ring at 10Gbps will hold $\sim 10,000$ 64Byte slots). N slots form a super-frame with slot# 1 of the super-frame being identified by a special indicator. Nodes maintain a slot counter that increments with each passing slot and resets at the beginning of the super-frame.

Once the nodes can read and update the control channel, it can be used to carry control bits. For example, within each frame a bit-vector can convey the wavelength availability information for the time slot. The vector will be of length *W*, where *W* is the number of wavelengths in the network. If bit *w* equals a '1,' then wavelength *w* is carrying a packet during the time period. A '0' bit indicates that the wavelength *w* is available in that time slot. Therefore by simply reading the control bits a node can immediately determine which wavelengths are busy and which are free. It can then choose a free wavelength, say *i* for transmission and update the corresponding control bit *i* to '1'. This ensures that the next node does not transmit on wavelength *i* in that particular time slot: collision avoidance.

In order for this scheme to work, it is important to align the packets transmitted on the data channels to the slot boundaries on the control channel: slot synchronization. Any amount of misalignment, either random or deterministic, has to be compensated by adding guard time (extra overhead) in between time slots. Figure 18.3 shows the portion of the node involved in the alignment process. First, each node transmits a data packet such that is time-aligned with the control frame at the node's output. This can be done by electrically controlling the transmission process. Second, the delay of the data channels passing transparently through the node has to be made equal to the electrical processing delay of the control channel. For that, the local clocks of the nodes are synchronized during the link establishment process, ensuring a deterministic electrical delay for the control channel. Then, a SMF (Single Mode Fiber) delay line (shown in Figure 18.3) is used to match the two delays to within a few 10s of nanoseconds. Furthermore, a calibration process is used fine-tune this delay difference to within one byte by adjusting the electrical delay (also shown in 18.3) seen by the control channel. This calibration technique is described in [6]. This can reduce the delay difference between data and control channel to within 1 byte.

In addition to the above, chromatic dispersion in SMF will cause slots on the wavelengths to drift away from each other. Figure 18.4 (a) plots the drift in ns/km for both SMF (single Mode Fiber) and DCF (Dispersion Compen-

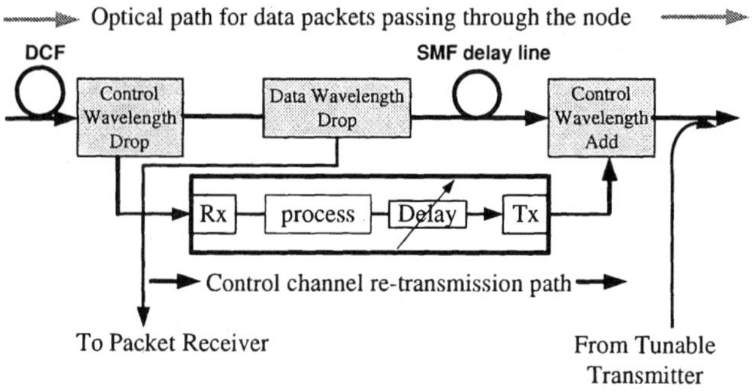


*Figure 18.3.* Control-channel and data-channel paths through a HORNET node.

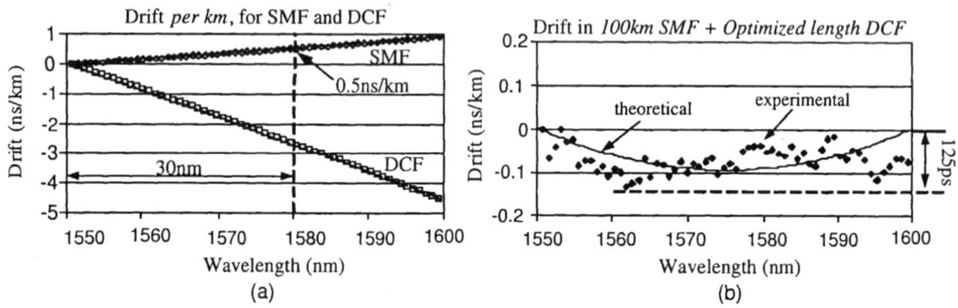


*Figure 18.4.* (a)Drift in ns/km measured in SMF and DCF as a function of wavelength. The drift is relative to the control wavelength of 1550.8nm. The data channels are at wavelengths greater than 1550.8nm. (b) Drift in ns after 100km of SMF + Optimized length of DCF is shown as a function of wavelength. Again, the control wavelength of 1550.8 nm is the reference.

sation Fiber) versus the wavelength. 1550.8nm is used as the control wavelength. Therefore as an example, for a 32-node, 100km SMF ring with 0.8nm WDM spacing and one wavelength per node design ( 30nm total bandwidth), the maximum drift = 50ns. This leads to a guard time of 50ns per time slot, i.e. 100-percent overhead for 64-byte slots at 10Gbps. Clearly this is a serious problem. We manage the dispersion on a link by link basis by using an optimal length of DCF per km of SMF such that the variance of the resulting drift at the output of the SMF+DCF span is minimized. The resulting drift for 100km SMF plus optimized length of DCF has a maximum value of only 125ps requiring a negligibly small overhead. This result is plotted in Figure 18.4 (b). We have confirmed that this technique works for different samples of SMF and DCF.

**Supporting best effort IP traffic.** In the case of best effort packet transport, there are two main challenges to overcome. The first challenge is the

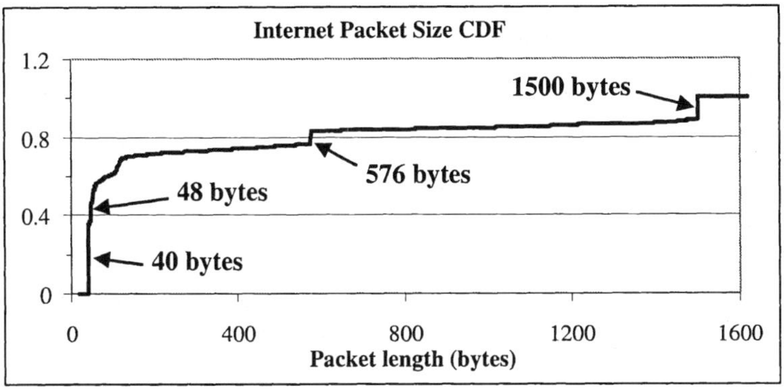


*Figure 18.5.* This CDF of IP packet sizes measured on a particular link by NLANR shows packets ranging from 40bytes to 1500bytes.

transport of variable size IP packets inside fixed sized HORNET frames, while the second is scheduling packets while maintaining fairness in the network. This section will cover our solutions to the two issues.

**Handling variable size IP packets.** Figure 18.5 shows a cumulative distribution function (CDF) of packet sizes measured on a typical IP link[7]. As shown in the figure, typical IP packets sizes range from 40 bytes to 1500 bytes making them incompatible with the fixed sized control channel format of HORNET. One possible solution is to segment all packets into 64byte frames, similar to IP-over-ATM transport. Although simple, the scheme results in an excessive amount of overhead owing to the addition of a header to each segment.

Adding only a small amount of intelligence into the MAC protocol can significantly reduce the overhead. Instead of segmenting all the packets into fixed size cells, the HORNET MAC protocol segments packets only when necessary: segmentation and re-assembly on demand (SAR-OD). In SAR-OD, if a packet is longer than the control frame (or slot) duration, the node transmits the packet across slot boundaries (without segmenting the packet and re-applying the header) until either the packet is complete or until the MAC protocol detects an imminent collision. If a collision is imminent, the node segments the packet by applying a byte that indicates that the segment is an incomplete packet and stops its transmission, avoiding the collision. The node is now free to send packets on different wavelengths while it waits for an opportunity to finish the packet it had begun. At the next opportunity, the node begins transmitting the segmented packet. When the final segment of a packet is completely transmitted, the node finishes the packet with a byte that indicates that the packet is complete. The receiver in a HORNET node has extra intelligence built into it

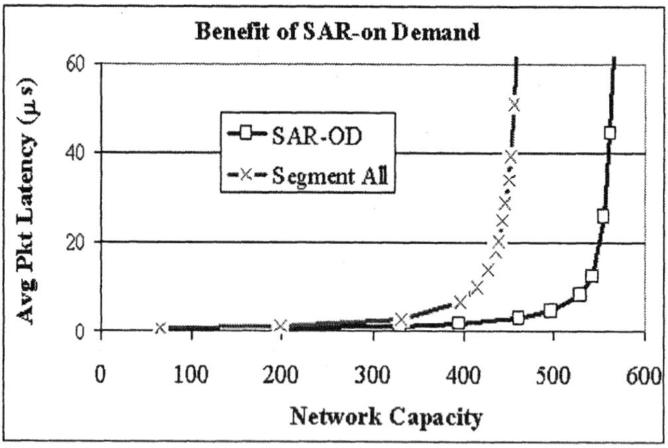


*Figure 18.6.* The simulated performance benefits of using SAR-OD. The network has 33 nodes, and the transmission rate is 10Gbps.

to work with the SAR-OD protocol. It queues the packet segments it receives in Virtual Input Queues (VIQs) based on the source of the segment (the HORNET packet header carries the source address of the HORNET node). Thus segments fall into the queues as the node receives them. It then reconstructs the original packet whenever all segments are received and passes the packets to the local area interface.

The performance advantage of SAR-OD is measured via simulations shown in Figure 18.6 for a 33-node HORNET network. The simulator uses a variable-sized packet CDF similar to the one shown in Figure 18.5. The graph in Figure 18.6 shows a significant performance advantage of approximately 15%, to a segment-all scheme similar to IP-over-ATM.

**The DQBR protocol for scheduling and fairness control.** In HORNET, the source node transmits a packet on wavelength $i$ and sets the control bit $i$ to a value = '1'. When the destination node drops the wavelength $i$, it clears the control bit $i$ to '0'. This leads to the commonly acknowledged unfairness problem where the node $j$ that is immediately downstream to a node $i$ will get first chance of using slots freed by node $i$. As a result, the VOQs for certain source-destination pairs will experience lower throughput than others. Fairness control is hence necessary for the HORNET MAC. To that end, we have developed a protocol called Distributed Queue Bi-directional Ring (DQBR) that attempts to transform HORNET ring architecture into a distributed FCFS queue (one queue per wavelength per direction). The protocol is an adaptation of an older protocol called Distributed Queue Dual Bus (DQDB)[8, 9], which was created for single channel dual-bus metro networks.

The operation of the DQBR fairness control protocol is shown in Figure 18.7. In each control channel frame, a bit stream of length $W$ called the request bit stream follows the wavelength availability information, where $W$ is the number of wavelengths. When a node on the network receives a packet into VOQ $w$, the node notifies the upstream nodes about the packet by setting bit $w$ in the request bit stream in the control channel that travels upstream with respect to the direction the packet will travel. For the case of variable-sized packets, the node places the number of requests corresponding to the length of the packet measured in frames. All upstream nodes take note of the requests by incrementing a counter called a request counter (RC). Each node maintains an RC corresponding to each wavelength. Thus, if bit $w$ in the incoming request bit stream is set, RC $w$ is incremented (Figure 18.7(a) and (b)). Each time a packet arrives to VOQ $w$, the node stamps the value in RCw onto the packet and then clears the RC (Figure 18.7 (c)). The stamp is called a wait counter (WC). After the packet reaches the front of the VOQ, if the WC equals $n$, the VOQ must allow $n$ frame availabilities to pass by for downstream packets that were generated earlier (Figure 18.7 (c) and (d)). When an availability passes by the node on wavelength $w$, the WC for the packet in the front of VOQ $w$ is decremented (if the WC equals zero, then RC $w$ is decremented). Not until the WC equals zero can the packet can be transmitted (Figure 18.7 (e)). The DQBR request-counting system attempts to ensure that if two packets arrive at two different nodes and desire the same wavelength, the one that arrived to the network first will be transmitted first, as if the network is one large distributed FCFS queue.

To demonstrate the fairness control of the protocol, we simulated the scheme when the network is saturated, i.e. the wavelengths are oversubscribed. Figure 18.8 shows one specific result: it plots the throughput for nodes sending packets to Node *18* on a 25-Node HORNET network. With DQBR, the throughput is equal for all nodes, whereas without DQBR, the nodes close to Node *18* have a very difficult time sending packets to Node *18*. Although not shown here, the throughput of the nodes to destinations other than Node *18* is also equal.

**Supporting guaranteed bit-rate circuit traffic.** The HORNET architecture is inherently best effort. Therefore the MAC should be extended to support guaranteed bit-rate services for applications requiring QoS and for carrying TDM (SONET for example) traffic across the network.

**Obtaining a constant bit-rate over HORNET.** As described previously the network is time slotted. Therefore, nodes can obtain a constant bit-rate by simply using a constant number of slots per unit time, to transmit on a particular destination wavelength. For example, if a node uses slot $i$ to transmit data on wavelength $w$, when slot $i$ passes through the node (once per super-

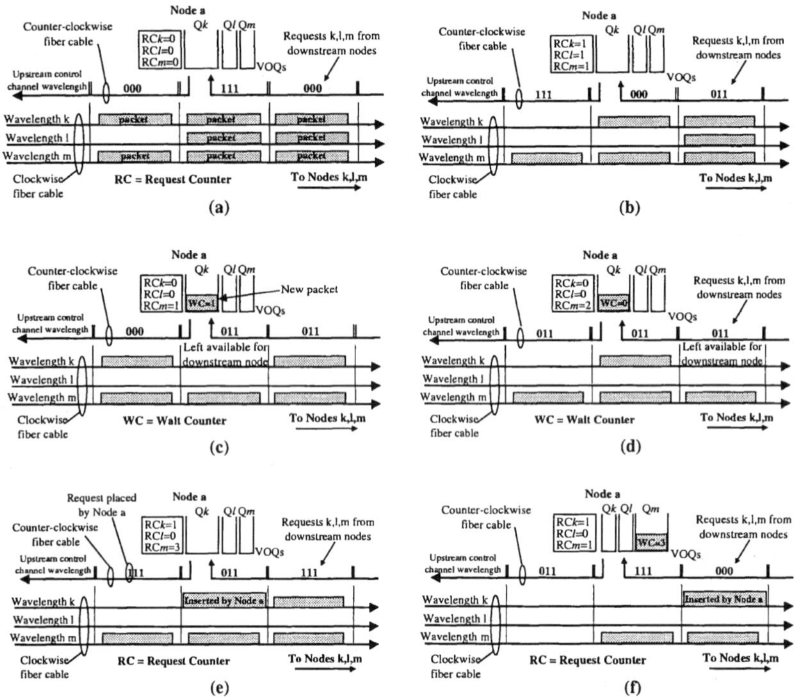


*Figure 18.7.* DQBR operation: (a) A node monitors the requests on the upstream control channel. (b) The node increments the RC counters for any requests it sees. (c) When a packet arrives in a VOQ, the value in the corresponding RC counter is stamped onto the packet as the WC. The packet cannot be inserted into the availability because the WC value is nonzero. (d) The WC counter is decremented for every availability that passes by on the corresponding wavelength. (e) The packet can now be transmitted. (f) When a packet arrives to VOQ *m*, the value from RC *m* is moved into the WC stamped onto the packet. The packet will have to allow three empty frames on Wavelength *m* to pass before it can be transmitted.

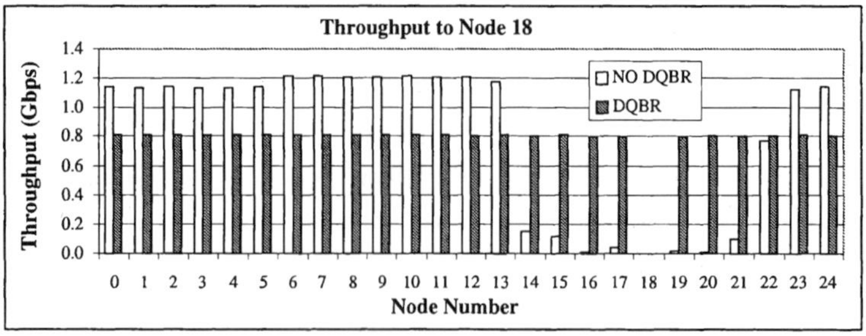


*Figure 18.8.* Throughput measured for VOQ *18* at each of the 25 nodes on a *HORNET* network is plotted. The total network load for wavelength *18* is 1.5 times its capacity. There is enough propagation delay between nodes to hold 50 control frames.

frame or round trip time RTT), then the node will obtain a constant bit-rate. The value of the bit-rate is equal to the number of bits transmitted divided by the round trip time (RTT) in seconds. For a 100km ring network with 64byte frame sizes this amounts to approximately 1Mbps, making it the granularity of circuit transport over HORNET. In order to obtain a circuit bit-rate of $R$ Mbps, the node can use $R$ slots per RTT. Therefore the MAC needs to a reservation and scheduling mechanism that can guarantee the availability of certain slots for certain nodes for the duration of their circuits. Additionally, the scheme should allow circuits of different bit-rates, fast setup and teardown times (millisecond) and fair access to all nodes.

Any reservation and scheduling (termed RS from hereon) protocol for HORNET needs to complete three steps in order to work within the bounds of the architecture. First, to begin servicing a circuit request, the mechanism needs to find a slot (or slots) that are free for both the source $s$ and destination $d$ of the circuit. This is because it is possible that a slot in which the wavelength $d$ is unreserved is being used by the source to transmit circuit data to some other destination $d'$. Second, once the required number of slots are found, these slots need to be reserved for the wavelength $d$ and the reservation status of the slots needs to be broadcast to all nodes. The broadcast is necessary so that all nodes are aware of this reservation and do not re-reserve the slot for the same wavelength or transmit best-effort packets in a reserved slot. Third, the source needs to schedule the transmission of data on the correct wavelength in the reserved slots. Different RS protocols can be designed based on how these three steps are handled [10]. A specific RS scheme developed for HORNET is described below.

**A source-based reservation and scheduling mechanism.** Similar to the wavelength availability vector for best effort packet transport, the control channel carries a reservation vector in each slot. If the reservation bit $d$ is '1' it means that wavelength $d$ is reserved in that time slot. Each node maintains a data structure in memory (SRAM) that is indexed by the slot# and has as many entries as there are slots in the ring network, $N$. At each entry, the structure carries three items: a reservation bit, a $log_2 W$ long bit-vector that holds the target wavelength where $W$ is the number of wavelengths, and a $log_2 N$ bit-vector to denote circuit ID. This implies that there can be $N$ independent circuits being serviced by a node simultaneously. This structure is called the control memory, since it controls the slot by slot functioning of a node similar to a control memory in a circuit switch.

To understand the RS process, assume that the node $s$ has to service a circuit request with ID *xyz* for $k$ slots to node $d$. Also assume that a slot $j$ is entering the node $s$. By reading the control channel reservation vector and its local reservation bit from the control memory, node $s$ immediately knows whether it can reserve wavelength *d* in slot *j*. If it can reserve the wavelength, it sets the reservation bit *d* to '1' on the control channel and enters *(1,d, xyz)* in its control memory at location *j*. That completes the slot reservation. Node *s* continues to do this until all *k* slots are reserved for the circuit *xyz* before moving on to the next request. Therefore the algorithm described above handles the three steps mentioned before very efficiently. A node finds the right slot by reading the destination's reservation status via the control channel and its status in the control memory. Setting the control channel bit to '1' reserves the slot for the particular wavelength and acts as a broadcast mechanism. The control channel acts like a single bulletin board to reserve wavelengths. It is thus not possible for two nodes to reserve the same slot simultaneously because of time lags or the large distances between nodes etc. Once a slot is reserved by node s, no other node can use it. This ensures that a reserved slot becomes available for use 1 RTT after it is reserved. The circuit setup time is thus 2 RTTs provided bandwidth is available: 1 RTT to find and reserve the slots and 1 RTT for the slots to become free for use. Similarly, a circuit can be torn within 1 RTT. This is because a node sees all slots in 1 RTT and can clear the control channel bits and its control memory for the slots to be cleared. The scheduling is easy since the nodes maintain a control memory. In the example, slot *j* is reserved by node *s*. When slot *j* passes by the node (once per RTT), the node reads the control memory and immediately knows that it has to tune the laser to wavelength *d* and transmit data belonging to circuit *xyz*.

Fairness control mechanisms are added to the core protocol described above. To obtain the results shown below we wrote a custom simulator. A Bernoulli arrival process is used to generate circuit requests. The destination of a request is randomly chosen amongst all nodes to simulate any-to-any traffic, or is some cases we use a single destination to simulate hot-spot traffic patterns. Requests are stored in VOQs as they arrive at a node. The arrival rate is kept high in order to induce congestion, i.e. wavelengths are over-subscribed.

The first fairness control mechanism is to make the source clear the slots during circuit tear down instead of the destination. This is because destination based clearing is unfair just like in the best-effort transport scenario. Figure 18.9 (a) depicts this phenomenon: the service rate of VOQ-10 at each node is plotted; node 9 suffers the most owing to the worse positional priority (node 10's service rate is 0 since a node does not transmit to itself). The intuition behind source based clearing is that slots are cleared at different points in the ring getting rid of unfairness. But source clearing can be unfair if a node is allowed to reserve a slot it has just cleared. In this scenario a node can hold-on to a slot forever (greedy protocol). This is avoided by not allowing the node to reserve a slot it has just cleared (non-greedy). Thus when a node clears the slot, the next downstream node that is free gets to use it first. This leads to a round-robin use of a particular slot and no node can hold onto a slot forever.

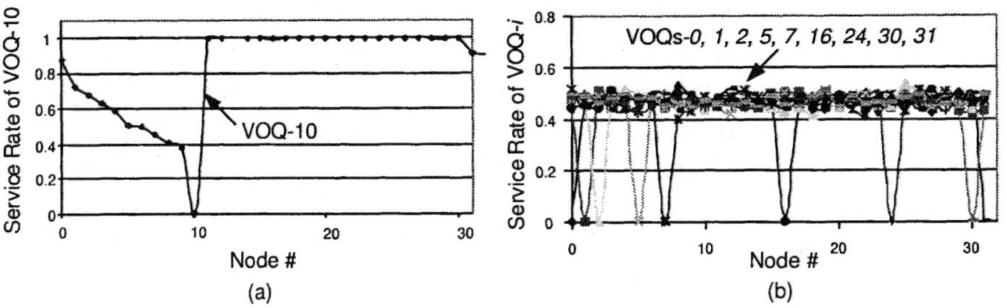


*Figure 18.9.* (a) The service rate for VOQ-10, plotted for all the nodes, shows the decreasing trend from nodes 31 to 9, with node 9 suffering the most while it picks up again from node 11 onwards to node 30. (b) Service rates for a traffic pattern consisting of a combination of small and large requests. The circuit size is distributed as a uniform random variable U(1, 32). Fair access is achieved and deadlock is avoided using random-timers.

We have found that a non-greedy version of source-based clearing is effective in rendering fair access to all wavelengths for single-slot circuits. For the more general case, i.e. circuits of varying sizes, additional challenges exist. It is possible for multiple nodes to get into a deadlock while reserving slots on the same wavelength. This happens because nodes try to reserve slots simultaneously. If the destination gets reserved in all slots without a single circuit finishing its reservation, the nodes will keep trying to reserve, a deadlock. To break possible deadlocks, the requests have a random timer that gets activated once they get to the head of the queue. If the node cannot finish its reservation in the random time indicated by the timer, it stops reserving and clears the slots it has reserved. The request then waits for a random amount of time before it can start reserving again. This basically breaks the deadlock with a random node winning and finishing its circuit reservation. The randomness ensures that no one node is favored in this process. We see that random timer technique offers excellent fairness and avoids deadlock when circuits of different sizes (small and big) are part of the traffic pattern. Figure 18.9 (b) plots the service rates when the size of the circuits is distributed U (1, 32). Note that the number of slots in this simulation N is 320. Similar fairness results are seen for hot-spot destination and unbalanced traffic scenarios (not shown here).

### 18.3.2 Survivability Protocols

A metro network must be survivable to a fiber cut or a node failure. SONET networks typically use a 2-fiber unidirectional path switched ring (2FUPSR) or a 2 or 4-fiber bi-directional line switched ring (2F or 4F BLSR) architecture [11]. We have previously developed a 2 fiber bi-directional path-switched ring (2FBPSR) survivable architecture for HORNET[6]. In this scheme, all of a node's transmission capacity can be used for working traffic, while the capacity drops to half for the nodes positioned adjacent to the cut (worse case) when a failure occurs. Therefore the amount of fully protected traffic that can be supported is half the maximum capacity of the network (similar to SONET), but the benefit over conventional SONET is that under normal working conditions, all of the bandwidth is used for working traffic, effectively doubling the capacity.

With the addition of circuit service over HORNET, the survivability mechanisms have to undergo a change. This is because a circuit is obtained by reserving slots. If the nodes were to re-reserve slots for all the active circuits when a failure occurs, it would take much longer than the 50ms restoration time specified by SONET. This is because to finish reserving a circuit, it takes at best 2 RTTs (1ms for 100km ring). Moreover, each node can have as many active circuits as there are slots in the ring. Hence, there can be no guarantees given about the restoration time.

So, if a circuit needs protection, the reservation protocol is made to reserve slots to the destination in both directions (fibers). One direction carries the working slots while the other carries the protection slots. The working direction can be chosen by a simple routing decision, such as shortest path to the destination. Another vector in the control channel is used, the survivability-vector, that denotes if the reserved wavelength $d$ in time slot $j$ is a working slot (bit $d$ = '1') or a protection slot (bit $d$ = '0'). This will be a $W$-bit vector, where $W$ is the number of wavelengths. When a node reserves a slot, it marks the corresponding survivability control bit to '1' or '0' and makes an entry in its control memory. For scheduling the node transmits data only in the working slots and leaves the protection slots empty. Thus under normal working conditions (no failures), any node can use the protection slots for carrying best effort traffic. When a fiber cut occurs, for a given source-destination pair, only one working direction exists. Nodes can compute the working direction by knowing the location of the failure (this can be broadcast by the nodes adjacent to the fiber cut or failed node) and their position in the ring (topology). Once the control channel is reestablished, the nodes can continue transmitting circuit data in the pre-reserved slots in whichever direction is still working. The restoration time is hence of the order of a few RTTs, the time it takes to repair the control channel, well within the 50ms specified for SONET equipment. Thus the pre-reservation of slots in a soft-state provides quick restoration while allowing protection slots to be used for best effort traffic during normal operation effectively doubling the capacity during normal operation.

## 18.4 HORNET Node and Sub-systems

### 18.4.1 Node design

Figure 18.10 shows a detailed block diagram of one half of the HORNET node (one transmission direction). At the input of the node is a DCF spool that aligns frames on data and control channels A wavelength drop (3-port thin-film filter) removes the control channel wavelength from the ring and drops it into the Data and Control Processor where the bit stream is analyzed for wavelength availability information, DQBR requests, reservation bits and any other control information. An optical amplifier boosts the power of the payload signals. It is important to drop the control channel wavelength before the amplifier so that it only provides gain to the data wavelengths. Because the loss through the node is relatively low ( 6dB), optical amplifiers are not necessarily contained in all nodes. Thus the amplifier is illustrated with a dotted line in Figure 18.10. After the payload signals receive the necessary boost, a thin film filter drops the data wavelength associated with the node into the burst-mode packet receiver. The data packets with the recovered clock are sent to the Data and Control Processor, where a lookup is performed and the packets are switched to the appropriate local area interface card. The data wavelength drop could be reconfigurable to drop a tunable sub-set (one or more) of the wavelengths. The payload wavelengths that are not dropped will then pass through a wavelength add that multiplexes the control channel wavelength onto the fiber. An SMF delay line is located just before the wavelength add. The delay line along with the frame synchronization protocol is used to perfectly match the delays of the payload signals to the control channel. Near the output of the node, the fast-tunable packet transmitter inserts packets onto the ring. The node could use more than one tunable transmitter to accommodate varying traffic demands. A variable optical attenuator (VOA) is placed at the output of the tunable transmitter to set the output power of the transmitter. The node must transmit its packets at a power level to match the power level of the packets passing through the node, as determined by the link budget. This power level is dependent upon the location of the nearest optical amplifier (recall that amplifiers are not necessarily located in all nodes). The link budget systems analysis is presented in detail in [12].

In the following subsections, we discuss the design and performance of the tunable transmitter and burst-mode receiver sub-systems of HORNET.

### 18.4.2 Fast tunable transmitter sub-system

The fast tunable transmitter is one of HORNET's main sub-systems. The transmitter consists of four components: the tunable laser, the tuning controller, the D/A converters, and the modulator. The tunable laser can be a

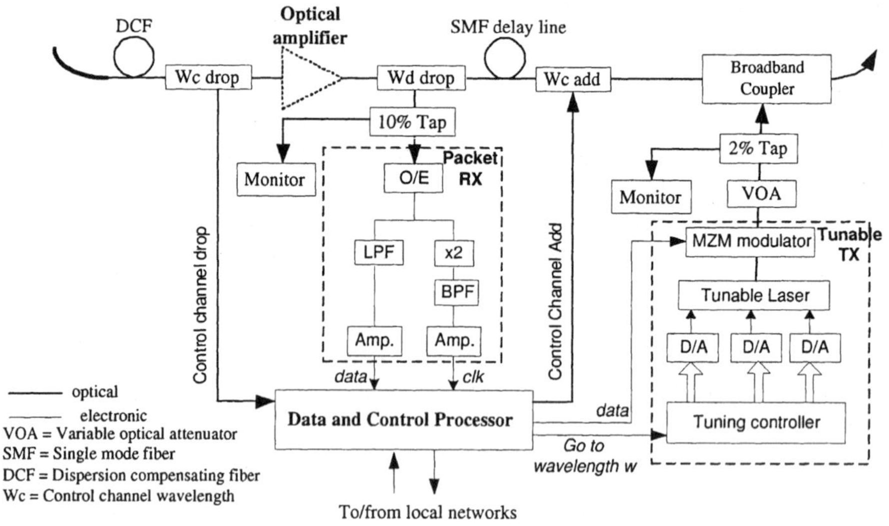


*Figure 18.10.* Detailed block diagram of the HORNET node. Sub-systems and components for only one transmission direction are shown.

Super-Structure Grating DBR (SSG-DBR) laser[13], a Sampled Grating DBR (SG-DBR) laser[14], or a Grating-Assisted Coupler with Sampled Reflector (GCSR) laser[15]. All three lasers exhibit a wide tuning range of approximately 30nm for commercially available devices. Moreover, all three are DBR lasers, relying on refractive index changes for tuning, an inherently fast physical effect. We have used both SG-DBR and GCSR lasers in our sub-systems. Our fast-tuning work is restricted to the GCSR laser.

Both SG-DBRs and GCSR lasers have three tuning sections. A change of current into these sections effectively tunes the laser to a different wavelength. Hence each wavelength is specified by a combination of three tuning currents. The accuracy of these currents is important. A block diagram of the transmitter is shown in the node figure itself (Figure 18.10. The tuning controller consists of a processor that holds three sets of 10-bit digital values per wavelength. There are three D/A converters (DACs) that convert the digital values to analog current. The DACs have a 0.02mA resolution (accurate) and their output current has a rise time $\sim 4ns$ in a 50-ohm load (fast). The output current of the three DACs drives the tuning sections of the laser. The optimum values of the tuning currents and hence the 10-bit digital values are obtained by a calibration procedure. To tune the laser from wavelength w1 to w2, the processor outputs a new set of digital values and clocks the DACs. The current output of the DACs changes and tunes the laser to the target wavelength.

To fast-tune the laser we used a tuning pulse with an overshoot or an under shoot to help maintain a fast rise or fall time of the current entering the laser chip. This was required because the laser packaging is not optimized to handle fast rise time signals. The overshoots and undershoots are created digitally by changing the digital inputs to the DACs and clocking the DACs at the appropriate times[16]. Using these techniques we have obtained less than 15ns tuning times for C-band tuning [17, 16].

Other researchers have also investigated the fast-tunable packet transmitter subsystem [18, 19, 20]. Similar tuning results have been achieved in each of those projects. The tunable laser technology will continue to improve and cost of these lasers will reduce even further, making the fast-tunable transmitters a viable sub-system for use in photonic packet switched networks.

### 18.4.3 Burst mode receiver sub-system

The HORNET project has experimented with two analog techniques for burst mode clock and data recovery. In the first method, called the Embedded Clock Tone (ECT) technique (first demonstrated in [21]), the transmitter frequency multiplexes its local clock with the payload of the data packet. In the packet receiver, the packet and embedded clock are separated using a low-pass filter for the data and a very narrow band-pass filter for the clock tone. The relationship between the clock phase and the data bit-phase is a design aspect of the transmitter, and can be made deterministic. Thus, the receiver can use the received clock to sample the payload data. The ECT technique is simple but the clock tone uses some modulation depth on MZ modulator that would have otherwise been used for data, thus reducing power in the transmitted payload data signal.

A technique that does not suffer from the modulation depth issue uses a nonlinear circuit element to re-create the clock signal from the incoming data signal. The nonlinear clock extraction subsystem implemented for HORNET [16] is shown inside the packet RX sub-system in Figure 18.10. A nonlinear element used could be a frequency doubler (shown in the figure) or a mixer. A narrow band-pass filter is required at the output of the nonlinear clock element. The filter rejects unwanted signal components and acts to average the clock cycles generated by the non-linear element. The filter must be narrow enough to withstand strings of bits in the payload bit stream with no transitions. This is dictated by the well-known relationship between filter response time and filter bandwidth. The resulting output clock tone phase has a deterministic relationship to the pay-load data bit phase, and thus the clock can be used to recover the data bits in the receiver.

However, the clock recovery time for both the nonlinear clock extraction and ECT techniques is limited by the rise time of the band-pass filter, which is inversely proportional to the bandwidth of the pass band. The rise time of the output signal for the band-pass filter used in HORNET is approximately 12ns, which equates to 4Bytes at 2.5Gbps and 16Bytes at 10Gbps. This is a fundamental limit for the analog techniques described above. For lower overhead, digital techniques are preferable.

### 18.4.4 Transmitter-Receiver performance tests

The transmitter-receiver sub-systems were setup back to back to measure the tuning time, recovery time and error-rate performance of the two sub-systems [16]. Figure 18.11 shows the setup. A PLD-based controller manages timing and controls the various devices in the set-up. The HP Pattern Generator (PPG) outputs $2^7 - 1$ PRBS data stream at 2.5Gbps. Data packets with a time duration approximately 1 microsecond are generated by controlling the switch placed at the PPG output. This is comparable to an average Internet packet at 2.5 Gbps. The timing is managed such that the packet arrives at the MZ modulator to modulate the optical carrier as soon as the laser is tuned. An optical filter at the output of the tunable transmitter drops a fixed wavelength. Hence, it selects packets only on the receiver's drop wavelength, to simulate a HORNET receiver. These packets are converted to electronic data by an O/E converter. The electrical 2.5Gbps data signal is split to feed the BERT (payload) and the non-linear clock recovery circuit (a mixer is used as the non-linear element). A strong tone at the exact frequency of the payload data's bit rate is present at the output of the mixer whenever there is a packet present at the inputs. A gating signal is applied to the BERT when clock and data are present at its inputs. The BERT counts errors when the gating pulse is high. One difficulty is that the BERT requires a continuous clock at its input. In our system, the recovered clock is present only when a packet is incident on the burst mode receiver. Thus, a continuous clock from the PPG is 'switched in', whenever the recovered clock is not present.

The results are shown in Figure 18.12. In Figure 18.12(a), a specific fast-tuning result is shown in which the benefit of using the overshoot-undershoot pulse shaping technique is clearly seen; a reduction of tuning time from 25 to 4 ns is achieved. In Figure 18.12(b), the data, clock and gating signal are shown just before they enter the BERT. The zoomed-in picture shows that the gating signal turns on 25 ns after the clock start rising. We were able to measure error-less transmission for this setup indicating that the clock is stable and recovered within 25 ns. We could not pull the gating signal within 25 ns because the switching-out of the BER clock causes an electrical *bounce* that renders the clock unusable. In (c) the BER plot is shown. The line to the extreme left represents conventional systems: the transmitter is not tuned but kept fixed such that its wavelength matches the drop wavelength, the PPG's clock is used

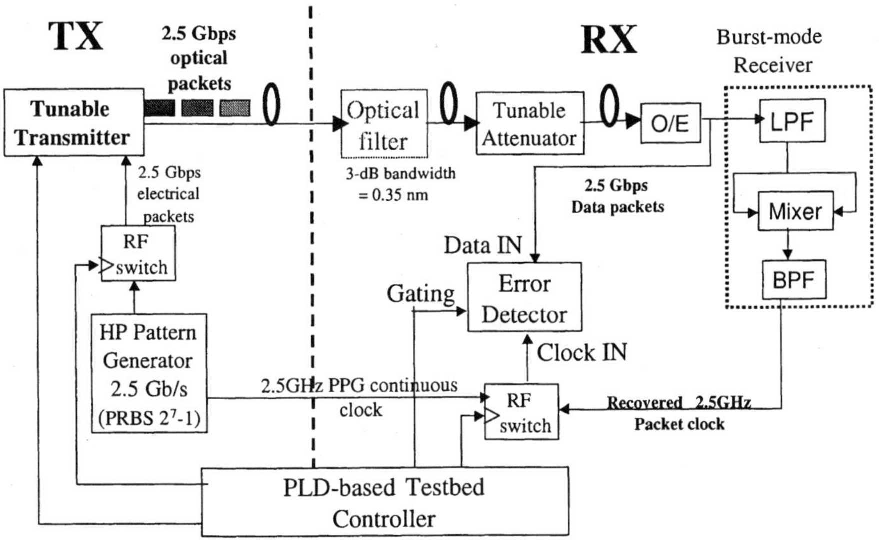


*Figure 18.11.* Experimental setup for the transmitter-receiver sub-system BER tests.

as the BERT input clock, and burst-gating mode is not used. The second line from the left is the case when data packets are generated using the RF switch at the output of the PPG, but the laser is not tuned, and the clock from the PPG is used. In the third line, the tunable transmitter is tuned between each packet transmission: it transmits on the receiver's wavelength, then on a channel 0.8nm away, then on a channel 5nm away and then back on the receiver's wavelength; but still the PPG clock is used. It can be seen that the first three lines are almost on top of each other showing negligible penalty when packets are created and when the laser is tuned. In the line on the extreme right, the clock recovery is added to the test such that the recovered clock is used as the input clock of the error detector. Extra power required for jitter-free clock creates a 0.7 dB penalty which is quite small.

## 18.5 Testbed

Over the duration of the HORNET project, we have built a comprehensive testbed that is used to demonstrate protocols, sub-systems and conduct system experiments. Key testbed results are presented in [5, 16, 22, 23]. In this article, we present our most recent testbed results that demonstrate the reservation protocol and other mechanisms that enable circuits over HORNET [24].

A four node unidirectional HORNET testbed is depicted in Figure 18.13. Each node monitors incoming traffic and outputs node statistics. For the purpose of illustrating the reservation protocol, we will focus on node 3. As Figure

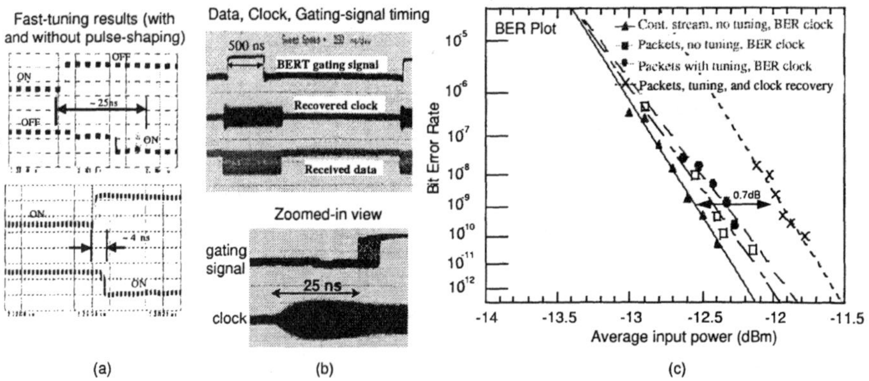


*Figure 18.12.* (a) A laser fast-tuning result: reduction in tuning time from 25 ns to 4 ns is achieved using a *shaped* tuning pulse. (b) Timing of the signals as they enter the BERT: data, clock and the gating-signal. Zoomed-in signals (clock, gating) shows that gating is on 25 ns after clock starts rising. (b) Burst mode BER curves for the transmitter-receiver sub-systems. Total penalty from baseline to final test is only 0.7dB.

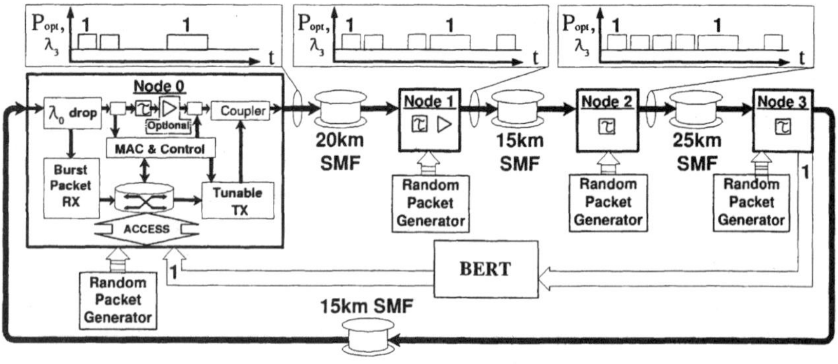


*Figure 18.13.* A 4-node testbed was built for demonstrating Circuits over HORNET.

18.13 shows, each node contains a random packet source. Node 0 serves as a circuit source with traffic generated by a Bit Error Rate Tester (BERT). Node 3 receives circuits and packets placed on wavelength w3. Packet arrivals into source nodes are generated with a pseudo-random number generator.

The first experiment uses a pre-configured pattern of circuit setups and tear-downs to illustrate the protocol capabilities. The statistics at node 3 presented in Figure 18.14 confirm proper operation. The x- and y-axis are normalized to round trip times (RTTs) and utilization, respectively. The figure shows a number of interesting properties. The plot begins with the circuit traffic increasing as was pre-configured for the circuit arrivals. As the circuit traffic increases more and more slots to node 3 get reserved resulting in a decrease in the best effort traffic. This exhibits that best effort packet traffic does not

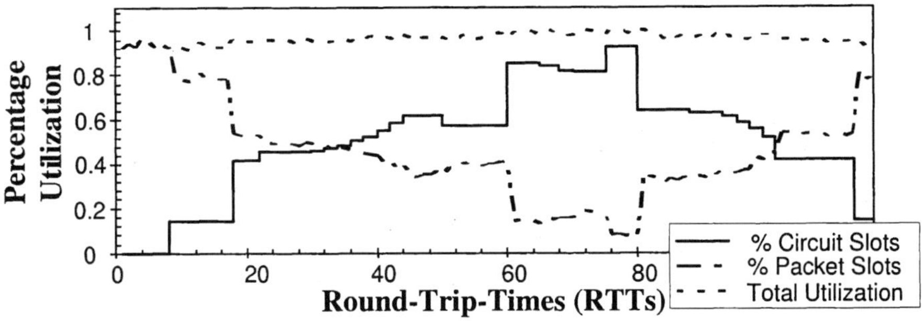


*Figure 18.14.* Circuit and Packet Statistics measured at node 3 show ability of the MAC to setup and teardown circuits and handle both circuits and best effort traffic simultaneously.

use slots that are reserved for circuit traffic. Second, circuits of several sizes are reserved ranging from 5Mbps (see the staircase in the circuit-plot starting at roughly the 32nd RTT) to approximately 310Mbps (at the 60th RTT). Each circuit has a pre-assigned time to live (TTL). The node maintains a timer for each circuit and when the timer for a circuit expires, it clears the slots reserved for the circuit. The circuit traffic therefore starts reducing (at 80th RTT) as pre-configured, allowing the best effort traffic to increase. Meanwhile, the combined utilization of node 3's bandwidth is 100% implying that almost every slot is full. This plot confirms the functioning of the control channel time slotting, link establishment and maintenance, the reservation protocol and the ability of the circuit and best effort scheduler to work hand-in-hand.

Bit Error Rate (BER) test constitutes the second test. These are continuous mode BER tests (not gated packet BER). As Figure 18.13 illustrates, the continuous streaming traffic incident at the access side of node 0 is transmitted across HORNET as bursts of data. The current testbed operates at a channel rate of 1.25Gbps, while the PRBS BER tests are taken at 155 Mbps (OC-3). Therefore, these tests include the entire electrical and optical path through both the source and destination nodes and across the network. The remaining channel bandwidth is still available for other circuits or best effort traffic. Therefore this is a realistic test of circuits over HORNET, in which an imaginary OC-3 circuit is established across HORNET from node 0 to 3 using the reservation protocol, to allow SONET network elements to communicate.

The BER plot is shown in Figure 18.15. The baseline case is actually continuous streaming data at 1.25Gbps from node 0 to node 3, back-to-back, representing the performance of a traditional circuit-based network. In the second line circuit data is packetized and transmitted from node 0 directly to 3 (not over the network) and node 3 reconstructs the circuit from packet data. Although the data channel rate is 1.25Gbps, the data in the plot corresponds to a 155Mbps circuit occupying only a portion of that bandwidth. The

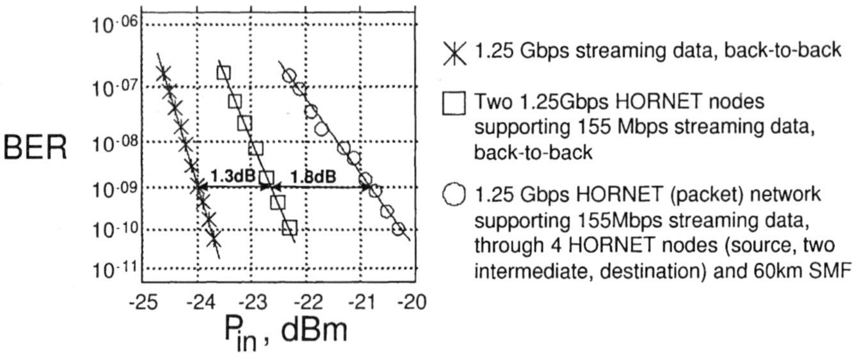


*Figure 18.15.* Continuous mode BER tests over HORNET's packet switched physical layer demonstrate the feasibility of Circuits over HORNET.

1.3dB penalty is due to the clock and data recovery employed in HORNET's packet receiver. The rightmost trend corresponds to the full HORNET testbed, as depicted in Figure 18.13. The data travels electrically through the source node's (node 0) queuing subsystem before entering HORNET where it travels at 1.25Gbps in reserved slots through two HORNET nodes, an amplifier, and approximately 60km of SMF to the destination node 3's packet receiver. The complete optical path causes a 1.8dB penalty at a BER of $10^{-9}$. The BER line exhibits a signal dependent penalty due to the presence of the in-line EDFA. This test confirms the concept of circuits over HORNET: the ability to transmit and maintain a synchronous OC-3 circuit over an OC-24 packet switched physical layer. It also demonstrates the functioning of the complete HORNET node, with tunable transmitter, packet receiver, data and control processor and LAN-side interfaces.

## 18.6 Summary

This article summarizes the accomplishments of the HORNET project. The primary achievement of the project is the development of a new architecture for next-generation metro networks. The use of tunable transmitters and wavelength routing enables HORNET to cost-effectively scale to high capacities compared to conventional networks.

The development of a new architecture created a new set of issues, requiring an entirely new suite of protocols and sub-systems. A control channel was used to design a time slotted MAC protocol. Time slotting simplifies the design of the control protocols allowing the protocols to work synchronously, on a slot by slot basis rather than asynchronously. The alignment of the slots is maintained by a combination of: clock synchronization, calibration procedures and dispersion compensation using optimal length of DCF. The MAC handles variable-sized IP packets using the SAR-OD protocol that segments on demand, and provides a significant improvement in throughput compared to a segment-all scheme. A fairness control protocol named DQBR was developed. DQBR operates in a distributed fashion using the control channel to effectively create a network-wide FIFO queue per destination. Simulations verified that DQBR guarantees equal opportunity to all network users. We designed a reservation protocol that reserves the desired number of slots that are used in each round trip time to create a fixed bit-rate channel. Circuits of different bit-rates can be established by reserving more or less slots. The protocol achieves fairness by employing a number of heuristics: source-clearing instead of destination-clearing, round robin service of VOQs inside nodes and random timers are used to avoid deadlock. The network is survivable to one node failure or fiber cut. In the survivability protocol, nodes changes the direction of transmission based on the location of the fiber cut, the destination of the data and the position of the node in the ring. For circuits that require protection, slots are reserved in both directions and path switching is used at the transmitter and receiver. The various sub-systems are described in Section 18.4. The tunable transmitter is shown to tune within 15ns and carry bits on the tuned optical carrier with a good bit error rate. Similarly two burst mode receiver designs are described. Both designs are analog and are simple to implement but the recovery time is limited to the rise time of the band pass filter 12ns. A complete node design depicting the various opto-electronic, optical, electrical sub-systems and components is described. The project benefited by the development of a 4-node testbed that was used for extensive demonstration of the protocol and sub-systems described above. In this article, key results demonstrating circuits over HORNET are presented. Namely a pre-configured pattern of circuits was established on the testbed. Statistics of circuit and best effort packet traffic were measured to prove the correct functioning of the protocols. A continuous mode bit-error rate test at 155Mbps over the packet switched 1.25Gbps HORNET network was also presented. The BER test mimics a synchronous circuit, such as an OC-3 SONET TDM channel, on top of HORNET.

In conclusion, the research completed in the HORNET project can help in the development of future packet-based metro and access networks.

## Acknowledgments

The authors thank DARPA for sponsoring the HORNET project under agreement number F30602-00-2-0544, and Sprint ATL, Burlingame, CA for their sponsorship. Special thanks go to Dr. Steven Gemelos and Dr. Duang-Rudee Wonglumsom for their work on the HORNET project in its earlier years. We also thank other PNRL students and visitors who contributed to the work, including Fu-Tai An, Moritz Avenarius, Hopil Bae, Yasuyuki Fukashiro, Yu-Li Hsueh, Eric Hu, Yoshiaki Ikoma, Dr. K. K. S. Kim, Akira Nakamura, Hiroshi Okagawa, Dr. Akira Okada, Dr. Takashi Ono, Carlo Tosetti and Scott Yam. We would also like to thank ADC Altitun for their generosity in providing GCSR laser samples and Genoa for donating their Linear Optical Amplifiers to this project.

## References

[1] Resilient Packet Ring Alliance. An introduction to resilient packet ring technology. White Paper, available at http://www.rpralliance.org, October 2001.

[2] The HORNET Project web-site. http://pnrl.stanford.edu/iphornet/iphornet.html.

[3] M. J. Spencer and M. A. Summerfield. WRAP: A medium access control protocol for wavelength-routed passive optical networks. *Journal of Lightwave Technology*, 18(12):1657–1676, December 2000.

[4] R. Gaudino, A. Carena, V. Ferrero, A. Pozzi, V. De Feo, P. Gigante, F. Neri, and P. Poggiolini. RINGO: A WDM ring optical packet network demonstrator. In *Proceedings of the 27th European Conference on Optical Communications*, September 2001.

[5] K. Shrikhande, I. M. White, D. Wonglumsom, S. M. Gemelos, M. S. Rogge, Y. Fukashiro, M. Avenarius, and L. G. Kazovsky. HORNET: A packet-over-WDM multiple access metropolitan area ring network. *IEEE Journal on Selected Areas in Communications*, 18(10):2004–2016, October 2000.

[6] I. M. White, M. S. Rogge, K. Shrikhande and L. G. Kazovsky. A summary of the HORNET project: a next-generation metropolitan area network *IEEE Journal of Selected Areas in Communications*, 21(9): 1478–1494, September 2003.

[7] National Laboratory for Applied Network Research, Measurement Operations and Analysis Team. http://pma.nlanr.net/Datacube/.

[8] IEEE Standard 802.6. Distributed queue dual bus (DQDB) subnetwork of a metropolitan area network (MAN), December 1990.

[9] E. L. Hahne, A. K. Choudhury, and N. F. Maxemchuk. Improving the fairness of Distributed-Queue-Dual-Bus networks. In *Infocom '90*, pages 175–184, 1990.

[10] M. S. Rogge, K. Shrikhande, C. Toseti, H. Bae, and L. G. Kazovsky. Circuits over HORNET (coho) - guaranteed bit rates over a packet-based metro network. In *Globecom Technical Digest*, Vol.5, pp. 2740–2744, December 2003.

[11] I. Haque, Wilhelm Kremer, and K. Raychaudhuri. Self-healing rings in a synchronous environment. *IEEE LTS*, pages 30–37, November 1991.

[12] I.M. White, Eric Hu, Yu-Li Hsueh, K. Shrikhande, M.S. Rogge, and L. G. Kazovsky. Demonstration and system analysis of the HORNET architecture. *Journal of Lightwave Technologies*, to appear in 2003.

[13] F. Kano and Y. Yoshikuni. Frequency control and stabilization of broadly tunable SSG-DBR lasers. In *Optical Fiber Communications Technical Digest*, pages 538–540, March 2002.

[14] Y. A. Akulova, C. Schow, A. Karim, S. Nakagawa, P. Kozodoy, G. A. Fish, J. DeFranco, A. Dahl, M. Larson, T. Wipiejewski, D. Pavinski, T. Butrie, and L. A. Coldren. Widely-tunable electroabsorption-modulated sampled grating DBR laser integrated with semi-

conductor optical amplifier. In *Optical Fiber Communications Technical Digest*, pages 536–537, March 2002.

[15] B. Broberg, P-J. Rigole, S. Nilsson, L. Andersson, and M. Renlund. Widely tunable semiconductor lasers. In *Optical Fiber Communications Technical Digest*, pages WH4:1–WH4:3, March 2002.

[16] K. Shrikhande, I. M. White, M. S. Rogge, F-T. An, E. S. Hu, S. S-H. Yam, and L. G. Kazovsky. Performance demonstration of a fast-tunable transmitter and burst-mode packet receiver for HORNET. In *Optical Fiber Communications Technical Digest*, pages ThG2:1–ThG2:3, March 2001.

[17] Y. Fukashiro, K. Shrikhande, M. Avenarius, M. S. Rogge, I. M. White, D. Wonglumsom, and L. G. Kazovsky. Fast and fine wavelength tuning of a GCSR laser using a digitally controlled driver. In *Optical Fiber Communications Technical Digest*, pages WM43:1–WM43:3, March 2000.

[18] O. A. Lavrova, G. Rossi, and D. J. Blumenthal. Rapid tunable transmitter with large number of ITU channels accessible in less than 5 ns. In *Proceedings of the 26th European Conference on Optical Communications*, pages 23–24 (Paper 6.3.5, September 2000.

[19] S. J. B. Yoo, Y. Bansal, Z. Pan, J. Cao, V. K. Tsui, S. K. H. Fong, Y. Zhang, J. Taylor, H. J. Lee, M. Jeon, and V. Akella. Optical-label based packet routing system with contention resolution in wavelength, time, and space domains. In *Optical Fiber Communications Technical Digest*, pages 280–282, March 2002.

[20] J. Gripp, M. Duelk, J. Simsarian, S. Chandrasekhar, P. Bernasconi, A. Bhardwaj, Y. Su, K. Sherman, L. Buhl, E. Laskowski, M. Cappuzzo, L. Stulz, M. Zirngibl, O. Laznicka, T. Link, R. Seitz, P. Mayer, and M. Berger. Demonstration of a 1.2 Tb/s optical packet switch fabric (32x40 Gb/s) based on 40 Gb/s burst-mode clock-data-recovery, fast tunable lasers, and high-performance NxN AWG. In *Proceedings of the 27th European Conference on Optical Communications, Postdeadline Session 3*, September 2001.

[21] I. Chlamtac, A. Fumagalli, and L. G. Kazovsky et. al. Cord: Contention resolution by delay lines. *IEEE Journal of Selected Areas in Communications*, 14(5):1014–1029, May 1996.

[22] I. M. White, M. S. Rogge, Y-L. Hsueh, K. Shrikhande, and L. G. Kazovsky. Experimental demonstration of the HORNET survivable bi-directional ring architecture. In *Optical Fiber Communications Technical Digest*, pages 346–349, March 2002.

[23] D. Wonglumsom, I. M. White, K. Shrikhande, M. S. Rogge, S. M. Gemelos, F-T. An, Y. Fukashiro, M. Avenarius, and L. G. Kazovsky. Experimental demonstration of an access point for HORNET - a packet-over-WDM multiple access MAN. *Journal of Lightwave Technology*, 18(12):1709–1717, December 2000.

[24] M. S. Rogge, K. Shrikhande, H. Bae, C. Toseti, and L. G. Kazovsky. Circuits over HORNET (coho) demonstration: guaranteed bit rates over a packet-based network. In *European Conference on Optical Communications (ECOC) Technical Digest*, Th1.4.2, September 2003.

# Index